Systems Engineering for Sustainable Development Goals

Systems Engineering for Sustainable Development Goals

Editor

Cecilia Haskins

MDPI • Basel • Beijing • Wuhan • Barcelona • Belgrade • Manchester • Tokyo • Cluj • Tianjin

Editor
Cecilia Haskins
Norwegian University of Science
and Technology
Norway

Editorial Office
MDPI
St. Alban-Anlage 66
4052 Basel, Switzerland

This is a reprint of articles from the Special Issue published online in the open access journal *Sustainability* (ISSN 2071-1050) (available at: https://www.mdpi.com/journal/sustainability/special_issues/SystemEngineering_SDGs).

For citation purposes, cite each article independently as indicated on the article page online and as indicated below:

LastName, A.A.; LastName, B.B.; LastName, C.C. Article Title. *Journal Name* **Year**, *Volume Number*, Page Range.

ISBN 978-3-0365-2221-0 (Hbk)
ISBN 978-3-0365-2222-7 (PDF)

Contents

About the Editor

Cecilia Haskins is an American living and working in Norway as a mentor, consultant, and author. She is an active volunteer with distinguished service to professional societies and homeowner associations. Her career includes 35 years as a practicing systems engineer, for which she has been recognized as an INCOSE-certified Systems Engineering Professional.

Cecilia spent 21 years in academia, where she taught organizational management, project management, and systems engineering courses, and supervised theses written at all levels of higher education.

Her educational background includes a BSc in Chemistry from Chestnut Hill College, an MBA from Wharton, University of Pennsylvania, and a PhD from the Norwegian University of Science and Technology (NTNU). She has emeritus status with both NTNU and USN. Her research interests are eclectic, but focused on the application of systems thinking and systems engineering to assorted problem domains.

sustainability

MDPI

Editorial

Systems Engineering for Sustainable Development Goals

Cecilia Haskins

Department of Mechanical and Industrial Engineering, Norwegian University of Science and Technology, Høgskoleringen 1, 7491 Trondheim, Norway; cecilia.haskins@ntnu.no

Citation: Haskins, C. Systems Engineering for Sustainable Development Goals. *Sustainability* **2021**, *13*, 10293. https://doi.org/10.3390/su131810293

Received: 20 August 2021
Accepted: 25 August 2021
Published: 15 September 2021

Sustainability is expanding the discipline and practice of systems engineering. The International Council on Systems Engineering (INCOSE) has foreseen the critical importance of addressing sustainability issues in their Systems Engineering Vision since 2007 [1,2]. As a concept, sustainability covers a broad range of issues ranging from the concern for rapidly depleting natural resources to the concern for the welfare of the earth's populations. Domains such as health care, transportation, natural resource management, social economics, and governance recognize the value of applying systems thinking and systems engineering practices to finding solutions that go beyond temporary Band-Aid solutions for symptoms and that instead address the root causes of these problems.

Traditionally, the discipline and practices of systems engineering are associated with solutions for technological systems and products. The term has been used since the late 1950s, and over the intervening decades, systems have been increasing in complexity. A realization that these practices, which have evolved to cope with technological complexity, are equally equipped to cope with political and society issues has gradually emerged [3]. Applying the concepts of systems thinking and systems engineering is both acceptable and expected [4]. Research applies these methods to various grand challenges and the UN Sustainability Development Goals (SDG) [5,6].

The 14 articles in this Special Issue are from researchers representing Brazil, Ireland, Germany, the Netherlands, Norway, Poland, Spain, and the USA. The contributions cover a range of industries. Of the articles in this Special Issue, four cover topics pertaining to the energy sector, three include work pertinent to the food, higher education, and regional planning sectors, respectively. Topics span sustainable business models, resilience and post-COVID recovery, models for working to achieve SDG targets, and one application of system dynamics to explore the peace nexus.

Seven of the contributors to this issue are young researchers who are also graduates of the NTNU industrial ecology program, the first such academic initiative of its kind, started in 1993 [7]. Systems engineering and industrial ecology both build on the strengths of interdisciplinary theory and practices [8]. This Special Issue of *Sustainability* features original papers by young and established researchers in systems engineering leading the advancement of this rapidly expanding discipline.

Addressing resilience in the energy sector, Neumann, van Erp, Steinhöfel, Sieckmann, and Kohl studied business model patterns for achieving corporate resilience. The identified patterns were analyzed and validated by expert interviews in the German electrical industry, confirming their usefulness in providing guidance for organizations to tackle industrial resilience by adapting their business models [9]. In the same sector, this time considering the oil and gas industry on the Norwegian continental shelf, Czachorowski also uses systems engineering to innovate business models for use in the exploration and production phases of a major operator. She applies morphological analysis to identify discrete components that can be applied in the industry as they "clean up their act" to address SDG targets [10]. Against the backdrop of energy transitions, Kirkels, Evers, and Muller contribute a position paper exploring the challenges facing systems engineering when it is applied to sociotechnical problems. They offer two approaches to address the limitations [11]. Muller also presents a case study of an energy transition in the Netherlands, applying roadmapping and conceptual modelling to support policy and decision makers who must cope with a large heterogeneous set of stakeholders. [12].

Agriculture, fisheries, and food waste in the hospitality arena provide examples of how applying methods from systems engineering can help formulate solutions to move the food sectors toward meeting SDG targets. Palmer, Burton, and Haskins propose a framework called MPAST: Mapping Problem Archetypes to Solutions for Transitions, then apply it to a Norwegian agricultural bioeconomy transition using data collected from a recent survey [13]. Building on research conducted around lost fishing gears and ghost fishing, Deshpande and Haskins apply a disciplined systems engineering process to examine ways to address the root causes of this problem. The study yields some interesting suggestions [14]. Finally, Buczacki, Gładysz, and Palmer propose a framework that is suitable for illustrating the dependencies between different micro actions and macro policies that impact the ability to meet SDG targets. They conclude that to resolve a pervasive issue, such as food waste, requires the concerted collaboration of consumers, businesses, and governing authorities [15].

Communities hoping to preserve their natural resources and regional planners also benefit from systems engineering research. Johansen, Aspen, Sparrevik and Æsøy developed a system-oriented sustainability scoring model that they apply to the port of Geiranger, Norway in an effort to balance the negative effects of cruise tourism [16]. A case study from a regional planning process in the Aalesund region in Norway was used by Aspen and Amundsen to demonstrate their participatory planning support system (PSS), derived using systems engineering methods, for integrating and operationalizing the SDGs [17].

Themes from industrial ecology inform two of the articles. The Capacity-building in Sustainability and Environmental Management model (the CapSEM-model), presents organizations a systemic way to transition to sustainability through the application of systemic methodologies that gradually help increase environmental and sustainability performance. Fet and Knudson map the SDGs onto the CapSEM model as an example of how they can be useful in the transition to sustainability [18]. A case of industrial symbiosis (IS) in Spain was analyzed by Ruiz-Puente and Jato-Espino. They concluded that the changes implemented in industrial parks resulted equated to contributions to nine SDGs and fourteen of their specific targets, proving the domino effect associated with the application of IS policies by governments and public entities [19].

In the domain of higher education, Lynch, Andersson, and Johansen address the potential for courses in entrepreneurship to establish a mindset in students toward a sustainable future. They assert that a systems perspective must be integrated into entrepreneurial education and suggest that merging systems thinking and entrepreneurship can be used to nudge students towards sustainability [20]. Yang and Cormican conducted a scoping study that analyzes extant evidence to uncover the contributions of systems engineering in advancing progress towards the SDGs. They conclude that systems engineering has been an active catalyst promoting the SDGs with the potential to encourage the transdisciplinary research necessary to achieve long-term transformational and sustainable change across sectors and disciplines [21].

The pièce de résistance is provided by Amadei, who addresses one of the quintessential SDG goals to promote peace. He examines the value proposition of adopting a systems approach to capture the linkages between the SDGs and peace sectors using basic system dynamics models to illustrate peace–development nexus dynamics [22].

Funding: This research received no external funding.

Institutional Review Board Statement: Not applicable.

Informed Consent Statement: Not applicable.

Data Availability Statement: Not applicable.

Conflicts of Interest: The author declares no conflict of interest.

References

1. INCOSE. *Systems Engineering Vision;* Crisp, H., Diego, S., Eds.; International Council on Systems Engineering: San Diego, CA, USA, 2007; INCOSE-TP-2004-004-02.
2. INCOSE. *A World in Motion-Systems Engineering Vision;* Diego, S., Ed.; International Council on Systems Engineering: San Diego, CA, USA, 2014; INCOSE-TP.
3. Sillitto, H.; Martin, J.; McKinney, D.; Griego, R.; Dori, D.; Krob, D.; Godfrey, P.; Arnold, E.; Jackson, S. *Systems Engineering and System Definitions;* Diego, S., Ed.; International Council on Systems Engineering: San Diego, CA, USA, 2019; INCOSE-TP-2004-004-INCOSE-TP.
4. Philbin, S.P. Driving Sustainability through Engineering Management and Systems Engineering. *Sustainability* **2021**, *13*, 6687. [CrossRef]
5. Brooks, I. The United Nations Sustainable Development goals in systems engineering: Eliciting sustainability requirements. In Proceedings of the 7th International Conference on ICT for Sustainability, Bristol, UK, 21–27 June 2020; pp. 196–199.
6. Madni, A.M. Transdisciplinary systems engineering: Exploiting disciplinary convergence to address grand challenges. *IEEE Syst. Man Cybern. Mag.* **2019**, *5*, 6–11. [CrossRef]
7. Ehrenfeld, J. Industrial ecology: A new field or only a metaphor? *J. Clean. Prod.* **2004**, *12*, 825–831. [CrossRef]
8. Sage, A.P. Systems engineering and management for industrial ecology and sustainable development. In Proceedings of the 1997 IEEE International Conference on Systems, Man, and Cybernetics, Computational Cybernetics and Simulation, Orlando, FL, USA, 12–15 October 1997; Volume 1, pp. 784–790.
9. Neumann, K.; van Erp, T.; Steinhöfel, E.; Sieckmann, F.; Kohl, H. Patterns for Resilient Value Creation: Perspective of the German Electrical Industry during the COVID-19 Pandemic. *Sustainability* **2021**, *13*, 6090. [CrossRef]
10. Czachorowski, K.V. Cleaning Up Our Act: Systems Engineering to Promote Business Model Innovation for the Offshore Exploration and Production Supply Chain Operations. *Sustainability* **2021**, *13*, 2113. [CrossRef]
11. Kirkels, A.; Evers, V.; Muller, G. Systems Engineering for the Energy Transition: Potential Contributions and Limitations. *Sustainability* **2021**, *13*, 5423. [CrossRef]
12. Muller, G. Applying Roadmapping and Conceptual Modeling to the Energy Transition: A Local Case Study. *Sustainability* **2021**, *13*, 3683. [CrossRef]
13. Palmer, E.; Burton, R.; Haskins, C. A Systems Engineering Framework for Bioeconomic Transitions in a Sustainable Development Goal Context. *Sustainability* **2020**, *12*, 6650. [CrossRef]
14. Deshpande, P.C.; Haskins, C. Application of Systems Engineering and Sustainable Development Goals towards Sustainable Management of Fishing Gear Resources in Norway. *Sustainability* **2021**, *13*, 4914. [CrossRef]
15. Buczacki, A.; Gładysz, B.; Palmer, E. HoReCa Food Waste and Sustainable Development Goals—A Systemic View. *Sustainability* **2021**, *13*, 5510. [CrossRef]
16. Johansen, B.H.; Aspen, D.M.; Sparrevik, M.; Æsøy, V. Applying System-Oriented Sustainability Scoring for Cruise Traffic Port Operators: A Case Study of Geiranger, Norway. *Sustainability* **2021**, *13*, 6046. [CrossRef]
17. Aspen, D.M.; Amundsen, A. Developing a Participatory Planning Support System for Sustainable Regional Planning—A Problem Structuring Case Study. *Sustainability* **2021**, *13*, 5723. [CrossRef]
18. Fet, A.M.; Knudson, H. An Approach to Sustainability Management across Systemic Levels: The Capacity-Building in Sustainability and Environmental Management Model (CapSEM-Model). *Sustainability* **2021**, *13*, 4910. [CrossRef]
19. Ruiz-Puente, C.; Jato-Espino, D. Systemic Analysis of the Contributions of Co-Located Industrial Symbiosis to Achieve Sustainable Development in an Industrial Park in Northern Spain. *Sustainability* **2020**, *12*, 5802. [CrossRef]
20. Lynch, M.; Andersson, G.; Johansen, F.R. Merging Systems Thinking with Entrepreneurship: Shifting Students' Mindsets towards Crafting a More Sustainable Future. *Sustainability* **2021**, *13*, 4946. [CrossRef]
21. Yang, L.; Cormican, K. The Crossovers and Connectivity between Systems Engineering and the Sustainable Development Goals: A Scoping Study. *Sustainability* **2021**, *13*, 3176. [CrossRef]
22. Amadei, B. Systemic Modeling of the Peace–Development Nexus. *Sustainability* **2021**, *13*, 2522. [CrossRef]

sustainability

MDPI

Article

Systemic Analysis of the Contributions of Co-Located Industrial Symbiosis to Achieve Sustainable Development in an Industrial Park in Northern Spain

Carmen Ruiz-Puente and Daniel Jato-Espino *

INGEPRO Research Group, Department of Transport and Projects and Processes Technology, University of Cantabria, 39005 Santander, Spain; mdelcarmen.ruiz@unican.es

* Correspondence: jatod@unican.es

Received: 16 June 2020; Accepted: 16 July 2020; Published: 19 July 2020

Abstract: Resource efficiency is a strategy with great potential to make progress towards the UN Sustainable Development Goals (SDGs), since it can contribute to meeting a variety of economic, environmental and social targets. In this context, this investigation developed a systemic analysis of co-located Industrial Symbiosis (IS) synergies in an industrial park formed of four companies. To this end, public data showing that the main activity in this park concerned materials, water and steam flows were supported with short visits to the companies for verification purposes. Then, the effects of nine exchange and twelve share synergies were analysed at different scales according to their impacts on sustainable development. The changes caused by these synergies in the flows in the industrial park enabled saving more than 10 k tonnes of raw materials and waste disposal and almost 10 Mm^3 of raw water per year, as well as six auxiliary service systems. In the end, these figures might be translated into more than 200 kt CO_2 eq. and EUR 6M saved per year, which in turn corresponds to 0.05% of the Gross Domestic Product (GDP) of the region in which the park is located. In terms of sustainable development, these modifications were translated into contributions to nine SDGs and 14 of their specific targets, proving the domino effect associated with the application of IS policies by governments and public entities.

Keywords: circular economy; systems thinking; industrial symbiosis; resource sharing; steam; sustainable development; synergy; industrial sustainability

1. Introduction

Resource efficiency has been proposed by the UNEP International Resource Panel (IRP) as a means to decouple economic growth from environmental deterioration while protecting human well-being, which is crucial for achieving the Sustainable Development Goals (SDGs) [1]. The worldwide consumption of primary materials was 90 billion tons by 2017, a figure which is expected to increase up to 186 billion tons by 2050 [2]. This trend has resulted in diverse negative impacts on soils, aquifers and biodiversity, whilst increasing food and water demands.

Industrial Ecology (IE) has been found to have huge potential to attenuate the effects of these harmful prospects [3]. IE is an interdisciplinary area whose goal is to reproduce the characteristics of natural ecosystems in the industry, closing material loops and searching for sustainable development [4]. Hence, IE requires industrial areas not to be isolated with respect to their surroundings, but considered alongside them to optimize the handling of resources from extraction to disposal [5]. Apart from this global and systemic vision of the industry as a natural ecosystem, IE must be based on the creation of dynamic networks of entities or companies within their area and the promotion of sustainable development [6]. In this vein, the retrofitting of industries by increasing resource efficiency has been identified as a means to contribute to the SDGs [7].

Industrial Symbiosis (IS) is a branch of IE aimed at creating cooperation networks whereby two or more industrial entities carry out mutually beneficial relationships, with emphasis on cases in which one entity makes productive use of streams that are considered as wastes by another [8]. The other main means of symbiotic association concerns the shared use of facilities, equipment and/or utilities for saving resources purposes [9]. As a result, there is a reduction in the amount of waste and emissions, as well as economic advantages for the industrial entities involved [10]. Geographical proximity is an essential factor in facilitating IS [11], since it increases the feasibility of opportunities through which to exchange products and/or share components [12]. Despite the additional complexity entailed by IS practices in relation to other alternatives (e.g., direct disposal), IS has been found to be capable of producing economic and environmental benefits at regional scales [13].

As a result, the principles of IS crystallize in Eco-Industrial Parks (EIP), which consist of communities of companies located in a common area that seek to improve economic, environmental and social performance through collaboration in managing resources [14]. EIPs differ from other forms of cooperation such as industrial clusters, whose main focus is to gain economic benefits and, therefore, disregards other aspects of sustainability [15]. They also differ from industrial districts, which are more specific and mostly refer to co-located small and medium companies devoted to light manufacturing sectors of the economy [16]. Although all these terms have limited geographic boundaries, they are closely linked to regional innovation systems, which in turn relate to the existence of learning experiences based on localized nodes of industrial activity. These nodes can lead to developing policies whereby a region is considered the most suitable scale to foster innovation-based economies [17].

Under these premises, the research community has produced a variety of scientific studies to explore the nexus between IE (including IS and EIPs), systems engineering and sustainability. More than two decades ago, Côté and Cohen (1998) [18] highlighted the need for widening the perspective on EIPs, which tended to be limited to waste exchange, and used a systems-approach to involve environmental, economic and social aspects within and outside of the park. In this vein, Haskins (2006) [19] studied existing eco-industrial developments and pointed out the adoption of regional approaches to achieve stable and sustainable ecosystems. This path was found to be hindered by geographic dispersion, as well as by additional efforts in terms of trust, coordination and data collection. As a contribution to deal with these barriers, the same author proposed [20] and applied [21] a systems framework to form and sustain industrial parks, including aspects such as stakeholder coordination, interdisciplinarity, unification and monitoring. The use of indicators to track sustainable development progress was identified as one of the main lines of additional work to develop. Similarly, Sopha et al. (2010) [22] built a systems framework for assessing IS, including its application through a case study. Although the approach served for stakeholder identification and reveals the need and means to solve the problem under consideration, facilitating communication was argued to be a field to explore in the future. The dynamic nature of EIPs was also addressed by Romero and Ruiz (2013) [23], who developed a framework expected to form the basis on which to build mathematical models for analyzing EIPs. Their objectives, surroundings and the relationships and decisions of stakeholders were underlined as key features to properly modelling these systems.

The link between territorial policies and sustainability was explored by Deshpande and Aspen (2018) [24], who provided a systems engineering framework to ensure sustainable resource management. The approach taken was found to facilitate the nexus between the SDGs and policymakers, such that the latter gained understanding of a complex process and improved their decision-making. In relation to the SDGs focused on zero poverty and hunger, Ginige (2018) [25] combined systems engineering with smart computing to support the coordination of agricultural production in Sri Lanka. The proposed framework was presented to provide a long-term solution to the complex multi-disciplinary nature of this problem. On a smaller scale, Moldavska and Welo (2019) [26] defined a corporate sustainability assessment method based on systems thinking to measure the links of manufacturing companies with the SDGs. The proposed approach proved to address the shortcomings of previous methods by using a set of sustainability criteria. This was in line with the perspective of van den Hoven (2019) [20], who emphasized the need for using a systemic and comprehensive engineering approach to couple technical issues with social and ethical aspects. Brooks (2020) [27] explored the link between systems engineering and the SDGs by adopting a design science research method to examine the elicitation of sustainability requirements in aerospace and healthcare facilities. The preliminary results achieved suggested that the application of the proposed method may lead to the enhanced coverage of the SDGs in most organizations.

The trend of these studies showed that previous research focused on developing theoretical frameworks to couple systems engineering with the monitoring of the SDGs. However, their use for cultivating symbiotic relationships in EIPs is still limited, especially under the consideration of how these potential processes of IS might contribute to achieving the SDGs. Therefore, this revealed a gap in what concerns the assessment of IS in industrial systems, whereby synergies are presented and analysed to assist public and private sector organizations in developing and implementing EIPs. This needs to be supported with an examination of the potential benefits that IS may have at meso (city and region) and macro (country) scales, including their evaluation using the SDGs as a benchmark. Consequently, this research concerned the identification of synergies in an industrial park, their analysis at different scales and the relationships of such synergies with the targets forming the SDGs, thereby providing an overview of the contributions of IS to sustainable development.

The rest of the article is structured as follows: Section 2 includes the methodology proposed to collect and process the data characterizing the main companies in the industrial park under study, as well as the framework used to value the contributions of synergies to be identified in terms of sustainable development. Section 3 starts by describing and illustrating the situation of the three main flows (materials, water and steam) in the industrial park. Then, potential direct and indirect synergies are overviewed according to their feasibility, such that a set of opportunities is selected and assessed based on their impacts of the three flows under consideration, The section ends with the estimation and discussion of the benefits to which these synergies may lead, emphasizing their associated economic and environmental savings. Finally, Section 4 contains the main findings of the study, highlighting both their implications for its research field and future lines of action to address its limitations.

2. Materials and Methods

The approach taken to conduct this research is depicted in Figure 1. It stemmed from a collaboration agreement between the city council to which the industrial park under study belongs and the university of the region. Therefore, the project involved the three main stakeholders required for promoting IS, namely public institutions, research centres and private companies, whose interests focused on welfare, knowledge and benefits, respectively. In this vein, the results to be obtained in this study were expected to work as the basis for developing future IS projects in the city. To this end, first was the acquisition of information from the public Integrated Environmental Authorisations (IEA) of the main companies (C1–C4) in the industrial park. Then, guided visits to their facilities and work sessions were arranged to corroborate these data and understand the processes leading to them. Finally, an analysis of potential IS opportunities in this industrial park was undertaken by taking into account both the contributions of the identified synergies to sustainable development and their scaling to regional and national levels.

2.1. Study Area and Industrial Park Description

The industrial park under study is located in the north of Spain and formed of more than 100 companies, most of which belong to a first subsystem of small and medium-sized enterprises (SMEs). Still, the four large companies forming the second subsystem are responsible for most of the material and energy resources in the area. Furthermore, since the park is an open industrial ecosystem, they can perform as a node to enable potential expansions involving other companies and/or industrial parks within the city or region. Consequently, this study focused on these dominant companies, since they might act as a catalyst in terms of IS due to their intense activity and turnover. Data for the characterization of these companies were acquired from their IEAs, which contain their global resource flows and can be publicly accessed on the official bulletin of their region. Nevertheless, in no case will details about the location and names of these companies be disclosed, since confidentiality has been maintained to prevent any leakage of sensitive information. Suffice it to state that the distance separating the companies is as follows: 2.6 km (C1–C2), 0.7 km (C1–C3), 2.7 km (C1–C4), 2.15 km (C2–C3), 4.0 km (C2–C4) and 2.3 km (C3–C4).

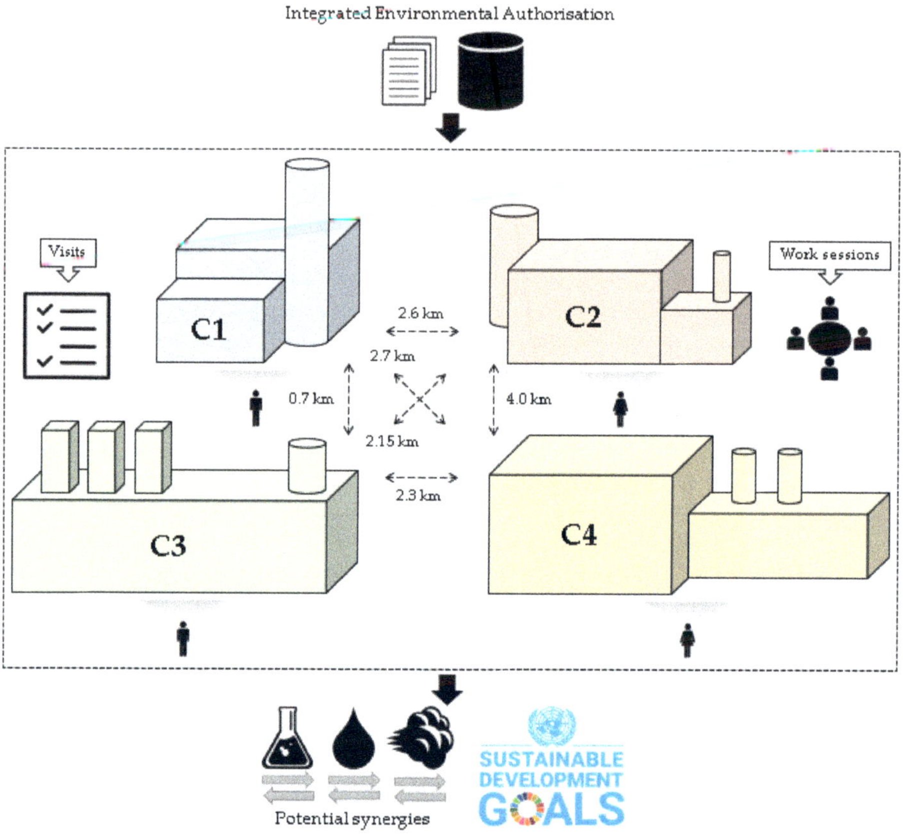

Figure 1. Graphical abstract of the steps taken to carry out a systemic analysis of Industrial Symbiosis (IS) synergies in an industrial park in the context of the Sustainable Development Goals (SDGs).

C1 was a company devoted to the conversion of plastic materials into flexible packaging and amounted to 490 employees on the date the study was conducted. With 466 workers, the area targeted by C2 was the manufacture of tires for private cars, trucks, tractor and industrial vehicles, pre-rolled treads and compound strips. C3 was the largest company, with 517 employees operating in facilities for the production, distribution and/or treatment of cellulose, inorganic chemical products (gas), electric energy, pressurized steam, viscose fibre, sodium sulphate organic chemical products (plastic materials and salts). Finally, C4 staff (463 people) were dedicated to several production processes with multiple applications, such as manufacturing of sodium carbonate, electrolysis, preparation of hydrogen peroxide solutions and cogeneration with thermal coal and cooling systems.

The input and output patterns observed in the companies forming the industrial park led to limiting the study to three flows: materials, water and steam. This particular trio of flows was selected because of their predominance in the activities carried out in the industrial park and their greater potential impacts with respect to others such as electricity, compressed air, nitrogen and fuels. Materials included raw materials, auxiliary materials and auxiliary production services, as well as the wastes derived from their production, such as hazardous and non-hazardous substances and emissions and discharges to air and water. Water included the consumption and waste flows required for its analysis as a resource, without taking into account temperature. Finally, steam concerned water thermal flows used in the companies as a process fluid or for power generation purposes.

2.2. Preparation of Questionnaires

The first step to undertake an inventory gathering information about the activity of the companies in the industrial park was the delimitation of the systems included in a company. According to Casals et al. (2008) [28], a company can be divided into five systems: management, operations, finance, marketing and human resources. Since the circulation of physical resources within the companies mainly concerns the operations system, this was the focus of the study hereinafter. This system refers to the set of activities oriented to the transformation of raw materials into products or the production of services demanded by the market, as well as to the auxiliary tasks required to carry out such processes. Material, water and steam flows circulate throughout several areas forming the operations system, such as production, auxiliary systems, general systems, waste treatment, storage and staff services.

With the support of these concepts, a questionnaire was designed to inventory data describing the different flows present in the industrial park. This questionnaire was structured according to three main parts: (1) an introduction showing the proposed workflow and a glossary of terms, (2) general information that must be provided by the companies regarding their organization and location maps and (3) specific data to be collected during the visits about the input, distribution and output flows in the companies. As proofs of the contents of these questionnaires, Supplementary Materials shows the data requested to compile information about the three flows analysed in this study: materials, water and steam. The questionnaires were filled out during short stays in the four companies forming the industrial park during 2011, in order to gather information about materials, water and steam flow inputs and outputs, as well as their distribution in the companies. The questionnaire was sent in advance to the technical coordinators of the companies to guarantee the effectiveness of the stays, which were divided into two main components: visits to facilities and office works sessions. The former were supported with explanations by technicians regarding the processes in their respective companies, whilst the latter consisted of putting together the information collected about the different flows, enabling the completion of the questionnaire.

2.3. Processing and Analysis of Data and Identification of Synergies

Once the stays were over, the data collected were processed and analysed according to the following sequence: inventory, flow schemes and floor plans. The inventory consisted of organising and simplifying the data collected during the stays independently for each company, and then combining them to represent the whole industrial park. Flow data were arranged to include quality specifications and details about the generation, distribution and treatment facilities, specifying their nominal and maximum capacities. Finally, the analysis was completed with the floor plans of the industrial park. Although this information was not strictly necessary for the identification of synergies, their consideration enabled the determination of the preliminary technical viability of some of them by locating consumption and waste generation spots.

The processing and analysis of data was the premise required for the identification of synergies. From a conceptual point of view, this research considered two types of synergies: substitution and mutuality. The former consists of replacing the consumption of input flows in a company by waste or output flows from another company, such that there is a reduction in resource needs and environmental impacts. Unlike this type of synergies in the form of exchanges, mutuality synergies refer to the shared use of services and/or infrastructures between different companies. In line with the most recent developments in the field of IS at the European level, the terminology used in the SCALER project [29] was adopted for this investigation. This project, which aimed at increasing the uptake of IS across

Europe through a systemic approach that considered all stakeholders, translated exchange and sharing practices into direct and indirect synergies, respectively. The criteria followed for the identification of materials, water and steam synergies in the industrial park are summarised in Table 1.

Table 1. Considerations for the identification of materials, water and steam synergies.

Flow	Synergy	Description	Parameters
Materials	Direct	Use of wastes as raw materials	Type and amount of materials and wastes
	Indirect	Shared supply	Type and amount of materials
		Shared storage	Type and amount of materials; Storage capacity
	Indirect	Joint management of waste	Type and amount of wastes; Current management
Water	Direct	Use of wastewater as raw water	Type and amount of raw water; Raw water uses; Water with similar quality; Type and amount of wastewater; Wastewater destination
	Indirect	Shared collection	Type and amount of raw water; Raw water uses; Water with similar quality
		Pre-treatment of water	Type and amount of water; Water uses; Ownership of facilities for water pre-treatment; Pre-treatment capacity
Steam	Direct	Use of non-recovered steam as consumption steam	Waste steam generation; Steam consumption; Physical characteristics of waste and consumption steam
	Indirect	Shared supply	Production and consumption
		Supply from one company to another	Production and consumption; Supply capacity

Due to their intrinsic characteristics, materials data could not be arranged in the form of a network using an origin-destination structure based on nodes and arrows. This was because these data consisted of separate lists of different materials and wastes, without sharing intermediate processes or treatments. Instead, water and steam uses were formed of a sequence of steps going from their collection to their disposal or recirculation. Consequently, their representation differed from that of materials, whose visualization was approached through bar charts detailing the variety of substances used in the industrial park.

Regarding water and steam, their graphical depiction was addressed with the support of Sankey diagrams, which display resource flows through directed arrows whose width is proportional to the quantity represented [30]. This concept was originally published by Riall Sankey at the end of the 19th century to analyze the efficiency of steam engines [31]. The need for improving material management in the steel production sector during the 1930s popularized the application of balances with the support of Sankey diagrams [32]. Since then, they have become particularly important in industrial ecology, due to their capacity to depict all relevant flows and their interdependences [33]. This trend provides evidence of the suitability of Sankey diagrams to represent both water and steam flows.

2.4. Industrial Symbiosis and Sustainable Development Goals (SDGs)

The broad implications of IS, whose consideration helps to promote economic growth and the safeguarding of the environment, have great potential to unlock mechanisms aimed at fostering sustainable development [3]. These potential contributions can be measured using the Sustainable Development Goals (SDGs) as a benchmark, which are 17 objectives that were set in the United Nations Conference on Sustainable Development held in Rio de Janeiro in 2012 for the period 2015–2030 [34]. In fact, the UN International Resource Panel highlighted the ripple effect of IS on sustainability, provided by its linked benefits across different SDGs [1].

Schroeder et al. (2019) [7] identified the extent to which circular economy practices may be relevant for achieving the SDGs. Overall, they found that a circular economy can contribute with greater or lesser intensity to all SDGs, either directly or indirectly. The strongest relationships were found to correspond to the targets of SDGs 6 (Clean Water and Sanitation), 7 (Affordable and Clean Energy), 8 (Decent Work and Economic Growth), 12 (Sustainable Consumption and Production) and

15 (Life on Land). These inferences provided an overview of the association between circular economy and the SDGs; however, further analysis is required to explore the relationships between sustainability and particular branches of circular economy, such as IS.

Hence, the synergies identified in the industrial park under study through the proposed methodology (Figure 1) were evaluated in terms of their potential contributions to achieving the 169 targets included in the SDGs. To the extent possible, this analysis was supported with coarse quantifications of the economic and environmental benefits derived from the implementation of the IS practices identified. In turn, this course of action was intended to enable making estimates about the scale effects of IS at regional and national levels.

3. Results and Discussion

In what concerns material flows, IS was aimed at (1) reducing the consumption of materials and the generation of wastes, (2) minimizing and centralizing the storage of input materials and wastes to increase space availability and (3) sharing supply services to save costs due to economies of scale, thereby improving the logistics in the industrial park. Overall, the consumption of virgin materials was distributed as follows: raw materials (2,906,119 t/yr., 96%), auxiliary materials (87,061 t/yr., 3%) and auxiliary production services (34,676 t/yr., 1%). Regarding the destination and type of wastes generated, 242,648 t/yr. (99.6%) were found to be non-hazardous, which means that only 1006 t/yr. (0.4%) were hazardous.

The breakdown of materials and wastes according to their types is depicted in Figure 2. C4 was the greatest contributor to the consumption of raw materials, since it was responsible for brine and limestone, which amounted to a yearly consumption of 1,576,974 and 930,646 t. Similarly, C3 was the dominant company in the consumption of auxiliary materials and auxiliary production services, with 17,280, 15,515 and 2191.6 t/yr. of sulphuric acid, sodium hydroxide and oxygen, respectively. Regarding hazardous wastes, sludge from ink distillation processes (253.8 t/yr.) and Industrial Wastewater Treatment Plants (IWTP) (179.3 t/yr.) were the main sources of generation. The fact that calcined calcium carbonate (205,684 t/yr.) and brine (110,000 t/yr.) were the main flows in terms of non-hazardous wastes and emissions and discharges supported the feasibility of some potential synergies, since both products were consumed as materials in the industrial park (Figure 2).

Apart from generic reduction and minimization considerations as mentioned for the materials, the analysis of water flows also required taking into account the following Spanish and European standards concerning the Hydraulic Public Domain: Royal Decree-Law 995/2000 (Quality objectives for certain polluting substances), Royal Decree-Law 60/2011 (Environmental quality standards in the field of water policy), Directive 91/271/EEC (Urban wastewater treatment) and Royal Decree-Law 1620/2007 (Legal framework for the reuse of purified water).

Overall, more than 94% of water (3739 m^3/h out of 3966 m^3/h) was captured from a river. Assuming an annual continuous work regime (8760 h) and according to the values represented in Figure 3, this volume amounted to 32,753,640 m^3/yr. Therefore, reducing the consumption of this source was a priority. From less to more quality, the types of water used by the companies are raw, filtered, clarified, decarbonated and demineralized (equivalent to osmotized). Since the most widely consumed type of water by all the companies in the industrial park was clarified (2746 m^3/h, 41% of the flow processed through either coarse filtering, osmosis, clarification, demineralization, decarbonation or chlorination), this was a flow of interest for the reduction in the amount of water used in the park. A small part of the flows derived from these treatment nodes was straightforwardly connected to output nodes, since it was associated with purges.

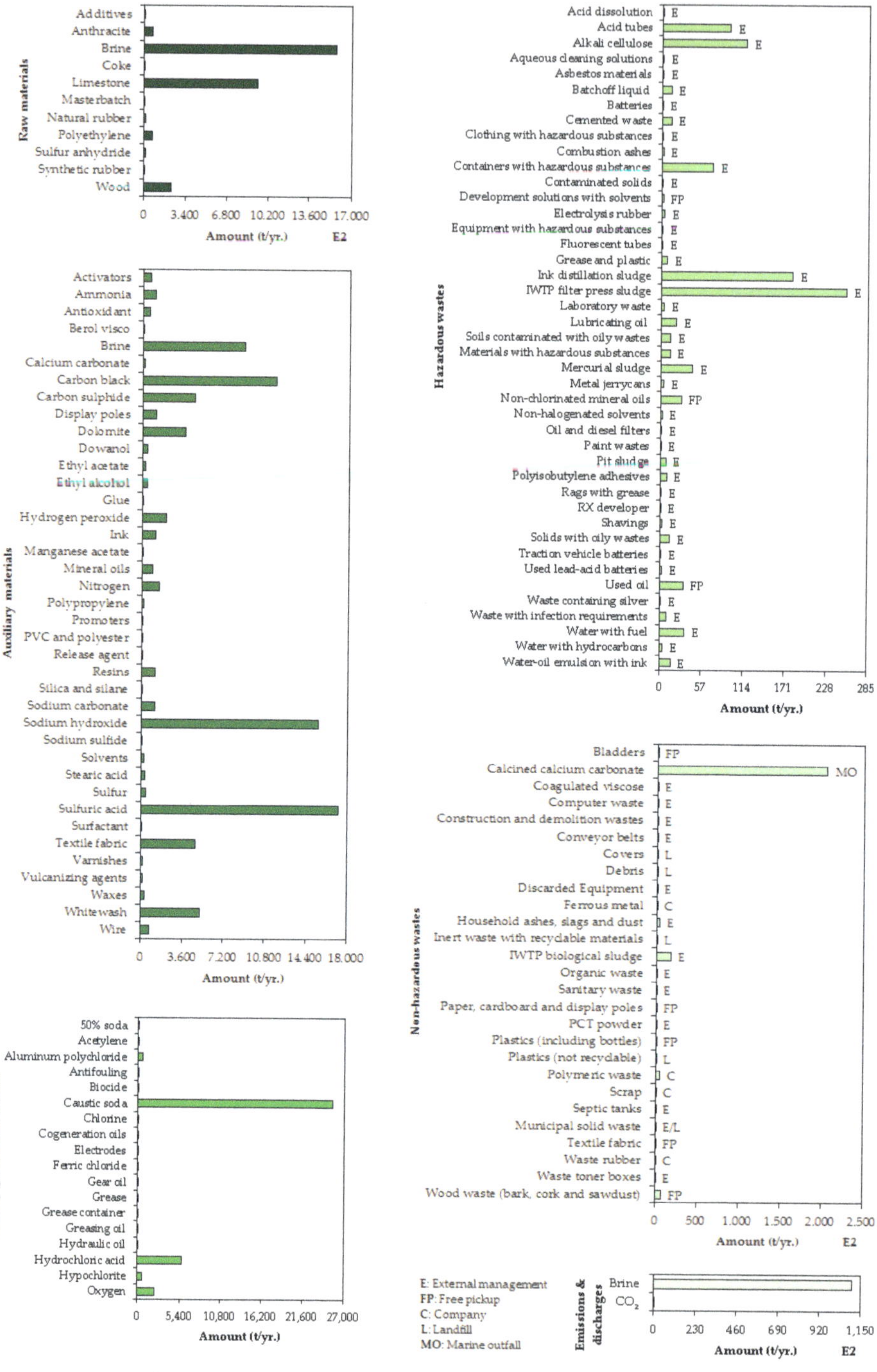

Figure 2. Breakdown of materials and wastes in the industrial park.

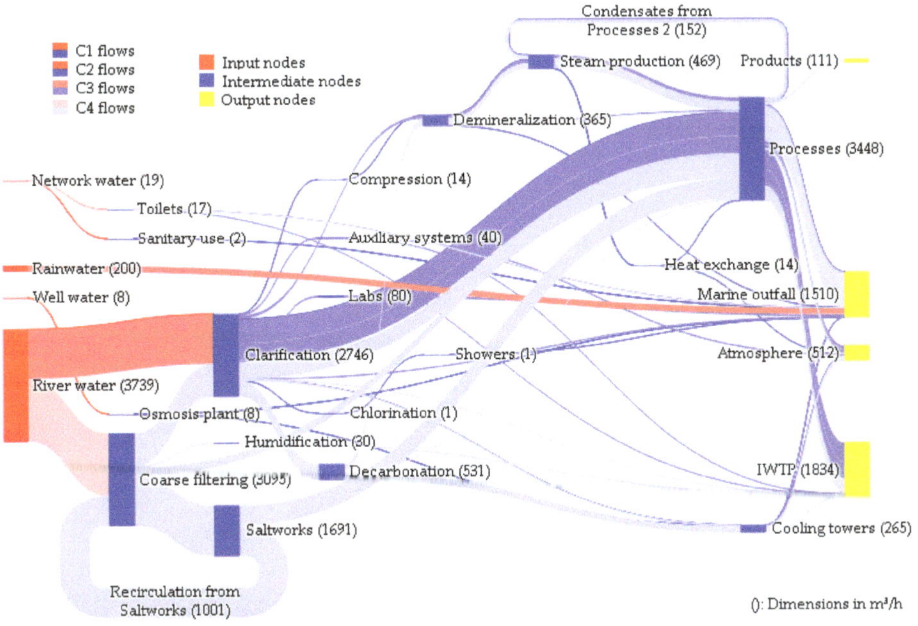

Figure 3. Sankey diagram of water flows in the industrial park.

The total flow consumed in intermediate nodes amounted to 12,811 m^3/h. In particular, three groups represented the main uses given to water: processes, steam production and cooling towers. These groups were taken as a criterion to determine whether any of their associated uses might admit lower quality water without compromising the performance of the companies. The highest consumption within this group corresponded to processes (3448 m^3/h, 82%), followed by steam production (469 m^3/h, 11%) and cooling towers (265 m^3/h, 7%).

Water destination plays an important role in this systemic analysis, since this concept is related to the minimization of discharge rates and potential reuse of water according to the implementing legislation. Most of the water used in the park was sent to an IWTP (1834 m^3/h, 46%), a marine outfall (1510 m^3/h, 38%) and the atmosphere (512 m^3/h, 13%). Instead, only 3% ended up being used as products (111 m^3/h).

The optimization of steam consumption in the industrial park aimed at achieving a higher efficiency in the companies and minimizing the presence of unnecessary infrastructures, while centralizing steam flows and achieving continuous operating regimes. The fulfilment of these targets was conditioned by the potential use of the steam produced in the park for generating electricity to be either sold externally or consumed in internal processes. Therefore, the approach taken to address the steam flow considered its link with electricity. The quality demands for the production of steam required demineralized water, such that 317 m^3/h out of the 365 m^3/h flow depicted in Figure 3 performed as an input (317 t/h) in the Sankey diagram represented in Figure 4.

The joint analysis of the four companies revealed that the steam and hot water generated in the industrial park amounted to 469 t/h (carbon boiling, gas boiling and gas cogeneration), which were produced in the range from 1.2 to 140 kg/cm^2. Most of this steam (420 t/h) was produced at high pressure (58–140 kg/cm^2) and then was turbined (Figure 4) for the generation of electric energy. The low and medium-pressure steam (5.5–33 kg/cm^2) derived from turbines, along with the remains of the production nodes (49 t/h) at 1.2–16.5 kg/cm^2, were used in different processes: Processes 1 involved input flows at medium pressure (9–33 kg/cm^2), such as steam recovery or drying, whilst Processes 2 (e.g., distillation) corresponded to low-pressure processes (1.2–5.5 kg/cm^2). Heat exchange was left apart as a single process to provide more details about C2, since the steam scheme of this company would be too simplified otherwise. Overall, the steam consumption in the industrial park was balanced in terms of pressure: 344 t/h were consumed at medium pressure (Processes 1 + Heat exchange) and 303 t/h at low pressure (Processes 2). In the end, the steam consumed in the processes was either recirculated as condensates (152 t/h) or converted into 21 t/h of non-recovered (atmosphere) or 296 t/h of condensate steam (marine outfall and IWTP), amounting to the 317 t/h that originally entered the system.

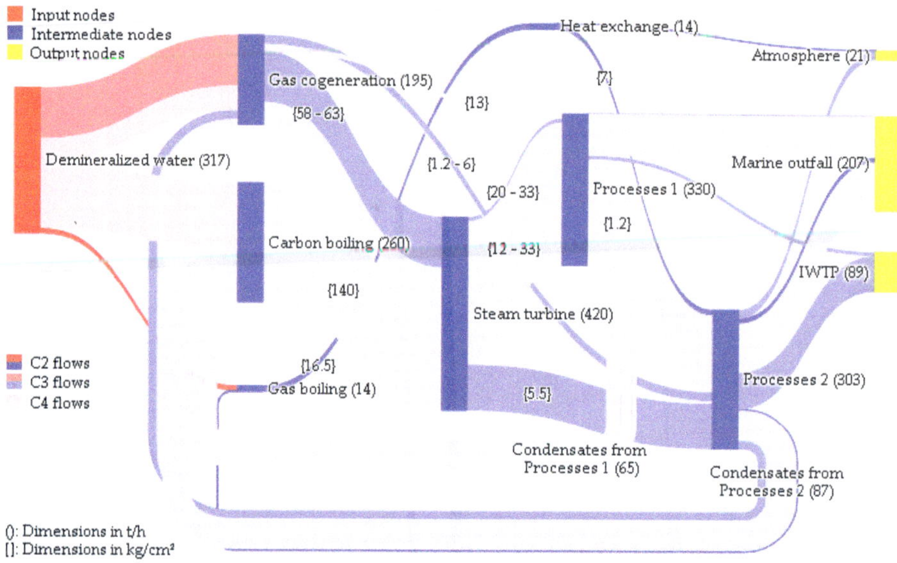

Figure 4. Sankey diagram of steam flows in the industrial park.

To quantify all these stages, Table 2 collects the path followed by steam in the companies that consumed it (C1's was a dry process), including those nodes involving production, pressure drops and consumption. Carbon was used only by C4, which also produced steam through cogeneration. This system was also used by C3, whilst C2 resorted to a gas boiler to generate steam. Table 2 specifies the amount of steam generated by each company using these systems, as well as their corresponding values of pressure. Once produced, high-pressure steam (>35 kg/cm^2) derived from carbon boiling and cogeneration in C3 and C4 was turbined in steam turbines at a medium (8–35 kg/cm^2) and low pressure (< 8 kg/cm^2), respectively (Table 2). Similarly, the flow generated in C2 through gas boiling was laminated at a medium and low pressure through heat contact and exchange, as indicated in Table 2. To support the consideration of potential synergies based on this scheme, Table 3 summarizes the situation of the companies involved in the steam flow in what concerns their use, capacity, needs and margin.

In addition, Table 2 also indicates the role played by steam at the scale of the whole system, specifying the electric production and consumption patterns of the companies. In this sense, the sole purpose sought by C2 with respect to its steam production was its subsequent consumption in processes associated with this flow. Thus, all the electricity consumed by this company (30,600 MWh/yr.) was supplied by the national electrical grid. Instead, C3 devoted part of its steam production to electric energy. This company sold the energy produced in its cogeneration system (719,300 MWh/yr.) to the national electrical grid, whilst the portion of high-pressure steam produced through steam turbines (81,600 MWh/yr.) was used for the generation of electric energy for self-consumption. The origin of C4 electricity stemmed from turbined steam derived from carbon boiling (208,139 MWh/yr.) and the national electrical grid (125,308 MWh/yr.).

Most of this electricity ended up being consumed in the processes (268,718 MWh/yr.), while a smaller part was returned to the national electrical grid (64,729 MWh/yr.).

Table 2. Scheme of the path followed by steam in the companies forming the industrial park.

ID	Flow	Production		Pressure drop 1		Pressure drop 2		Consumption	
		System	Amount *	Node	Amount *	Node	Amount *	System	Amount
C2	Steam	Gas boiling	14 {16.5}	Heat exchange	14 {7}	–	–	Processes 2	13
	Electricity	National electrical grid	30,600	–	–	–	–	Processes	30,600
C3	Steam	Gas cogeneration	110 {63}	Steam turbine	6 {33}	–	–	Processes 2	130
					4 {20}	–	–		
					100 {5.5}	–	–		
			20 {6}	–	–	–	–		
	Electricity	Gas cogeneration	719,300	–	–	–	–	National electrical grid	719,300
				Steam turbine **	81,600	–	–	Processes	81,600
C4	Steam	Gas cogeneration	50 {58}	Steam turbine	50 {13}	–	–	Processes 2	50
			10 {13}	Processes 1	10 {1.2}	–	–	Processes 2	10
			5 {1.2}	–	–	–	–	Processes 2	5
		Carbon boiling	260 {140}	Steam turbine	50 {33}	–	–	Processes 2	50
					210 {13}	Processes 1	155 {1.2}	Processes 2	155
								Processes 2	55
	Electricity	Gas cogeneration	NA					Private company	NA
		Carbon boiling		Steam turbine ***	208,139			Processes	268,718
		National electrical grid	125,308					National electrical grid	64,729

* Values in t/h (pressure in kg/cm^2); ** From gas cogeneration; *** From carbon boiling.

Table 3. Situation of steam flow use, capacity, needs and margin in the companies forming the industrial park.

ID	System	Status	Pressure (kg/cm^2)	Use (t/h)	Nominal Capacity (t/h)	Needs (%)	Available Margin (%)	Available Margin (t/h)
C2	Global	Installed	–	14	65	21	79	51
	Production	Active *	16.5	14	40	35	65	26
		Backup **	16.5	0	25	0	100	25
	Pressure drops	Active ***	7	14	40	35	65	26
C3	Global	Installed	–	130	295	44	56	165
	Production	Active *	63	110	125	88	12	15
			6	20	35	57	43	15
		Backup **	63	0	150	0	100	150
	Pressure drops	Active ***	33	6	21	28	72	15
			20	4	19	21	79	15
			5.5	100	115	87	13	15
C4	Global	Installed	-	325	530	61	39	205
	Production	Active*	140	260	270	96	4	10
			58	50	60	83	17	10
			13	10	20	50	50	10
			1.2	5	15	33	67	10
		Backup **	58	0	185	0	100	185
	Pressure drops	Active ***	33	50	60	83	17	10
			13	270	290	93	7	20
			1.2	170	190	89	11	20
		Backup ****	13	0	185	0	100	185

* C2: Gas boiling; C3: Gas cogeneration; C4: Gas cogeneration (140 kg/cm^2) and Carbon boiling (58/13/1.2 kg/cm^2); ** C2: Backup gas boiling; C3: Backup carbon boiling; C4: Backup gas boiling *** C2: Heat exchange; C3: Steam turbine; C4: Steam turbine (33/13 kg/cm^2) and Processes 1 (1.2 kg/cm^2); **** C4: Steam turbine.

3.1. Direct Synergies

Table 4 summarises the purpose, participating companies and characteristics (Ch.) of the proposed direct synergies, as well as their identification through a three character code indicating type (S—Substitution), resource (M—Materials; W—Water; S—Steam) and number. The amount of limestone waste (calcined $CaCO_3$) generated by C4 is currently poured into the sea through marine outfall, since it is not harmful for the environment as long as long the spillage is restricted. This was a priority due to its great proportion in the wastes generated by C4 (99%). C2 consumes this material as an additive in the production of tyres, whilst C3 uses dolomite that can be partially replaced by $CaCO_3$ in the production of cellulose (SM1). According to the BAT (Best Available Techniques) documents [35], about 70% of dolomite might be substituted by $CaCO_3$ (2650 t/yr.). Overall, the residual mass flow of $CaCO_3$ produced by C4 (205,684 t/yr.) would completely meet the demands of C2 and C3 (2826.4 t/yr.).

Table 4. Summary of direct synergies identified in the industrial park.

ID	Purpose	Donor				Recipient			
		Company	Type	Ch. 1 *	Ch. 2 **	Company	Type	Ch. 1 ***	Ch. 2 ****
SM1	$CaCO_3$ exploitation	C4	Calcined $CaCO_3$	205,684	Suspended	C2	$CaCO_3$	176.4	Powder
						C3	Dolomite	2650	Solid
SM2	CO_2 distribution	C1	CO_2 (VOCs combustion)	435	Adequate quality	C4	CO_2 (limestone & anthracite calcination)	666,000	95% pure
		C2	CO_2 (VOCs combustion)	NA	Adequate quality				
SM3	Oil exploitation	C2	Oil	20.7	Used as a lubricant	C1	Reused oil	NA *****	Impurities are allowed
SM4	Brine exploitation	C4	Used brine	110,000	NA	C3	Brine	8965	Low salt. Mercury
SW1	Purified water reuse	C3	IWTP (purified water)	1230	Suspended solids	C4	River (filtered water)	993	<35 mg/L suspended solids
SW2	Purified water reuse	C3	IWTP (purified water)	1230	Suspended solids	C1	Well (raw water)	8	<5 mg/L suspended solids
SW3	Rainwater reuse	C3	Rainwater	200	Suspended solids	C4	River (filtered water)	993	<35 mg/L suspended solids
SW4	Purified water reuse	C4	IWTP (purified water)	604	Mercury (5×10^{-5} ppm)	C3	River (raw water)	NA	$<5 \times 10^{-5}$ ppm mercury
SS1	Unrecovered steam reuse	C3	Waste steam	NA	NA	C1	Hot steam	150	95 °C

* Amount in t/yr. (SM1–SM4)/Amount in m^3/h (SW1–SS1); ** Qualitative feature (SM1–SM4, SS1)/Outflow condition (SW1–SW4); *** Amount in t/yr. (SM1–SM4)/Amount in m^3/h (SW1–SW4, SS1); **** Qualitative feature (SM1–SM4); Limit values for water reuse (SW1–SW4); Temperature in °C (SS1); ***** NA: not specified.

Both C2 and C1 generate controlled CO_2 emissions through Regenerative Thermal Oxidizers (RTO) for Volatile Organic Compounds (VOCs) that may fit the interests of C4 (SM2). Along with calcium oxide (CaO), CO_2 was the main compound of $CaCO_3$ calcination, which is a reactant involved in C4's main chemical process for manufacturing sodium carbonate. However, C4 was found to have no CO_2 deficit, to the extent that part of its CO_2 exceeded the amount necessary for the reaction and was released to the atmosphere. Hence, a complementary option may consist of creating some infrastructure for capturing the CO_2 emissions produced by all these companies.

In a similar vein, C1 consumes oil previously used in its own combustion engines for subsequent reuse in the removal of polyethylene pigments. Consequently, this company has no deficit either, thereby not requiring the oil produced by C2 (SM3). C4 generates brine waste from electrolysis processes with an initial concentration of 250 g/kg of NaCl. Then, this flow is purified in the IWTP to reduce the amount of mercury to the allowed limits. Although C3 consumes brine in its production processes, the characteristics of this waste should be analysed more in detail to guarantee the compatibility of this synergy (SM4).

Two potential synergies (SW1 and SW2) were identified to reuse water from C3 IWTP in C4 and C1. The amount of suspended solids in the IWTP would not be a problem, since it is within the limits to reuse water for industrial purposes. Therefore, it might be used by C4, since most of its consumption is filtered water from a river, whose quality requirements meet those of the IWTP outflow. On the contrary, the amount of suspended solids in C3 water exceeds that of the raw water collected from the well by C1. Although C1 facilities have the capacity for reducing dissolved solids, chlorine and organic matter, this treatment would produce new wastes that were not previously generated by this company.

The collection of rainwater in C3 is discharged into the river. Thus, its reuse by C4 as filtered or raw water (SW3) would require some treatment to ensure it meets current standards of quality [36]. Furthermore, since most of the water consumed by C4 is filtered, rainwater could be recirculated and reused to moisten ashes in generators and refrigerate the electrolysis unit.

Some of the processes carried out by C3 do not require high quality water, such as wood barking. Hence, wastewater from C4 might be used as a replacement in these situations (SW4), since its mercury contents meet the standards of environmental quality [37]. However, C3 water supply is undertaken through a pressurized circuit, whose flexibility should be studied to evaluate more in depth the suitability of this synergy. Additionally, it is also necessary to detail the use and conditions of water quality in C3 to assess the adequacy of this substitution.

The steam-related direct synergy (SS1) involved reusing non-recovered steam from C3 in C1. Hot water consumption in C1 was carried out at 95 °C; however, there was no certainty about whether this temperature may be reached by some non-recovered steam flow from C3. Hence, although this substitution might be potentially feasible, the lack of data regarding the flow and characteristics of non-recovered steam in C3 precluded its implementation (Table 4).

3.2. Indirect Synergies

Table 5 is analogous to Table 4 but applies to the indirect synergies identified, such that the first character in their codification denotes mutuality (M). The analysis of auxiliary material consumption in the industrial park revealed that some companies shared input flows. In particular, C3 and C4 were found to consume brine in their respective production processes. Hence, improvements in economies of scale might be obtained in case both companies had the same brine supplier (MM1). Since C4 is self-dependent in the extraction of brine from salt surveys, it may directly supply C3, whose stock stems from a company belonging to the same corporate group.

In the same vein, there were a list of synergies based on sharing the same supplier of auxiliary production services (MM2). These implementations would not only yield economic improvements, but also environmental benefits in distribution logistics. The role played by C4 was particularly relevant in this synergy, since it owns the facilities required to provide C1 with hypochlorite and C1–C2–C3 with hydrochloric acid, chlorine (own internal generation) and caustic soda. Data availability hindered the proposal of specific actions for other cases involving aluminum polychloride, oxygen and hydraulic oils, which were limited in providing details about suppliers and/or amounts.

The common storage of these same auxiliary production services (MM3) would result in the optimization of logistics, since the supply of materials would be limited to only one location, and an increase in room availability. C4 proved to be capable of storing the whole amount of hypochlorite, hydrochloric acid and hydraulics oils in the industrial park. C2 was found to have space for storing hydraulic oils too, as well as oxygen. Caustic soda was storable by C3 and C4, whilst data scarcity precluded determining the situation of aluminium polychloride.

MM4 and MM5 sought to produce economic and environmental gains in the industrial park by jointly managing both hazardous and non-hazardous wastes. C1 was the company producing a wider variety of shared wastes (8 hazardous and 6 non-hazardous), followed by C3 (4 hazardous and 6 non-hazardous) and both C2 (6 hazardous and 3 non-hazardous) and C4 (9 hazardous). C4 was the greater contributor to generating hazardous wastes with the potential for cooperation (MM4), especially with regards to the amount produced (82.3 t/yr.), followed by C1 (67.8 t/yr.), C3 (38.7 t/yr.) and C2 (37.7 t/yr.). Used oils (81.1 t/yr.) and plastic and metal containers (73.2 t/yr.) were the most shared hazardous waste in the industrial park. The same external company was in charge of the free pickup of used oils for C1, C2 and C4; instead, this information was unspecified for C3. The remaining 8 shared hazardous wastes (Table 5) were handled by different external managers (unknown in some cases). Therefore, an improvement may consist of homogenizing the management of these wastes and reducing the number of external companies involved. However, this option should be considered carefully, since it may eventually lead to a monopolistic situation characterized by rising collection prices.

Table 5. Summary of indirect synergies identified in the industrial park.

ID	Purpose	Type	Characteristics	C1	C2	C3	C4
MM1	Shared supply of auxiliary materials	Virgin brine	Amount (t/yr.)	–	–	8965	1,579,974
MM2	Shared supply of auxiliary production services	Hypochlorite	Amount (t/yr.)	3.4	–	–	589
		Aluminum polychloride	Amount (t/yr.)	–	NA *	510	–
		Hydrochloric acid	Amount (t/yr.)	–	180	637	1850
		Chlorine	Amount (t/yr.)	0.5	3.8	3	Internal
		Caustic soda	Amount (t/yr.)	1.2	25	15,515	25,552
		Oxygen	Amount (t/yr.)	1	3.6	3.2	External
		Hydraulic oils	Amount (t/yr.)	4	6.6	NA	26
MM3	Shared storage of auxiliary production services	Hypochlorite	Storage capacity	N **	–	–	Y
		Aluminum polychloride	Storage capacity	–	N	NA	–
		Hydrochloric acid	Storage capacity	–	N	NA	Y
		Chlorine	Storage capacity	N	Y	N	–
		Caustic soda	Storage capacity	N	N	Y	Y
		Oxygen	Storage capacity	N	Y	NA	–
		Hydraulic oils	Storage capacity	N	Y	Y	Y
MM4	Joint management of hazardous waste	Water with hydrocarbons	Amount (t/yr.)	0.0 (E) ***	–	–	33.5 (E)
		Metal/plastic containers	Amount (t/yr.)	60.2 (E)	–	9.5 (E)	3.5 (E)
		Solids with oily waste	Amount (t/yr.)	–	7.8 (E)	0.6 (E)	5.1 (E)
		Oil filters	Amount (t/yr.)	0.1 (E)	0.2 (E)	–	13.2 (E)
		Equipment	Amount (t/yr.)	0.5 (E)	–	0.6 (E)	1.6 (E)
		Non-halogenated solvents	Amount (t/yr.)	0.4 (E)	0.9 (E)	–	0.1 (E)
		Batteries	Amount (t/yr.)	0.4 (E)	2 (E)	–	0.1 (E)
		Fluorescent tubes	Amount (t/yr.)	0.2 (E)	0.6 (E)	–	0.5 (E)
		Used oils	Amount (t/yr.)	2.2 (E)	26.2 (E/FP)	28 (FP)	24.7 (FP)
MM5	Joint management of non-hazardous waste	Solid urban waste	Amount (t/yr.)	8.2 (L)	–	111.3 (E)	–
		Debris	Amount (t/yr.)	97.8 (L)	–	249 (E)	–
		Wood waste	Amount (t/yr.)	86.1 (FP)	–	7445 (FP)	–
		Paper and paperboard	Amount (t/yr.)	227.1 (FP)	42.5 (FP)	22.4 (FP)	–
		Scrap	Amount (t/yr.)	85.6 (L)	249.6 (C)	324.7 (E)	–
		Plastics	Amount (t/yr.)	5046 (C)	126.6 (FP)	60.7 (FP)	–
MW1	Pre-treatment of demineralized water	Demineralized water	Amount (m^3/h)	–	13.75	110	241
			Nominal capacity (m^3/h)	–	30	280	480
			Available margin (m^3/h)	–	16.25	170	69 ****
MW2	Pre-treatment of clarified water	Clarified water	Amount (m^3/h)	–	17.3	1628	1101
			Nominal capacity (m^3/h)	–	120	7000	1500
			Available margin (m^3/h)	–	102.7	5395	399
MW3	Joint collection of river water	River water	Amount (m^3/h)	8.3	17.3	1628	2094
			Nominal capacity (m^3/h)	10	120	7000	7000
			Available margin (m^3/h)	1.7	102.7	5018	4906
MS1	Steam supply from C3 to C2	C3 supply options to meet C2 needs	Amount in t/yr. {Pressure in kg/cm^2}}	–	14 {7}	15 {20}	–
MS2	Steam supply from C4 to C2	C4 supply options to meet C2 needs	Amount in t/yr. {Pressure in kg/cm^2}	–	14 {7}	–	205 {13}
MS3	Steam supply from C3 to C4	C3 supply options to meet C4 needs	Amount in t/yr. {Pressure in kg/cm^2}	–	–	15 {33/20} 15 {6/5.5}	270 {13} 170 {1.2}
MS4	Steam supply from C4 to C3	C4 supply options to meet C3 needs	Amount in t/yr. {Pressure in kg/cm^2}	–	–	110 {63} 20 {6}	195 {58} 205 {13}

* NA: not specified; ** N: there is not enough capacity to store the flow of auxiliary production services/Y: there is enough capacity to store the flow of auxiliary production services; *** E: external management/FP: free pickup/C: sale to company/L: landfill; **** restricted capacity due to the decarbonation step previous to water demineralization.

The management of non-hazardous wastes (MM5) was even more difficult to characterize due to the lack of detailed data. C3 was the main producer of non-hazardous wastes suitable for synergies, generating 8213 t/yr. C1 and C2 were responsible for 5524 and 419 t/yr., respectively. Although C4 produced 205,715.9 t/yr. (mostly calcined calcium carbonate), none of its residues were suitable for sharing. In those cases in which the same approach was used to manage waste (wood, scrap and paper and paperboard), there was at least one company that did not provide any details about the destination of either their companies or landfills. Potential alternatives to better handle hazardous wastes might focus on the coordination of their management methods, such that transportation costs and emissions may be reduced. These actions should be especially oriented to wood waste (7531 t/yr.)

and plastics (5233 t/yr.), which were the shared non-hazardous waste generated in a greater quantity in the industrial park.

Two of the three water-related synergies consisted of the pretreatment of either demineralized (MW1) or clarified water (MW2). The use of demineralized water was common for C2, C3 and C4. Both C3 and C4 had the capacity for supplying the needs of C2 (13.75 m^3/h). Water in C4 went through previous treatments before demineralization, whereby the amount available was reduced to 69 m^3/h, which would only cover 55.25 of the 110 m^3/h required by C3 (50.23%). Instead, this figure increased up to 64.73% (156 of 241 m^3/h) when the supply was from C3 to C4. Similarly, C3 was found to be capable of supplying C2 (17.3 m^3/h) and C4 (1101 m^3/h) in terms of clarified water (MW2), due to its 77% available margin (5390 m^3/h). Finally, the last water synergy (MW3) concerned the collection of river water from the same point. In this case, three companies (C2–C4) collected river water, whilst C1 used water of similar quality from a well. Therefore, this synergy would involve C1 replacing well water by river water from a common collection point, in order to obtain economic savings in terms of collection fees.

The remaining indirect synergies consisted of the supply of high and medium pressure steam from one company to another. To enable these types of synergies, the companies involved must meet certain conditions. First, recipient companies should be inefficient in terms of steam use with respect to their nominal capacity; otherwise, the supply of steam from another company might not be beneficial at the scale of the whole industrial park. Additionally, recipient companies should have steam needs that were replaceable without compromising its linkage to the generation of electric energy. From the point of view of suppliers, they must have the available margin to provide recipients with steam. In addition, the levels of pressure at which such margin was available should not involve large leaps with respect to the uses of recipient companies, in order to make optimal use of steam.

The first opportunity identified concerned the supply from either C3 or C4 to C2, since the latter was found to be very inefficient in using steam. It only used 21% of its nominal capacity (Table 3) and operated under a discontinuous working regime with weekly breaks, whereby a synergy for the steam supply to C2 might increase the productivity in the industrial park. C2 was also suitable in terms of electricity, since its production was null in this sense (Table 2) and a potential steam-related synergy would not interfere in its energy flow. C2 needs included the production of 14 t/h at 16.5 kg/cm^2 to be used for heat exchange at 7 kg/cm^2, as represented in Tables 5 and 6.

MS1 involved C3 as a supplier, which had 56% of the margin with respect to its nominal capacity to supply C2. Due to the values of pressure at which flow was available in C3 (Tables 5 and 6), the first option would consist of supplying 14 t/h at 63 kg/cm^2 to be used at 7 kg/cm^2; however, this was discarded too due to the large pressure difference it involved. Instead, C3 may shorten that leap by supplying the 14 t/h required by C2 at 20 kg/cm^2, once the steam produced through cogeneration was turbined (Table 3).

In the case of MS2, the options were rather similar. The margin available in C4 was 39% of its nominal capacity (Table 3). Again, the pressure drops caused by turbines made available 10 t/h at 33 kg/cm^2 and 205 t/h at 13 kg/cm^2 that would be sufficient to cover C2's requirements. Thus, C2 may use 14 t/h at 13 kg/cm^2 to replace C2 uses at 7 kg/cm^2.

The two remaining synergies implied the supply from C3 to C4 (MS3), and vice versa (MS4). Again, the aim of these synergies was to improve the efficiency of the suppliers and reduce the equipment used by the recipient companies. In this sense, both companies proved to be more efficient than C2. C4 was found to operate at 61% of their nominal capacity, whilst this figure decreased to 44% in the case of C3 (Table 3).

The applicability of MS3 would be limited to supply part of the uses derived from C4 carbon boiling, since the flow generated by C4 through cogeneration was linked to the sale of electricity to a private company (Table 2). Since the steam produced in C4 through carbon boiling was at a much higher pressure (140 kg/cm^2) than any flow available in C3 (Table 3), the feasible options associated with MS3 were restricted to subsequent lamination steps (Table 5). Hence, C3 may substitute part of

C4 uses at 33, 13 and 1.2 kg/cm^2 through the supply of flow at 33, 20 and 6/5.5 kg/cm^2, respectively, as collected in Table 3.

Table 6. Sustainable Development Goals (SDGs) and targets to which the selected Industrial Symbiosis (IS) synergies can contribute.

SDG	Target
3. Good health and well-being	3.9. By 2030, [...] reduce [...] deaths and illnesses from hazardous chemicals and air, water and soil pollution [...]
6. Clean water and sanitation	6.3. By 2030, improve water quality by reducing pollution, eliminating dumping and minimizing [...] hazardous chemicals and materials [...]
	6.4. By 2030, [...] increase water-use efficiency across all sectors and ensure sustainable withdrawals and supply of freshwater [...]
7. Affordable and clean energy	7.3. By 2030, double the global rate of improvement in energy efficiency
8. Decent work and economic growth	8.4. Improve [...] resource efficiency in consumption and production and [...] decouple economic growth from environmental degradation [...]
9. Industry, innovation and infrastructure	9.4. By 2030, [...] retrofit industries to make them sustainable, with increased resource-use efficiency and [...] clean [...] technologies [...]
11. Sustainable cities and communities	11.6. By 2030, reduce the [...] environmental impact of cities [...] by paying special attention to air quality and [...] waste management
12. Responsible consumption and production	12.2. By 2030, achieve the sustainable management and efficient use of natural resources
	12.4. By 2020, achieve the environmentally sound management of chemicals and all wastes [...] and [...] reduce their release to air, water and soil [...]
	12.5. By 2030, substantially reduce waste generation through prevention, reduction, recycling and reuse
	12.6. Encourage companies, especially large and transnational companies, to adopt sustainable practices and to integrate sustainability information [...]
14. Life below water	14.1. By 2025, prevent and significantly reduce marine pollution of all kinds, in particular from land-based activities [...]
17. Partnerships for the goals	17.13. Enhance global macroeconomic stability, including through policy coordination and policy coherence
	17.14. Enhance policy coherence for sustainable development

The viability of MS4 (from C4 to C3) was limited too, in this case by the double role played by the C3 turbine: it not only served to laminate steam, but also to generate the 81,600 MWh/yr. that was consumed by this company in its own processes (Table 2). Taking into account this link with the electricity flow, the only feasible option in synergy MS4 was the supply of steam from C4 at 58 and 13 kg/cm^2 to replace the whole flows produced in C3 through gas cogeneration, which amounted to 110 t/h at 63 kg/cm^2 and 20 t/h at 6 kg/cm^2, respectively (Table 2).

3.3. Implementation of Selected Synergies

Figure 5 summarises the synergies selected for subsequent analyses, which were shortlisted because of their feasibility and potential benefits for the industrial park as a whole. Most of these synergies involved C3 and C4, which was logical due to the larger size and capacity of these two companies. Still, although some of the selected synergies might be feasible based on the information collected throughout the study, the existence of data for their numeric characterization was limited due to their especially indirect nature. This was the case of MM1, MM2, MM3, MM4, MM5 and MW3.

In the case of materials, both feasible and selected synergies coincided. This was because of the absence of overlaps, whereby the same synergy in conceptual terms might be posed using different companies as donors and/or recipients. Instead, the water-related synergies required selecting among different feasible options according to their potential positive impacts on the industrial park. This was the case of SW1, which was chosen over SW4 due to the larger capacity of C3 for supplying purified water meeting the required water quality standards for reuse.

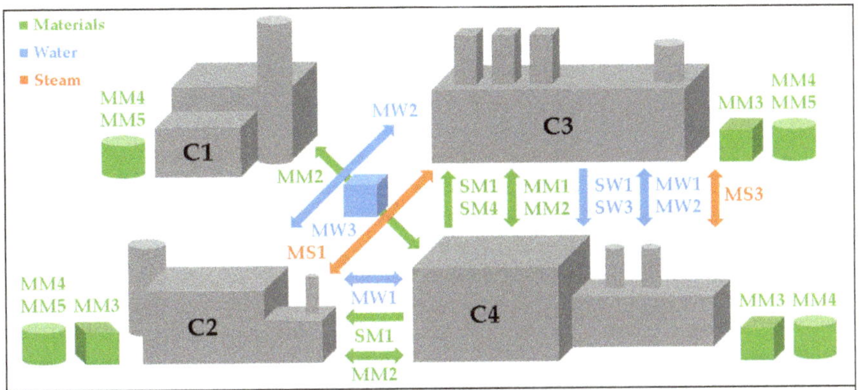

Synergy	Companies	Effect
SM1	C4 → C2, C3	−176.4 t/yr. $CaCO_3$ (C2); −2,650 t/yr. dolomite (C3) −2,826.4 t/yr. calcined $CaCO_3$ wasted (C4)
SM4	C4 → C3	−8,965 t/yr. brine (C3) −8,965 t/yr. used brine wasted (C4)
MM1	C3, C4	Shared supply of brine
MM2	C1, C4	Shared supply of hypochlorite
	C1, C2, C3, C4	Shared supply of hypochloric acid, chlorine and caustic soda
MM3	C4	Shared storage of hypochlorite, hydrochloric acid and hydraulic oils
	C2	Shared storage of hydraulic oils and oxygen
	C3, C4	Shared storage of caustic soda
MM4	C1, C4	Joint management of water with hydrocarbons
	C1, C3, C4	Joint management of metal/plastic containers and equipment
	C2, C3, C4	Joint management of solids with oily waste and used oils
	C1, C2, C4	Joint management of oil filters, non-halogenated solvents, batteries and fluorescent tubes
MM5	C1, C3	Joint management of wood waste
	C1, C2, C3	Joint management of paper and paperboard
	C2, C3	Joint management of plastics
SW1	C3 → C4	−793 m^3/h river water (C4)
SW3	C3 → C4	−200 m^3/h river water (C4)
MW1	C3 → C2, C4	−1 demineralizer (C2)
MW2	C3 → C2, C4	−1 clarifier (C2); −1 clarifier (C4)
MW3	C1, C2, C3, C4	−8 m^3/h well water (C1); + 8 m^3/h river water (C1)
MS2	C4 → C2	−1 gas boiler (C2)
MS4	C4 → C3	−1 gas cogeneration (C3); + 1 backup gas boiler (C4)

Figure 5. Summary of the selected synergies identified in the industrial park under study.

Similarly, C3 was found to be the best company to perform as a supplier of demineralized and clarified water (MW1 and MW2). Although both C3 and C4 could provide C2 with their needs in terms of demineralized water, their available margin to supply each other was larger if the synergy was posed from C3 to C4. In a similar vein, C3 proved to be capable of completely supplying both C2 and C4 with clarified water. Instead, C4 would only have capacity for supplying C2 and a small part of C3 needs. The remaining water-related synergies (SW3 and MW3) were only addressable in one manner, thereby not requiring any choice among companies.

The actions to be taken in the steam-related synergies were restricted by their potential impact on the energetic strategy of the companies. Under these two premises, the best combo of indirect synergies corresponded to MS2 and MS4 (Table 3). On the one hand, these two synergies involved the same supplier (C4) and made using two infrastructures in C2 and C3 unnecessary (Figure 5), thereby meeting the first criterion related to global efficiency. On the other hand, these synergies were also aligned with the requirement of not interfering in the electricity flow of the companies. C2 did not produce electricity through steam, whilst C3 sold it to the national electrical grid. Thus, C3 was subject to the uncertainty of demand and supply, such that the productivity of its energy sales would not be solid enough as to discourage the implementation of synergies involving benefits at a larger scale.

Proposing C4 as a supplier was the best alternative, even when considering that part of its production stemmed from carbon boiling, which is environmentally harmful in terms of greenhouse gas emissions. In fact, MS2 and MS4 involved C4 backup gas boiler for supplying, thereby not contributing to carbon burning. In this vein, the alternative of using C3 as a supplier would not cause a significant variation in C4 carbon-related production, since only a small part of the steam required by C4 could be supplied through MS3 (Table 5).

Hence, with respect to the diagrams in Figure 2, the consideration of the selected materials synergies only affected three groups: auxiliary materials, non-hazardous wastes and emissions and discharges. SM1 and SM4 were responsible for these changes, whereby both brine and calcium carbonate wastes were recirculated from some companies for reuse as auxiliary materials in others (Figure 5). As a result, the auxiliary materials' bars representing brine and calcium carbonate production in Figure 6 were reduced to zero, whilst that corresponding to dolomite was shortened by 70%. Consistent with these changes, calcined calcium carbonate and brine non-hazardous wastes and discharges were diminished too.

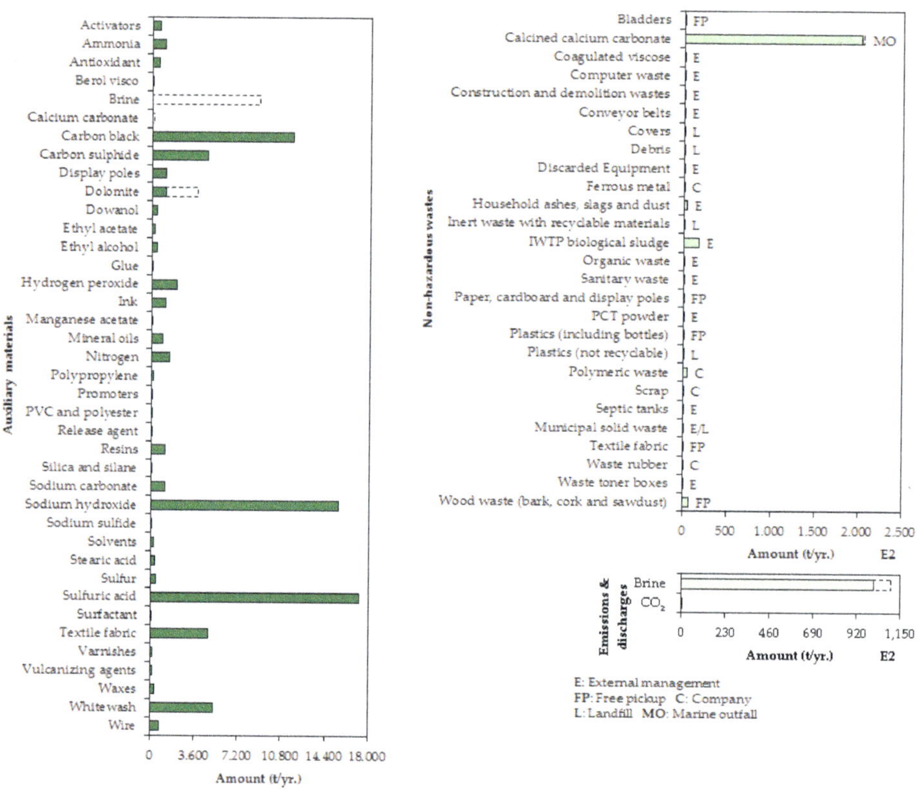

Figure 6. Breakdown of materials and wastes in the industrial park after implementing the synergies.

The remaining cooperation opportunities identified in Table 5 (MM1–MM5) consisted of either the shared use of supply or storage facilities for auxiliary production services or the joint management of hazardous and non-hazardous wastes, which had no visual repercussion on the charts included in

Figure 2. Although no data were found to quantify the contribution of these synergies, they would produce positive impacts on the economic and environmental response of the park, optimizing its performance by improving its logistics through the centralization of materials flows, as well as by increasing its spatial availability in anticipation of potential future expansions.

Figure 7 depicts the evolution of the Sankey diagram for water flows once the synergies proposed in Figure 5 were implemented. Direct synergies SW1 and SW3 enabled a reduction of 793 and 200 m^3/h in the amount of river water required in the industrial park, respectively. Both synergies involved C4 as a recipient, whose river water flow was completely directed to the coarse filtering node. Since the water quality associated with these synergies was equivalent to that of filtered water, rainwater and IWTP changed their previous roles, whereby the former flowed to marine outfall and the latter was an output node exclusively, to flow to the clarification node. Additionally, IWTP maintained its output role, whereby it received flows from the following nodes: cooling towers, toilets, processes, auxiliary systems and generation.

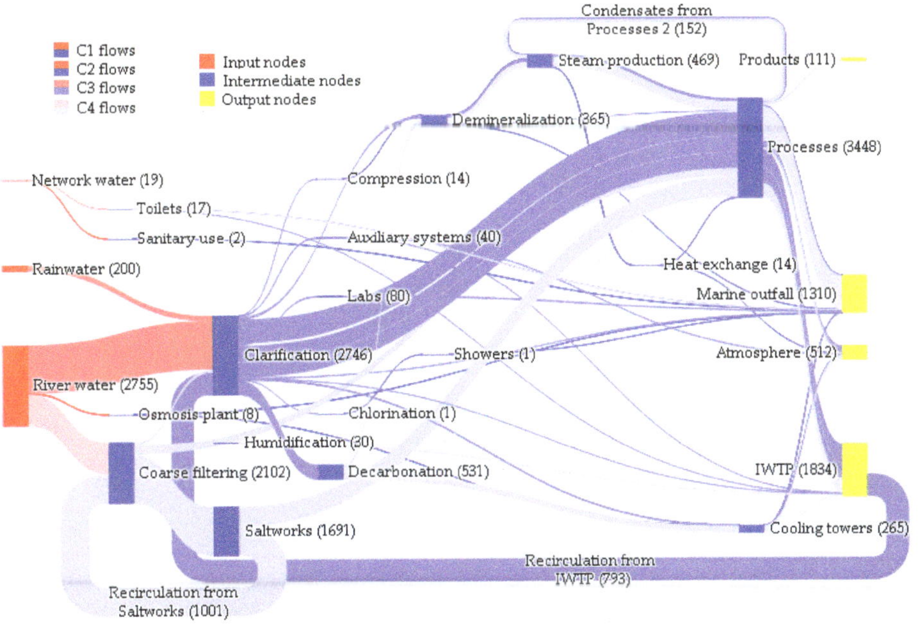

Figure 7. Sankey diagram of water flows in the industrial park after implementing the synergies.

The river water node was also affected by MW3, whereby C1 stopped collecting 8 m^3/h of well water due to its similarity to the river water collected by the remaining companies in terms of water quality. As a result of these synergies, the river water flow decreased from 3739 to 2755 m^3/h ($3739 - 793 - 200 + 8$). In the end, this caused a reduction in the economic costs associated with water catchment fees. These variations also produce alterations in the overall water flows in the industrial park. After the implementation of synergies, the input and output flows continued to be balanced but decreased to 2974 m^3/h. The breakdown of this figure in the case of the inputs was as follows: 2755 m^3/h from river water, 200 m^3/h from rainwater and 19 m^3/h from network water. As for the outputs, the distribution involved 1310 m^3/h to marine outfall (200 m^3/h less than before due to the change in rainwater), 512 m^3/h to the atmosphere and 1041 m^3/h to IWTP (793 m^3/h of the total 1834 m^3/h were recirculated to C4 coarse filtering).

The remaining indirect synergies related to the supply of either demineralized (MW1) or clarified (MW2) water from one company to others. Although the changes caused by these synergies were not noticeable at the scale of the whole system (they would be in the individual diagrams of the companies), they involved 2 demineralizers (C2 and C3) and 2 clarifiers (C2 and C4) that were no longer necessary for managing water flows in the industrial park, since other companies had extra capacity to satisfy their needs.

The updated version of the steam Sankey diagram after implementing the synergies identified in Figure 5 resulted in a variety of modifications in relation to Figure 4, whose main impacts were found in the production nodes. On the one hand, MS2 involved C2 that was supplied using steam from the C4 backup gas boiler, once turbined at 13 kg/cm^2. On the other hand, MS4 consisted of C4 supplying C3 needs with steam from backup gas boiling (58 kg/cm^2) and turbined flows (13 kg/cm^2). This caused an increase of 130 t/h in the gas boiling node, resulting in a final value of 144 t/h (Figure 8). The gas cogeneration node changed in the same proportion, decreasing to 65 t/h. Hence, the global value of steam produced in the industrial park remained at 469 t/h. The breakdown of this figure into carbon boiling, gas boiling and gas cogeneration evolved from 260, 14 and 195 t/h to 260, 144 and 65 t/h, respectively.

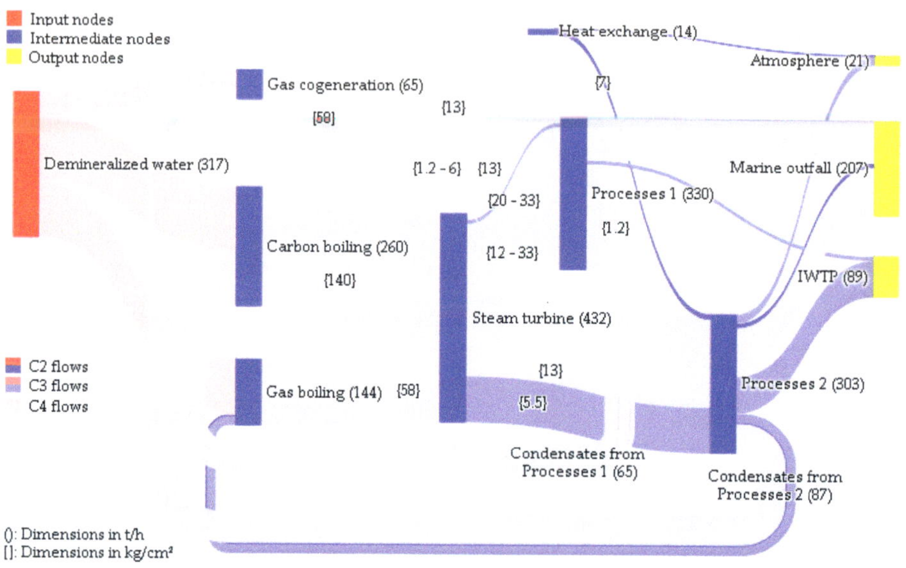

Figure 8. Sankey diagram of steam flows in the industrial park after implementing the synergies.

Apart from their implications in terms of the Sankey diagram, the application of these practices might contribute to optimizing steam consumption in the industrial park. MS2 and MS4 would result in an increase of 144 t/h in the production of C4, such that its efficiency would raise up to 88% (27% more than in the original situation). Since the efficiency of C4 cogeneration was already high in the starting situation (87%), both synergies were approached using its backup gas boiler, which would work at 78% of its nominal capacity (144 t/h out of 185 t/h). In terms of equipment, the impacts of these synergies would make the gas boiler used by C2 and the cogeneration system operated by C3 unnecessary. In turn, this would help to achieve a centralization of steam flows and continuous working regimes, avoiding unnecessary starts and stops that would raise operating costs.

The proposed synergies would also be feasible in what concerns the electric energy strategies taken by C2 and C3. As mentioned before, C2 only focused on the generation of steam, whilst C3 sold its energy to the national electrical grid. The convenience of these synergies would be justified by the dependence of C3 action plan on the fluctuations in demand and supply, whereby the cost effectiveness of its production would not be stable enough as to dissuade the consideration of alternatives involving global benefits such as these. The energy consumed by C3 in its own processes would remain constant, since the steam turbine from which it stemmed (Table 2) continued to produce electricity after the consideration of synergies.

3.4. Contribution to Achieving the UN Sustainable Development Goals

The consideration of the synergies compiled in Figure 5 resulted in a series of modifications in the material, water and steam flows of the industrial park under study. The implications of such changes in terms of sustainability can be addressed using the SDGs as a benchmark. Table 6 provides a summary of the specific targets to which the implementation of IS synergies as those described above might contribute.

Some of these targets were conceptually similar, such as 3.9, 6.3, 11.6 and 12.4, but differed in line with the perspective from which they were considered. Thus, the amount of waste and air and water emissions avoided due to the application of the proposed synergies can contribute to both protecting the welfare of living beings, controlling water quality, reducing environmental impacts on cities and optimizing consumption and production patterns.

Another group of targets associated with generic benefits derived from the synergies identified were 7.3, 8.4, 9.4, 12.2 and 12.5, since they all concerned resource efficiency, either in terms of energy, raw materials, water or waste management. As a result, the IS opportunities found in the industrial park might strengthen sustainable and modern energy, economic growth, industry and infrastructure innovation and conscientious consumption and production. In the same vein, synergies SW1, SW3, MW1, MW2 and MW3 contributed to achieving water-use efficiency in the form of sustainable withdrawals and the supply of freshwater (target 6.4). The recirculation processes involved in these water-related synergies also helped to mitigate marine pollution, thereby having a positive impact on life below water through target 14.1.

Finally, there is a last group of targets focusing on the main strengths of the approach taken, consisting of the systemic assessment of entire industrial parks. The results collected in this research can inspire other companies to adopt these kinds of strategies (target 12.6), thereby contributing to extending the benefits of IS at local and regional scales. In such cases, the broader implementation of these practices might favour the stability of industry-based economies (target 17.13), serving as an exemplar for other regions in the adoption of policies oriented to promote sustainable development (target 17.14).

The impacts of these links on economic and environmental terms were quantifiable in some cases, depending on data availability regarding costs, emissions and the processes in the industrial park. Table 7 summarises the approximate economic and/or environmental savings associated with these cases, which corresponded to the following synergies: SM4 (brine), SW3 (river water), MS3 (cogeneration system), SM1 (dolomite) and MW2 (clarifiers). These values were combined using 2011 as a baseline, in accordance with the year when data were collected. In the case of economic values, first was their conversion into 2011 figures according to the inflation rates of the corresponding currency. Then, the 2011 values were transformed into EUR and multiplied by the amount of materials, water or steam saved, as indicated in Table 7. Finally, these results were updated again based on inflation rates to obtain 2020 savings.

Table 7. Estimates of the economic and environmental savings associated with computable synergies.

Concept	Amount (Year)	Savings	2020 Savings (SI)	References
Brine	8965 t	0.21 (2015) $/gallon (962 gallons/t)	195 €/t	[38]
River water	8,619,840 m^3	0.02 (2007) €/m^3	0.024 €/m^3	[39]
Cogeneration	719,300 MWh	40 (2010) £/kW; 0.26 kg CO_2 eq./kWh	57.96 €/kW; 0.26 t CO_2 eq./MWh	[40,41]
Dolomite	2650 t	0.91 t CO_2 eq./t	0.91 t CO_2 eq./t	[42]
Clarifiers (2)	–	21,140 t CO_2 eq./yr.	21,140 t CO_2 eq./yr.	[43]

The production costs associated with brine were estimated at EUR 0.21/gallon (2015), which involves 962 gallons/t [38]. Since the amount of brine saved during a year was 8965 t, the economic savings associated with this exchange (Figure 6) were EUR 1,755,670/yr. (2020). As for the second flow, the amount of river water required after the implementation of the synergies was reduced to 984 m^3/h due to the recirculation of wastewater. Considering a collection cost of EUR 0.02/m^3 (2007) [39] and a continuous operating regime throughout the year (8760 h), this would result in EUR 208,332/yr. (2020).

Regarding the steam flow, data in the industrial park were limited in what concerns the gas boiler used by C2, such that the economic calculations only applied to the gas cogeneration system in C3, which was no longer necessary after implementing the synergies depicted in Figure 5. The cogeneration system, which produced 719,300 MWh/yr., involved GBP 40/kW (2010) [40] in terms of operation and maintenance. Consequently, this resulted in GBP 4,499,035 (2020) saved per year with a continuous work regime (8760 h/yr.). Overall, the implementation of these synergies would result in an annual economic improvement in the industrial park of EUR 6,463,037/yr. (2020).

Similar calculations were carried out to determine the emissions saved due to the synergies, expressed in t CO_2 eq. Dolomite involved 0.91 t CO_2 eq. per t saved [42], which yielded 2411 t CO_2 eq./yr. when considering the annual amounts used in the industrial park (Figure 6). Based on the study carried out by Heffernan et al. (2012) [43], the annual emissions associated with the two clarifiers avoided thanks to MW2 could be estimated in 21,140 t CO_2 eq./yr.

Since the cogeneration system in C3 produced 719,300 MWh/yr. and the emissions associated with this process were 0.26 kg CO_2 eq./kWh [41], the environmental savings in the steam flow were 187,010 t CO_2 eq./yr. Again, the effects of the remaining synergies were not taken into account because of the lack of data to determine their corresponding emissions. Hence, the calculable environmental savings in the industrial park amounted to 210,561 t CO_2 eq./yr.

The financial balance (incomes–expenses) and consumption expenses (raw materials, supplies and commodities) of the industrial activity in the city where the park was located were EUR 33,292,281 and EUR 868,798,157 in 2011, respectively [44]. Thus, the economic savings derived from the subset of the synergies quantified above (EUR 5,850,672/yr. (2011)) would mean 17.57% and 0.67% of these values at that time. Considering the limited number of synergies from which they stem, these rates provide evidence of the capacity of these companies to be the cornerstone for future IS developments in the city. To illustrate how the consideration of plausible IS practices might result in important benefits if applied at larger scales, the savings associated with this subset of synergies, expressed as a percentage, were applied to regional and national scales. Hence, the figures calculated above would reduce to 3.65% and 0.14% when referred to the region (autonomous community in Spanish), since 20.77% of its industrial production corresponded to the city of the industrial park [44]. In turn, the economic savings associated with the proposed synergies would mean 0.05% of the Gross Domestic Product (GDP) of the region, which was EUR 12,622,706k in 2011 [45]. Finally, taking into account that this region was responsible for 1.1% of the total GDP of Spain in 2011 (EUR 1,069,323M) [46], if the effects of the subset of quantifiable synergies were applied to the scale of the whole country, the economic savings would amount to approximately EUR 500M.

In the case of emissions, there were no available records of the environmental performance of the companies in the industrial park. Instead, the Spanish Inventory System provided data about the region where the park was located, which released 6,310,000 t CO_2 eq. during 2011 [47]. Hence, the environmental savings calculated for the subset of quantified synergies would amount to 3.34% of the Greenhouse Gas (GHG) emissions of the region, since they were equivalent to 190,610 t CO_2 eq./yr. in 2011. Again, if the subset of actions calculated above as a percentage was proportionally applied to the size of the country, which emitted 358 million t CO_2 eq. in 2011, the emissions avoided in Spain may amount to 11,953,958 t CO_2 eq.

The Spanish Government has recently approved its Circular Economy Strategy to reduce the generation of wastes and improve resource efficiency, whereby a series of targets were established to be achieved in 2030, taking the 2010 data as the baseline. These targets included a 30% reduction in the consumption of materials, a 15% reduction in the generation of waste and a 10% improvement in water efficiency [48]. At the scale of the industrial park studied, the amount of materials saved and waste diverted from disposal (i.e., including discharged brine) through the proposed synergies was 11,791.4 t (Figure 5). This figure represented 0.39% of the 3,027,856 t of materials consumed (raw materials, auxiliary materials and auxiliary production services) and 3.33% of the 353,654 t of wastes generated (hazardous and non-hazardous) (Figure 2). Therefore, more detailed information about potential means of either reducing and reusing wastes as raw materials or extending the lifetime of some products in the park should be collected and analysed to contribute further to meeting the first two objectives. Instead, since the use of water in the park would be reduced by 993 m^3/h out of the initial 3,739m^3/h (26.55%) when implementing the synergies related to the replacement of river water (Figure 5), the water-related target would be amply met.

4. Conclusions

This research took the activity of an industrial park located in the north of Spain to undertake a systemic analysis of co-located Industrial Symbiosis (IS) opportunities. To this end, information on the three main flows observed in the park (materials, water and steam) was collected from public data of its main companies and guided visits to their facilities. Next was the proposal and analysis of direct (exchange or substitution) and indirect (share or mutuality) synergies to improve the sustainability of the industrial park and explore its potential for future scaling-up.

As a result, 9 direct (4 materials, 4 water and 1 steam) and 9 indirect (5 materials, 3 water and 1 steam) synergies were identified during the study. Since 6 of them were discarded because of the lack of some data and/or their incompatibility with other cooperation opportunities, the industrial park was subject to a series of synergies that produced several changes in its flows. These changes involved the following savings per year: 2826.4 t of virgin materials replaced by compatible wastes (176.4 t of calcium carbonate and 2650 t of dolomite), 8965 t of brine replaced by used brine that otherwise would be discharged into water bodies, 8,619,840 m^3 of river water, 2 clarifiers, 1 demineralizer, 1 gas boiler and 1 gas cogeneration system.

Overall, these figures would entail a reduction in the production and operation economic costs of the companies, which in turn would crystallize in environmental savings due to the centralization of flows and the increase in resource efficiency. In this vein, most of the water and steam-related synergies would result in one company (C3 and C4, respectively) maximizing its capacity to act as a supply node for the whole industrial park. In turn, this increased efficiency would result in a reduction in the production unitary costs in this company, thereby putting it in a position to achieve competitive selling prices. This would justify the remaining companies in the park putting aside some of their equipment, such that purchasing from those supply nodes might be more profitable for them than generating their own production services. The importance of these synergies would be especially remarkable due to the connection among flows, whereby water performed as an input in steam, which in turn was partially used for electric energy purposes.

Regarding the potential contributions of the selected synergies, only some of them could be turned into preliminary economic and environmental benefits. Still, this limited quantification of synergies would imply saving EUR 6,463,037 and 210,561 t CO_2 eq. per year in the area where the industrial park is located. The proportional application of these actions to the region in which this area is located would result in 0.05% of its Gross Domestic Product (GDP) and 3.34% of its annual Greenhouse Gas (GGE) emissions. Furthermore, when valuing their potential benefits at the scale of the whole country, the implementation of these practices may mean about EUR 500M and more than 10M t CO_2 eq. saved per year.

Although approximate, the figures determined for Spain provide insight into the large annual economic gains that might be obtained by companies involved in IS practices, as well as the associated important benefits for the environment. In this vein, the selected synergies would result in reductions of 0.39%, 3.33% and 26.55% in the materials consumed, waste generated and water used in the industrial park, respectively, thereby contributing to meeting the targets established in the Spanish Circular Economy Strategy for 2030, especially concerning water efficiency. Apart from this quantification, the applied synergies were also analysed in terms of their contributions to achieving the Sustainable Development Goals (SDGs). In total, they proved to be in line with 9 SDGs and 14 of their more specific targets. The cascade effect inherent to these synergies explained the extension of their obvious benefits for the industry to other areas, such as water pollution, clean energy, economic growth, responsible consumption and production, sustainable cities and policy coherence.

The results achieved in this investigation are presented as a proof of the potential benefits of IS at different scales and under the perspective of the three pillars of sustainability: economy, environment and society. In this sense, they also served to prove the main assumption of the study, whereby the city council requesting it expected that the analysis of these large companies may lay the foundations for designing future IS strategies to strengthen the economy, environment and people's welfare in the area. Hence, new schemes of industrial production systems and retrofitting of industrial parks based on the collaboration and partnerships of companies are highly recommended as the keys to future business success. In addition, these approaches can improve both private and public decision-making and support funding allocation, as well as raise awareness and provide reputational benefits and marketing advantages. In the end, these outputs are expected to contribute to shedding light on the multiple positive impacts associated with the promotion of IS policies by governments and public entities. Despite its contributions with respect to recent related literature, this research still needs further development in what concerns the automation of the systemic identification and analysis of IS opportunities, in order to facilitate their future implementation.

Supplementary Materials: The following are available online at http://www.mdpi.com/2071-1050/12/14/5802/s1, Table S1: Summary of the questionnaire distributed in the industrial park to collect materials data. Table S2: Summary of the questionnaire distributed in the industrial park to collect water data. Table S3: Summary of the questionnaire distributed in the industrial park to collect steam data.

Author Contributions: Conceptualization, C.R.-P.; data curation, D.J.-E.; formal analysis, D.J.-E. and C.R.-P.; funding acquisition, C.R.-P.; methodology, C.R.-P.; project administration, C.R.-P.; software, D.J.-E.; supervision, C.R.-P.; visualization, D.J.-E.; writing—original draft preparation, D.J.-E.; writing—review and editing, D.J.-E. and C.R.-P. All authors have read and agreed to the published version of the manuscript.

Funding: This research was funded by the Spanish Ministry of Science, Innovation and Universities, grant number DPI2017-88127-R (AEI/FEDER, UE).

Conflicts of Interest: The authors declare no conflict of interest.

References

1. Potocnik, J.; Teixeira, I. *Resource Efficiency for Sustainable Development: Key Messages for the Group of 20*; UN Environment, International Resource Panel: Paris, France, 2018.
2. Hatfield-Dodds, S.; Schandl, H.; Newth, D.; Obersteiner, M.; Cai, Y.; Baynes, T.; West, J.; Havlik, P. Assessing global resource use and greenhouse emissions to 2050, with ambitious resource efficiency and climate mitigation policies. *J. Clean. Prod.* **2017**, *144*, 403–414. [CrossRef]
3. Kusch, S. *Industrial Symbiosis: Powerful Mechanisms for Sustainable Use of Environmental Resources*; The Global Sustainable Development Report (GSDR): Ulm, Germany, 2015.
4. El-Haggar, S.M. Chapter 10-Sustainability of industrial waste management. In *Sustainable Industrial Design and Waste Management*, 1st ed.; Academic Press: Oxford, UK, 2007; pp. 307–369, ISBN 978-0-12-373623-9.
5. Kapur, A.; Graedel, T.E. Industrial ecology. In *Encyclopedia of Energy*; Elsevier: New York, NY, USA, 2004; pp. 373–382, ISBN 978-0-12-176480-7.
6. Cervantes, G. A methodology for teaching industrial ecology. *Int. J. Sustain. High. Educ.* **2007**, *8*, 131–141. [CrossRef]
7. Schroeder, P.; Anggraeni, K.; Weber, U. The relevance of circular economy practices to the sustainable development goals. *J. Ind. Ecol.* **2019**, *23*, 77–95. [CrossRef]
8. Betts, K.S. Environmental news: Quantifying industrial symbiosis. *Environ. Sci. Technol.* **2005**, *39*, 354–360. [CrossRef] [PubMed]

9. Taddeo, R.; Simboli, A.; Ioppolo, G.; Morgante, A. Industrial symbiosis, networking and innovation: The potential role of innovation poles. *Sustainablity* **2017**, *9*, 169. [CrossRef]
10. Neves, A.; Godina, R.G.; Azevedo, S.; Pimentel, C.C.O.; Matias, J. The potential of industrial symbiosis: Case analysis and main drivers and barriers to its implementation. *Sustainability* **2019**, *11*, 7095. [CrossRef]
11. Velenturf, A.P.M.; Jensen, P.D. Promoting industrial symbiosis: Using the concept of proximity to explore social network development. *J. Ind. Ecol.* **2016**, *20*, 700–709. [CrossRef]
12. Chertow, M.R. Industrial symbiosis: Literature and taxonomy. *Annu. Rev. Energy Environ.* **2000**, *25*, 313–337. [CrossRef]
13. Deutz, P.; Lyons, D.I. Editorial: Industrial symbiosis-an environmental perspective on regional development. *Reg. Stud.* **2008**, *42*, 1295–1298. [CrossRef]
14. Lowe, E.A. Creating by-product resource exchanges: Strategies for eco-industrial parks. *J. Clean. Prod.* **1997**, *5*, 57–65. [CrossRef]
15. Le Tellier, M.; Berrah, L.; Stutz, B.; Audy, J.-F.; Barnabé, S. Towards sustainable business parks: A literature review and a systemic model. *J. Clean. Prod.* **2019**, *216*, 129–138. [CrossRef]
16. Porter, M.; Ketels, C. Clusters and industrial districts: Common roots, different perspectives. In *A Handbook of Industrial Districts*; Edward Elgar Publishing: Cheltenham, UK, 2009.
17. Asheim, B.T.; Isaksen, A. Location, agglomeration and innovation: Towards regional innovation systems in Norway? *Eur. Plan. Stud.* **1997**, *5*, 299–330. [CrossRef]
18. Côté, R.P.; Cohen-Rosenthal, E. Designing eco-industrial parks: A synthesis of some experiences. *J. Clean. Prod.* **1998**, *6*, 181–188. [CrossRef]
19. Haskins, C. Multidisciplinary investigation of eco-industrial parks. *Syst. Eng.* **2006**, *9*, 313–330. [CrossRef]
20. Haskins, C. A systems engineering framework for eco-industrial park formation. *Syst. Eng.* **2007**, *10*, 83–97. [CrossRef]
21. Haskins, C. The story of verdal-how one intelligent community uses systems engineering to enable sustainable development. In Proceedings of the 17th Annual International Symposium of the International Council on Systems Engineering, INCOSE 2007-Systems Engineering: Key to Intelligent Enterprises, San Diego, CA, USA, 24–28 June 2007; Volume 3, pp. 1465–1475.
22. Sopha, B.M.; Fet, A.M.; Keitsch, M.M.; Haskins, C. Using systems engineering to create a framework for evaluating industrial symbiosis options. *Syst. Eng.* **2010**, *13*, 149–160.
23. Romero, E.; Ruiz, M.C. Framework for applying a complex adaptive system approach to model the operation of eco-industrial parks. *J. Ind. Ecol.* **2013**, *17*, 731–741. [CrossRef]
24. Deshpande, P.C.; Aspen, D.M. A framework to conceptualize sustainable development goals for fishing gear resource management. In *World Sustainability Series*; Springer: Berlin, Germany, 2018; pp. 727–744.
25. Ginige, A. Systems Engineering approach to smart computing: From farmer empowerment to achieving sustainable development goals. In Proceedings of the International Research Conference on Smart Computing and Systems Engineering-SCSE 2018, Kiribathgoda, Sri Lanka, 29 March 2018; pp. 1–4.
26. Moldavska, A.; Welo, T. A Holistic approach to corporate sustainability assessment: Incorporating sustainable development goals into sustainable manufacturing performance evaluation. *J. Manuf. Syst.* **2019**, *50*, 53–68. [CrossRef]
27. Brooks, I. The United Nations Sustainable Development goals in systems engineering: Eliciting sustainability requirements. In Proceedings of the International Conference on ICT for Sustainability (ICT4S), Bristol, UK, 21–27 June 2020; Association for Computing Machinery (ACM): New York, NY, USA, 2020; Volume 7, pp. 1–8.
28. Casals, M.; Roca, X.; Forcada, N. *Diseño de complejos industriales*, 1st ed.; Iniciativa Digital Politecnica, Ed.; Universitat Politecnica de Catalunya (UPC): Barcelona, Spain, 2008; ISBN 9788476537428.
29. Stéphane, O.; Jean-Baptiste, Q.; Charles-Xavier, S.; Gwenaël, L.M.; Mouad, M.; Alexandre, B. A cross-sectorial synergies identification methodology for industrial symbiosis. In Proceedings of the Smart Innovation, Systems and Technologies; Springer Science and Business Media Deutschland GmbH: Berlin, Germany, 2019; Volume 155, pp. 229–240.
30. Soundararajan, K.; Ho, H.K.; Su, B. Sankey diagram framework for energy and exergy flows. *Appl. Energy* **2014**, *136*, 1035–1042. [CrossRef]

31. Sankey, H.R. The thermal efficiency of steam engines. Report of the committee appointed to the council upon the subject of the definition of a standard of standars of thermal efficiency for steam engines. *Minutes Proc. Inst. Civ. Eng.* **1898**, *134*, 278–312.
32. Schmidt, M. The Sankey diagram in energy and material flow management: Part I: History. *J. Ind. Ecol.* **2008**, *12*, 82–94. [CrossRef]
33. Schmidt, M. The Sankey diagram in energy and material flow management-Part II: Methodology and current applications. *J. Ind. Ecol.* **2008**, *12*, 173–185. [CrossRef]
34. Griggs, D.; Stafford-Smith, M.; Gaffney, O.; Rockström, J.; Öhman, M.C.; Shyamsundar, P.; Steffen, W.; Glaser, G.; Kanie, N.; Noble, I. Policy: Sustainable development goals for people and planet. *Nature* **2013**, *495*, 305–307. [CrossRef] [PubMed]
35. European Commission. *Industrial Emissions (Integrated Pollution Prevention and Control)*; European Parliament and Council of the European Union: Brussels, Belgium, 2010; pp. 17–119.
36. Official State Gazette. *Legal Regime for the Reuse of Purified Water*; Ministry of the Presidency: Madrid, Spain, 2007; pp. 50639–50661.
37. Official State Gazette. *Criteria for Monitoring and Evaluating the State of Surface Sater and Environmental Quality Standards*; Ministry of Agriculture, Food and Environment: Madrid, Spain, 2015; pp. 80582–80677.
38. SIMA Calculating True Costs of Salt Brine. Available online: https://www.sima.org/news2/2015/08/01/calculating-true-costs-of-salt-brine (accessed on 24 March 2020).
39. Hispagua ¿Cuánto Cuesta Al Agua? Available online: http://hispagua.cedex.es/sites/default/files/especiales/Tarifas_agua/introduccion.html (accessed on 24 March 2020).
40. *Electricity Generation Costs Update*; Mott MacDonald UK: Brighton, UK, 2010.
41. *POSTnote Carbon Footprint of Heat Generation*; POSTnote: London, UK, 2016.
42. IPCC. Industrial processes. In *Revised 1996 IPCC Guidelines for National Greenhouse Gas Inventories*; Houghton, J., Filho, L.M., Lim, B., Treanton, K., Mamaty, I., Bonduki, Y., Griggs, D., Callender, B., Eds.; Intergovernmental Panel on Climate Change: Ginebra, Switzerland, 2003; p. 2.5.
43. Heffernan, B.; Blanc, J.; Spanjers, H. Evaluation of greenhouse gas emissions from municipal UASB sewage treatment plants. *J. Integr. Environ. Sci.* **2012**, *9*, 127–137. [CrossRef]
44. *Encuesta Industrial de Empresas-Cantabria 2012*; ICANE: Santander, Spain, 2013.
45. *Informe del Mercado de Trabajo de Cantabria-Datos 2012*; Observatorio de las Ocupaciones: Santander, Spain, 2013.
46. INE Contabilidad Regional de España. Available online: https://www.ine.es/dyngs/INEbase/es/operacion.htm?c=Estadistica_C&cid=1254736167628&menu=resultados&idp=1254735576581#!tabs-1254736158133 (accessed on 4 June 2020).
47. *Emisiones de GEI por Comunidades Autónomas a Partir del Inventario Español-Serie 1990-2018*; MITECO: Madrid, Spain, 2020.
48. *España Circular 2030-Estrategia Española de Economía Circular*; MITECO: Madrid, Spain, 2020.

sustainability

MDPI

Article

A Systems Engineering Framework for Bioeconomic Transitions in a Sustainable Development Goal Context

Erika Palmer [1,*], Robert Burton [1] and Cecilia Haskins [2]

[1] Ruralis—Institute for Rural and Regional Research, N-7049 Trondheim, Norway; rob.burton@ruralis.no
[2] Department of Mechanical and Industrial Engineering, Norwegian University of Science and Technology, 7491 Trondheim, Norway; cecilia.haskins@ntnu.no
* Correspondence: erika.palmer@ruralis.no

Received: 3 July 2020; Accepted: 13 August 2020; Published: 17 August 2020

Abstract: To address sustainable development goals (SDGs), national and international strategies have been increasingly interested in the bioeconomy. SDGs have been criticized for lacking stakeholder perspectives and agency, and for requiring too little of business. There is also a lack of both systematic and systemic frameworks for the strategic planning of bioeconomy transitions. Using a systems engineering approach, we seek to address this with a process framework to bridge bioeconomy transitions by addressing SDGs. In this methodology paper, we develop a systems archetype mapping framework for sustainable bioeconomy transitions, called MPAST: Mapping Problem Archetypes to Solutions for Transitions. Using this framework with sector-specific stakeholder data facilitates the establishment of the start (problem state) and end (solution state) to understand and analyze sectorial transitions to the bioeconomy. We apply the MPAST framework to the case of a Norwegian agricultural bioeconomy transition, using data from a survey of the Norwegian agricultural sector on transitioning to a bioeconomy. The results of using this framework illustrate how visual mapping methods can be combined as a process, which we then discuss in the context of SDG implementation.

Keywords: systems engineering; bioeconomy; Sustainable Development Goals (SDGs)

---

1. Introduction

A central motivation for developing a bioeconomy is to foster "a continuing evolutionary process of transition from systems of mining non-renewable resources to farming renewable ones" [1] (p. 1)—an objective that ties the transition closely to the UN Sustainable Development Goals (SDGs) [2,3]. The SDGs are a network of interconnected goals used as a reference for the international community in working towards sustainable development [4–6]. Yet, an analysis of the SDGs has found them to be weak on agency, while requiring little from governments and nothing from business or consumers [7]. The SDGs focus on the state and impact, ignoring conflict and stakeholder knowledge of the issues [8,9] and the pressures and drivers that counteract each other in competing impact categories in the same or different geographic contexts [7]. This is of particular importance in bioeconomic transitions, as agriculture has an important role in the bioeconomy and land use has a significant impact on the environment [10], with Norway as no exception [11]. In addition, the economic development of rural areas in a bioeconomy is unclear [12]. As the ambition of individual states varies, the focus on state implementation of policies for increasing SDG impact means that targets will not be realized or there must be new strategies for implementation. If bioeconomy transitions are to be used to reach SDGs, new methods are needed to incorporate stakeholder perspectives and understand counteracting drivers and pressures in the SDG network for successful policy implementation. In the bioeconomy, there is a specific concern that policy implementation will be outside cultural norms in farming communities [13] and will require collaborative action within these communities to be successful [14]. In addition, there is an established need to consider citizens and other stakeholders disrupted by bioeconomy transitions [15].

Within this context, the conceptualization of bioeconomy transitions together with meeting SDGs naturally lends itself towards to systems engineering because SDGs are easily framed in a systems theoretical context [16], and capacity building to meet SDGs has been argued to require systems thinking [17]. System archetypes [18,19] can be helpful to gain insight into patterns of behavior, especially in contexts of interconnected and competing goals [20]. In addition, SDGs intersect many societal domains, and a systems approach to investigating persistent social and environmental problems has been shown to help policymakers tackle the complexity associated with many intersected societal domains [21]. Concerning using systems archetypes as part of a methodological design, however, there needs to be a systematic method for mapping archetypes rather than an arbitrary assignment of a generic system structure to a specific situation [22]. This systematic mapping requires an identification of a problem archetype to a solution archetype (the start and end points in a transition) with an identified path between the problem and solution. For mapping bioeconomy transitions, this mapping needs to be nested in the SDG network in order to complement national and international strategies that seek to use the bioeconomy as a means of reaching SDGs, and which can be complementary to SDG mapping [23].

We use systems engineering as the approach for developing a framework for a systematic and systemic mapping of bioeconomic transitions in an SDG context because bioeconomic transitions are transitions within and between socio-technical systems [24], and systems engineering has the potential to significantly contribute to the evaluation of socio-technical systems [22]. Systems engineering is developing ways to evaluate systems in social domains [25], and this paper provides a new arena in which to apply systems engineering methods: the bioeconomy. In this paper, we develop a framework for applying a qualitative systems engineering method: visual systems mapping. Qualitative methods in systems engineering have been cited as being useful in the evaluation of socio-technical systems because of their high degree of "messy complexity", making quantitative methods much more difficult to apply [26]. Szajnfarber & Gralla (2017) specifically mention the use of visual mapping techniques for systems engineers for process-orientated analysis (in this case, transition processes).

Figure 1 illustrates the connection between systems engineering, bioeconomy transitions and SDGs. Why are SDGs part of this process? The context of SDGs provides feedback for the success of bioeconomic transitions. Since the bioeconomy is being used as a means to meet SDGs, a bioeconomy transition should be planned and evaluated in an SDG context. The SDGs can then be used as performance indicators of the transition. This then leads to the question of: how do we plan and evaluate a bioeconomy transition in an SDG context? Using a systems engineering approach, this paper addresses this question by developing framework for applying visual systems mapping techniques. In doing so, we address the lack of social science inclusion in the study of bioeconomic transitions [9].

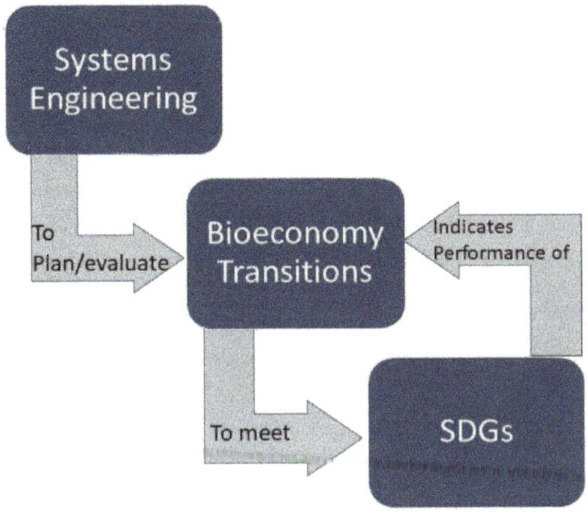

Figure 1. Relationship between Systems Engineering, Bioeconomic Transitions and Sustainable Development Goals (SDGs).

This is a methodological paper, where we use empirical data to develop the methodological framework. Although we are working within systems theory to develop this framework, we do aim for a theoretical contribution. The paper's aim is two-fold. First, it seeks to build upon previous archetype mapping methods in sustainability science to provide a system archetype process framework for use in bioeconomy transitions in an SDG context [18,19,27–29]. We do not provide a new framework for SDG modeling and scenarios (e.g., [30]), but rather build a process framework to bridge bioeconomy transitions while addressing SDGs. Second, this paper applies this framework on sustainable bioeconomy transitions in Norwegian agriculture. Given the brief background provided in this introduction, the next section outlines the process framework (called MPAST: Mapping Problem Archetypes to Solutions for Transitions) and the background on the framework application in the Norwegian bioeconomy transition. The following section provides the results of the mapping framework application, which is then discussed in the context of the SDGs. Lastly, this paper provides areas for future research and for framework development in a more general context.

2. Systems Mapping for Bioeconomy Transitions

Systems mapping for the purposes of this study is the qualitative modeling of a system that describes problematic and solution-orientated system conditions. We do not develop new system mapping techniques but instead develop a process framework for applying them in bioeconomic transitions. We do this because there are competing visions (solution states) of the bioeconomy, and different sectors in the bioeconomy will have different pathways for reaching solution states [31,32]. The application of system archetypes and mapping them in this context requires a more rigorous sector specific methodology than a simple selection of established system archetypes from those established by systems thinking practitioners (e.g., [19]), which is why we developed a process framework for applying them (within the agricultural sector). The background to the case application of this method is Norwegian bioeconomy transitions. To ground the reader in the description of the methodological design, we begin with a short background on the Norwegian bioeconomy.

2.1. Bioeconomy Transitions in Norway

One of the problems with transitioning to a bioeconomy is that while the development of a bioeconomy in Norway *necessitates* integrated development across sectors, multiple actors and institutions [33,34], industrial economies are siloed and sectorial. Each works on separate problems, has different stakeholders and is guided by independent governance/policy-regimes (i.e., different ministries). To achieve integration, Norway's National Bioeconomy Strategy prioritizes the development of cross-sectoral value chains in four areas, namely; cross-sectoral cooperation, markets for bio-based products, processing of biological products and sustainable extraction of bio-based products [35].

Within this framework, agriculture plays an economically limited, but politically important role and is a key sector in the bioeconomy. Since 1972, the maintenance of rural communities has been a key policy goal in Norway, with considerable amounts of public money flowing into decentralizing funds and initiatives [36]. Agricultural bioeconomy developments also feature in government policies to enhance food security and reduce climate emissions. Here, its main sectoral integrations are as consumers of bioeconomic solutions for feed production (high-protein biosynthesized animal feed from forestry to replace imported soy meal) and utilization of residual materials and waste (including biogas) [35]. Overall, Norway's vision of a bioeconomy transition is a fully integrated one and "encompasses all sectors and related services, which produce, process or use biological resources in whatever form" [37] (p. 15). In Norway, agriculture is a key part of this vision, as "it produces for a protected domestic market and is expected to achieve such policy goals as food security, food sovereignty, rural settlement and employment, and the maintenance of agricultural land and cultural landscapes" [32] (p. 11).

2.2. *MPAST: Mapping Problem Archetypes to Solutions for Transitions*

The systems mapping framework we developed for bioeconomy transitions is called MPAST: Mapping Problem Archetypes to Solutions for Transitions. Essentially, the MPAST framework identifies the start and finish lines for evaluating bioeconomy transitions by sector. MPAST is a "framework of frameworks", where it facilitates the systematic application and validation of established system archetypes to bioeconomy transitions. We do not develop established frameworks but put them together in a new way, and there are many different methods for analysis that can be built into each of the phases. There are three phases to this framework: Phase 1—start (problem state); Phase 2—end (solution state); and Phase 3—pathway. There are start and end points for transitions (Phases 1 and 2): we transfer from one (problematic) system condition/state to a new (solution) system condition/state. In a systems approach, these start and end points can be conceptualized as problem archetypes and solution archetypes. Between the start and ends points and their mapped archetypes is the pathway from problem to solution (Phase 3)—see Figure 2 for an overview.

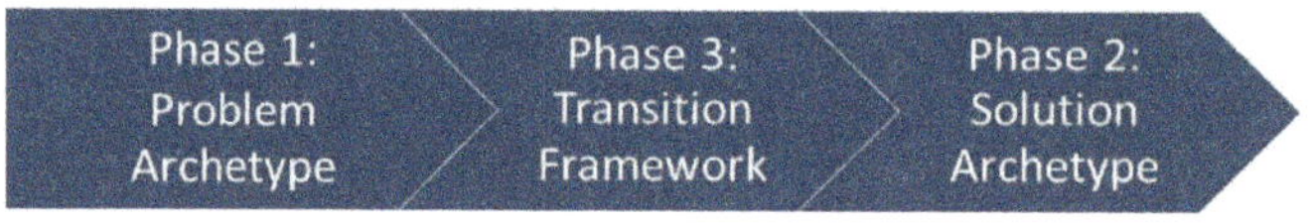

Figure 2. Mapping Problem Archetypes to Solutions for Transitions (MPAST) Process Framework. The phases are the methodological application sequence; the order in the figure is the transition process sequence.

2.2.1. Problem Archetypes (MPAST)–Phase 1

A system archetype is a generic qualitative description of a system structure that leads to problematic behavior or unintended consequences over time [18]. There are ten system archetypes that are generally acknowledged by systems thinking practitioners [19]. Although established as "archetypes" in academic literature, this does not have to be the only list that those using archetype mapping can look for to identify a system structure that describes a problematic system structure. Any system model that describes a problematic behavior can be applied in the MPAST framework. System archetypes in systems thinking are generic but not comprehensive for all problematic system behavior. Another example beyond the traditional system archetypes is the list of syndromes. Syndromes are a scaled-approach that identifies qualitative sub-models on an intermediate level that work towards global change [28,38]. As described in Section 2.1, the Norwegian bioeconomy and its competing stakeholder perspectives by sector lends itself well to the use of syndromes in the application of MPAST. The requirement of Phase 1 of MPAST application is that the problematic archetypes chosen to describe the initial system condition in a transition. In a participatory mapping session, this would require the stakeholders to acknowledge that the chosen problem archetype describes the system well. To validate the chosen problem archetype without stakeholders requires validation through consultation with subject matter experts (SMEs).

2.2.2. Solutions (MPAST)–Phase 2

The solution phase in MPAST may or may not depend on the set of problem archetypes chosen in Phase 1. If, for example, system archetypes outlined by Braun (2002) [19] are chosen to describe the problem, these archetypes also provide the solution archetype as well. However, when mapped onto the specific bioeconomy transition problem, the solution may have to be adapted. There are independent solution archetype sets that can be applied in Phase 2; for example, specifically for climate change, see Eisenack (2012) [39]. A helpful, extensive example of sustainable business model archetypes as solutions that can be applied to the bioeconomy is by Bocken et al. (2014) [27]. Validation of the solution archetype—i.e., knowing whether the solution of the bioeconomy end goal for a specific sector in a bioeconomy transition is a solution to the problem state (phase 1)—requires empirical evidence in scientific literature. For example, in the bioeconomy, creating value through waste with industrial symbiosis as a business model would be an empirically validated solution to a waste-based bioeconomy transition [40].

2.2.3. Transitions (MPAST)–Phase 3

For the bioeconomy, transitions require the society to change from one that mines non-renewable resources to a society that farms renewable resources [1]. A shift in system conditions has been cited as a requirement of bioeconomy transitions [32]. For transitions in MPAST, it is possible to use many pathway frameworks that work systematically towards a change in system states. The requirement of Phase 3 is that there is a systematic process for transitioning from Phase 1 to Phase 2. Methodological approaches to mapping transitions are found in fields such as strategic planning and systems engineering. For example, backcasting is a well-established strategic planning method often used in sustainability science [41]. Validation of Phase 3 depends on the transitioning framework chosen.

As we will explain in Section 3, we have chosen a framework for bioeconomy transitions from systems engineering—PADE. SPADE is an acronym, which stands for: Stakeholders, Problem Formulation, Analysis, Decision-Making and Evaluation [22]. SPADE is a non-sequential research process, each part can be taken individually or several simultaneously, and iteratively.

In the following section, we apply the MPAST framework to a sector-specific bioeconomy transition evaluation.

3. Case Analysis: Application of the MPAST Framework

For each of the phases in MPAST, we were informed by data from a survey given as part of a project that aimed to promote a smart transition to the bioeconomy [42]. The aim of the survey was to identify how different relevant sectors in the bioeconomy believe that the bioeconomy will look like in 2030, and what role they envisage playing in this economy. The survey was given to businesses and other stakeholders (e.g., R&D, advisory services) in sectors that have a connection to the bioeconomy. This included sectors such as industry, education and research, but also primary industries such as agriculture, forestry and fisheries. In total, there were 1313 survey respondents. We have chosen agriculture for the sector-specific bioeconomy transition evaluation using the MPAST framework. Norwegian agriculture is an interesting case for examining bioeconomy transitions because Norway's approach to developing a bioeconomy is a particularly integrated one [37], and as mentioned in Section 2.1 is meant to serve multiple goals: "food security, food sovereignty, rural settlement and employment, and the maintenance of agricultural land and cultural landscapes" [32] (p. 11). The survey data and analysis can be accessed through the link provided in the reference list [42].

3.1. Application: Problem Archetypes (MPAST)–Phase 1

When we began this analysis, the intention was to use traditional system archetypes summarized by Braun (2002) [19] for application in Norwegian agriculture bioeconomy transitions. However, we felt that these were not specific enough to environmental challenges in Norwegian agriculture, and we found that using syndromes of global change [28,38] was a better fit. Syndromes are sets of underlying actions that have consequences (symptoms). For example, the overexploitation of natural ecosystems is a syndrome that leads to a host of environmental problems that are symptoms (e.g., loss of biodiversity) of the underlying syndrome. In Phase 1, we identified each of the syndromes that were applicable to Norwegian agriculture bioeconomy transitions based on the survey data. Table 1 lists these syndromes and their meaning.

Table 1. Applied Syndrome in MPAST Phase 1 from Schellnhuber et al. (2002) [28].

Syndrome Name	Syndrome Meaning
Sahel Syndrome	Overcultivation of marginal land
Overexploitation Syndrome	Overexploitation of natural ecosystems
Dust Bowl Syndrome	Non-sustainable agro-industrial use of soils and bodies of water
Aral Sea Syndrome	Environmental damage of natural landscapes as a result of large-scale projects
Green Revolution Syndrome	Environmental degradation through the introduction of inappropriate farming methods

The syndromes in Table 1 act as the initial starting point in bioeconomy transitions for Norwegian agriculture. These issues must be solved by the solution phase of MPAST.

3.2. Application: Solutions (MPAST)–Phase 2

In Phase 2 of MPAST, we map solution archetypes to the problem archetypes identified in Phase 1. Syndromes as problem archetypes do not have prescribed solution archetypes. Bocken et al. (2014) [27] provide sustainable business model archetypes that fit well in a Norwegian agricultural bioeconomy transition context. The reason for this is because the archetypes are specific to business models, and business stakeholders by sector was the focus of the survey used in this analysis. Two sustainable business archetypes address the syndromes listed in Table 1 (see Table 2).

Table 2. Mapping solution archetypes to problem archetypes for Norwegian agriculture bioeconomy transitions.

Problem Archetypes-Syndromes [28]	Solutions-Sustainable Business Archetypes [27]
Sahel Overexploitation Dust Bowl Aral Sea Green Revolution	Adopt a Stewardship Role Repurpose for Society and the Environment

The sustainable business archetypes by Bocken et al. (2014) [27] give examples of how these solutions can be achieved. For example, in adopting a stewardship role, a business can create a business model that focuses on biodiversity protection. However, there is a pathway to reaching these end goals, and development of the specific business model is an iterative process, which we discuss in the next phase of MPAST.

3.3. Application: Transitions (MPAST)–Phase 3

As mentioned, we used the SPADE, which is "a streamlined methodology that is visually representative of the intrinsically iterative nature of systems engineering" [29] (p. i). We chose one solution in Table 2 to evaluate further based on the opinions of farmers towards bioeconomy goals in the survey: Adopt a Stewardship Role. We organized the results of the SPADE analysis for MPAST Phase 3 in Table 3.

Table 3. Overview of SPADE Analysis.

SPADE Analysis				
Stakeholders	**Problem**	**Analysis**	**Decision/Tradeoffs**	**Evaluation**
Farmers Agri-Business Local/Rural Communities Consumers Cooperative Agricultural Supplier Academia/Research Politician Environment	Energy Sources Waste Soil depletion Markets	Resource Stewardship	Income and Financing Heritage/Cultural Identity	Iterative evaluation of the S,P,A,D parts of SPADE framework

When conceptualizing the role of the SPADE framework and how to use it, we need to clarify its role in MPAST. Table 4 shows this organization. Phases 1 and 2 set the goal posts of where a sector is and where it needs to go in terms of bioeconomy transitions for a specific identified problem archetype in one sector. For the SPADE analysis, the overexploitation syndrome with the "adopt a stewardship role" was the best fit from the survey results as a solution for agriculture. The Phase 3 analysis of MPAST organizes elements to begin transitioning from overexploitation to stewardship.

Table 4. Overview of the MPAST Application to Norwegian Agriculture Bioeconomy Transitions.

MPAST	Phase 1: Problem Archetype	Phase 3: Transition to	Phase 2: Solution Archetype
Archetypes Set/Framework	Syndromes	SPADE	Sustainable Business Model Archetypes
Analysis	Overexploitation	Resource Stewardship Iterative Pathway	Adopt a Stewardship Role

Referring to Table 3, the *stakeholders* part of SPADE is necessary to make sure that the problem archetype (Phase 1) and solution archetype (Phase 2) includes all the relevant stakeholders. This may have already been systematically covered in Phase 1 and 2 depending on which archetypes were chosen, but Phase 3 ensures that they have been identified. The stakeholders listed in Table 3 were provided in the application of MPAST by the survey (in addition to the environment, which must be included as a stakeholder in sustainability analysis). The *problem* part of SPADE lists the problems (symptoms related to syndromes in this application) that were identified in Phase 1. The *analysis* part of SPADE lists the solution archetype identified in Phase 2. The *decision/tradeoffs* part of SPADE identifies structural issues (in this case identified by the stakeholders from the survey) that inhibit the transition from Phase 1 to Phase 2. The *evaluation* part of SPADE is an iterative evaluation of the stakeholders, problems, analysis and decisions/tradeoffs over time. This is a non-sequential transition analysis that can be updated as new data sources are used, new stakeholders are included and problem and solution states (Phases 1 and 2) shift. We used survey data to begin the process of analyzing a bioeconomy transition for Norwegian agriculture, but once started, this is an agile process, and no pathway analysis is set in stone. What we mean by this is that no analysis is final, and future work makes transition pathways clearer over time.

In the bioeconomy, we are transitioning from non-renewable to renewable resources. This can lead to the overexploitation of renewable resources [43], and a sustainable business model must exercise a stewardship role of the renewable resources. For this transition, it is critical to identify all relevant stakeholders to increase policy implementation success (in addition to reasons of social justice). In agriculture, the problems in the transition are (among others) energy sources, waste, soil depletion and markets. To begin a transition to resource stewardship, decisions/tradeoffs must be made. Agricultural stakeholders in the survey had many concerns related to income and financing and cultural heritage in bioeconomy transitions, which were identified as structural issues that inhibit the transition in the SPADE analysis.

In a more concrete example, "laboratory meat," despite its controversial status, offers a strong illustrates how bioeconomic development (protein synthesis) could both improve food security and address climate change as new industrial and scientific applications create entirely new platforms of industrial production, eradicating the need for existing farming systems in particular product sectors [44]. However, the energy consumption of the necessary bioreactors may be considerable, necessitating a move to renewable energy sources if sustainability is to be assured [45]. We can identify in this example competing SDG pressures (which we discuss in the next section), where a bioeconomy transition is a solution (farming renewable energy sources). The new market for laboratory meat and the energy sources are both identified problems in the SPADE analysis. The sustainable business model needs a resource stewardship role, which needs to be iteratively designed in consultation with stakeholders. Designing the transition with stakeholders must address the structural issues identified in decisions/tradeoffs: income/financing and cultural heritage/identity. For laboratory meat, this means the design must address the questions of: how new markets are going to affect farmer income; how they are going to finance transitioning to new markets; how will farming as part of cultural heritage and farmers' cultural identity affect or be affected by this transition; and how will renewable energy resources be sustainably harvested?

MPAST does not answer these questions, but it does identify, organize and gives power to stakeholder voices in setting the goal posts of bioeconomy transitions and in the iterative process of designing the pathway between the goal posts. In the next section, we discuss this in the context of SDGs.

4. Discussion: MPAST and SDGs

Using a systems engineering approach, the methodological framework developed in this paper seeks to build upon previous archetype mapping methods in sustainability science for use in in bioeconomy transitions. In addition, MPAST seeks to strengthen the use of qualitative methods in systems engineering for the evaluation of socio-technical systems [26]. The MPAST framework provides a process for applying visual systems mapping to plan and evaluate bioeconomy transitions to meet SDGs (see Figure 1).

Because the bioeconomy has garnered attention as a way of addressing SDGs, this section will discuss how the application of MPAST can work in an SDG context. SDG mapping requires continual improvement [23], and although this paper does not attempt a contribution to SDG mapping itself, MPAST can act as a bridge between SDG mapping and concrete goal setting for the bioeconomy by sector. To discuss how this bridge is built, we explain how the application of MPAST in Norwegian agricultural bioeconomy transitions is nested in several SDGs. We highlight three SDGs for discussion [4]: SDG 8: Decent work and economic growth; SDG 13: Climate Action; and SDG 15: Life on Land, which aims to protect endangered animal species and terrestrial ecosystems.

The critical assessment by Spangenberg (2017) [7] of the SDGs regarding requiring little to nothing from business and consumers, the counteracting drivers and pressures, and the lack of agency represented in the SDGs is of particular relevance for applying MPAST for bioeconomy transitions. MPAST provides a systematic framework for incorporating elements of agency and the identification of counteracting pressures and drivers over time in one geographic context. Because MPAST is sector-specific and has one geographic context per application, it does not address the criticism regarding pressures and drivers counteracting each other in different geographic contexts. For example, in Figure 3, the case application given in section three relates to SDGs 8, 13, and 15. Accomplishing these SDGs requires economic growth while addressing environmental issues related to climate and land use. These have counteracting pressures and drivers and require little to nothing from business, and agency is not well represented. At its core, sustainable development is aiming to address the counteracting pressure between economic development and climate action (in addition to all environmental impact categories). The bioeconomy is meant to help address this, but as bioeconomic industries develop, this puts pressure on life on land as this will require the farming of bio-resources. In addition, although the bioeconomy is politically assumed to be a win-win for rural areas and the environment, Burton and Fuglestad (2020) [12] indicate that this is unlikely as biotechnologies improve. Life on land also has a counteracting pressure with climate action as greenhouse gas emissions from agriculture, and the changes in land use necessitated by a bioeconomy transition, have a significant effect on mitigating climate change [10].

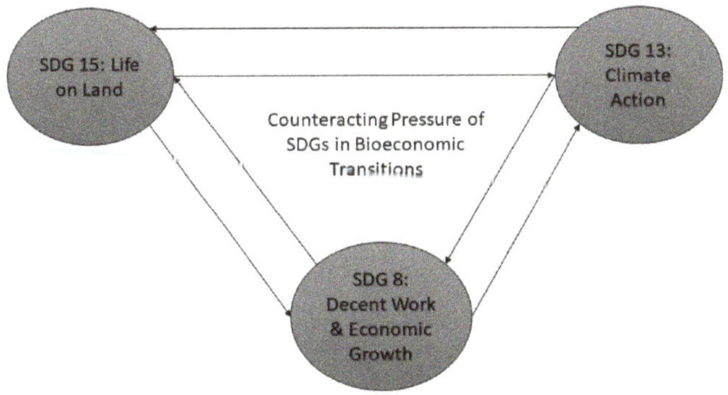

Figure 3. SDGs examples highlighting counteracting pressure.

MPAST facilitates pathways to address these criticisms when bioeconomy transitions are the means to reach SDGs. The SPADE analysis requires a stakeholder assessment. It does not provide the agency for SDGs, but it provides a tool for the incorporation of agency in the planning of SDG implementation with bioeconomy transitions. MPAST is centered on planning bioeconomy transitions by sector using established frameworks and works as a "framework of frameworks". The focus on sectors and stakeholders, rather than specifically the state, requires more of business in particular sectors. Business stakeholders become part of the process of setting the goal posts for bioeconomy transitions (MPAST Phase 1 and 2) and developing the transition pathway for meeting bioeconomy (and hence SDG) goals (MPAST Phase 3). Counteracting drivers and pressure are more easily identified with the application of system archetypes to problems and solutions, as illustrated with the laboratory meat example in Section 3.3.

The economics of bioeconomy transitions are complex, and the boundaries of traditional sectors may stretch or change shape in transitions as value-chains in the bioeconomy are not yet well-defined [46]. Social considerations such as the development of social and cultural capital also contribute strongly to the complexity of transitions—for example, by inhibiting adoption of agri-environmental policies that fall outside cultural norms [13] or facilitating collaborative actions within farming communities [14]. However, the social science perspective is not well developed in bioeconomy transitions literature [9]. This is especially problematic in the case of Norwegian agriculture because, as noted above, agriculture's value in the bioeconomy is as much (or more) its contribution to wider society as the economic value of production.

Stakeholder perspectives contributing to shared visions of bioeconomy solutions and transitions through iterative strategic planning to shape new sectors and value-chains can facilitate the bioeconomy as a means of addressing SDGs. The development and use of MPAST seems straightforward and clear as a process of using archetype mapping for transitions, but this was not clear at the beginning of MPAST development. There is an array of archetypes/frameworks that can be applied to sustainable bioeconomy transitions. There is no systematic method for applying these frameworks. At the beginning of the analysis of the Norwegian agricultural bioeconomy case, we identified many relevant archetypes and mapping frameworks that were relevant for different stages of bioeconomy transitions.

We knew that we needed to use systems archetypes in the analysis because of the systemic nature of SDGs. However, the arbitrary and unsystematic process of using system archetypes for bioeconomy transitions necessitated the development of MPAST.

There are limitations of the MPAST framework that require it to be developed for a more general context. As stated, we developed this framework to analyze the Norwegian agricultural case because of a lack of a mapping methodology that met the needs of our research design. While we see the potential to develop this framework to other contexts, further empirical work is required. Although we used Norwegian agricultural data in this study, this could also be applied in other geographic contexts using country specific data (or in other industries with industry specific data). Further applications in other contexts beyond Norwegian agriculture will allow further development of the MPAST framework. In addition, as new SDG mapping evolves, as seen in Dalampira & Nastis (2019) [23], MPAST will need to develop to match this evolution.

5. Conclusions

MPAST is a methodological framework using a systems engineering approach that maps processes for bioeconomy transitions; it can more easily be described as a "framework of frameworks". MPAST is designed to make a clear identification of start and finish lines in transitions and the pathway between using a systems approach. One could argue that MPAST could be used in any transition, but it is particularly useful for the bioeconomy. The bioeconomy has either weak/nonexistent or differing visions (solution states) of what the bioeconomy should be [31,32], and using MPAST is a way to systematically incorporate stakeholder perspectives into the analysis of bioeconomy development with a focus on sectors. Stakeholder perspectives are of particular importance given the focus the bioeconomy has had as a solution to addressing SDGs, and criticism of SDGs has shown them to be weak on agency. In addition to agency, SDGs have been criticized for their counteracting pressures and drivers in the same or different contexts. The systematic use of systemic frameworks helps identify what these drivers and pressures are, which is required for strategic planning to address them. Although there are important criticisms of the SDGs, the criticism itself can help provide the solutions.

Author Contributions: Writing and methodology, E.P., R.B. and C.H.; review, C.H. All authors have read and agreed to the published version of the manuscript.

Funding: This research was funded by the Norwegian Research Council, BioSmart and Protein 2.0 projects; grant numbers 244608 and 294777. The APC was funded by Ruralis.

Conflicts of Interest: The authors declare no conflict of interest.

References

1. Zilberman, D.; Kim, E.; Kirschner, S.; Kaplan, S.; Reeves, J. Technology and the future bioeconomy. *Agric. Econ.* **2013**, *44*, 95–102. [CrossRef]
2. Heimann, T. Bioeconomy and SDGs: Does the bioeconomy support the achievement of the SDGs? *Earth's Future* **2019**, *7*, 43–57. [CrossRef]
3. Ramcilovic-Suominen, S.; Pülzl, H. Sustainable development–a 'selling point' of the emerging EU bioeconomy policy framework? *J. Clean. Prod.* **2018**, *172*, 4170–4180. [CrossRef]
4. United Nations General Assembly (UN). *Transforming our World: The 2030 Agenda for Sustainable Development*; United Nations: New York, NY, USA, 2015.
5. Le Blanc, D. Towards integration at last? The sustainable development goals as a network of targets. *Sustain. Dev.* **2015**, *23*, 176–187. [CrossRef]
6. Pradhan, P.; Costa, L.; Rybski, D.; Lucht, W.; Kropp, J.P. A systematic study of Sustainable Development Goal (SDG) interactions. *Earth's Future* **2017**, *5*, 1169–1179. [CrossRef]
7. Spangenberg, J.H. Hot air or comprehensive progress? A critical assessment of the SDGs. *Sustain. Dev.* **2017**, *25*, 311–321. [CrossRef]
8. Wong, R.; Heijden, J. Avoidance of conflicts and trade-offs: A challenge for the policy integration of the United Nations Sustainable Development Goals. *Sustain. Dev.* **2019**, *27*, 1–8. [CrossRef]

9. Sanz-Hernández, A.; Esteban, E.; Garrido, P. Transition to a bioeconomy: Perspectives from social sciences. *J. Clean. Prod.* **2019**, *224*, 107–119. [CrossRef]
10. Parry, M.L. *Climate Change and World Agriculture*; Routledge: London, UK, 2019.
11. Olsson, E.G.A.; Rönningen, K.; Hanssen, S.K.; Wehn, S. The interrelationship of biodiversity and rural viability: Sustainability assessment, land use scenarios and Norwegian mountains in a European context. *J. Environ. Assess. Policy Manag.* **2011**, *13*, 251–284. [CrossRef]
12. Burton, R.J.F.; Fuglestad, E. *Etter Oljen: Vår bioøkonomiske fremtid. Bortenfor bioøkonomien. ("Beyond the Bioeconomy")*; Burton, R.J.F., Forbord, M., Ellingsted, M.B., Fuglestad, E., Eds.; Cappelen Damm: Oslo, Norway, 2020.
13. Burton, R.J.F.; Kuczera, C.; Schwarz, G. Exploring farmers' cultural resistance to voluntary agri-environmental schemes. *Sociol. Rural.* **2008**, *48*, 16–37. [CrossRef]
14. Sutherland, L.-A.; Burton, R.J.F. Good Farmers, Good Neighbours? The role of cultural capital in social capital development a Scottish farming community. *Sociol. Rural.* **2012**, *51*, 238–255. [CrossRef]
15. Mustalahti, I. The responsive bioeconomy: The need for inclusion of citizens and environmental capability in the forest based bioeconomy. *J. Clean. Prod.* **2018**, *172*, 3781–3790. [CrossRef]
16. Skene, K.R.; Malcolm, J. Using the SDGs to nurture connectivity and promote change. *Des. J.* **2019**, *22*, 1629–1646. [CrossRef]
17. Stafford-Smith, M.; Griggs, D.; Gaffney, O.; Ullah, F.; Reyers, B.; O'Connell, D. Integration: The key to implementing the Sustainable Development Goals. *Sustain. Sci.* **2017**, *12*, 911. [CrossRef] [PubMed]
18. Senge, P.M. *The Fifth Discipline Fieldbook: Strategies and Tools for Building a Learning Organization*; Doubleday: New York, NY, USA, 1995. [CrossRef]
19. Braun, W. The System Archetypes. The Systems Modeling Workbook. 2002. Available online: http://www.albany.edu/faculty/gpr/PAD724/724WebArticles/sys_archetypes.pdf (accessed on 17 August 2020).
20. Wolstenholme, E. Using generic system archetypes to support thinking and modelling. *Syst. Dyn. Rev.* **2004**, *20*, 341–356. [CrossRef]
21. Probst, G.; Bassi, A. *Tackling Complexity: A Systemic Approach for Decision Maker*; Routledge: New York, NY, USA, 2017. [CrossRef]
22. Haskins, C. Using Patterns to Transition Systems Engineering from a Technological to Social Context. *Syst. Eng.* **2008**, *11*, 147–155. [CrossRef]
23. Dalampira, E.-S.; Nastis, S.A. Mapping Sustainable Development Goals: A network analysis framework. *Sustain. Dev.* **2020**, *28*, 46–55. [CrossRef]
24. Bauer, F. Narratives of biorefinery innovation for the bioeconomy: Conflict, consensus or confusion? *Env. Inno. Soc. Trans.* **2018**, *28*, 96–107. [CrossRef]
25. Palmer, E. *Systems Engineering Applied to Evaluate Social Systems: Analyzing Systemic Challenges to the Norwegian Welfare State*; The University of Bergen: Bergen, Norway, 2017.
26. Szajnfarber, Z.; Gralla, E. Qualitative methods for engineering systems: Why we need them and how to use them. *Syst. Eng.* **2017**, *20*, 497–511. [CrossRef]
27. Bocken, N.M.; Short, S.W.; Rana, P.; Evans, S. A literature and practice review to develop sustainable business model archetypes. *J. Clean. Prod.* **2014**, *65*, 42–56. [CrossRef]
28. Schellnhuber, H.J.; Lüdeke, M.K.B.; Petschel-Held, G. The syndromes approach to scaling describing global change on an intermediate functional scale. *Integr. Assess.* **2002**, 3, 201–219. [CrossRef]
29. Haskins, C. *Systems Engineering Analyzed, Synthesized, and Applied to Sustainable Industrial Park Development*; Norwegian University of Science and Technology: Trondheim, Norway, 2008.
30. Allen, C.; Metternicht, G.; Wiedmann, T. An Iterative Framework for National Scenario Modelling for the Sustainable Development Goals (SDGs). *Sustain. Dev.* **2017**, *25*, 372–385. [CrossRef]
31. Meyer, R. Bioeconomy strategies: Contexts, visions, guiding implementation principles and resulting debates. *Sustainability* **2017**, *9*, 1031. [CrossRef]
32. Hansen, L.; Bjørkhaug, H. Visions and expectations for the Norwegian bioeconomy. *Sustainability* **2017**, *9*, 341. [CrossRef]
33. Valseth, M.S. Innovation Norway supporting the development of an algae industry in Norway. In Proceedings of the Workshop, Nordic Algae Network and Blue Bio, Ås, Norway, 15 November 2012.
34. Kleinschmit, D.; Lindstad, B.H.; Thorsen, B.J.; Toppinen, A.; Roos, A.; Baardsen, S. Shades of green: A social scientific view on bioeconomy in the forest sector. *Scand. J. For. Res.* **2014**, *29*, 402–410. [CrossRef]

35. Norwegian Government. *Known Resources, Limitless Possibilities*; The Government's Bioeconomy Strategy; The Ministry of Trade, Industry and Fisheries: Oslo, Norway, 2016.
36. Ellingson, W. Rural Second Homes: A Narrative of De-Centralisation. *Sociol. Rural.* **2017**, *57*, 229–244. [CrossRef]
37. Organisation for Economic Co-operation and Development (OECD). *Bio-Economy and the Sustainability of the Agricultural and Food System: Opportunities and Policy Challenges*; OECD: Joint Working Party on Agriculture and the Environment: Paris, France, 2018.
38. Schellnhuber, H.J.; Block, A.; Cassel-Gintz, M.; Kropp, J.; Lammel, G.; Petschel-Held, G. Syndromes of global change. *GAIA-Ecol. Perspect. Sci. Soc.* **1997**, *6*, 18–33. [CrossRef]
39. Eisenack, K. Archetypes of Adaptation to Climate Change. In *Human-Nature Interactions in the Anthropocene*; Routledge: New York, NY, USA, 2012. [CrossRef]
40. Cavallo, M.; Gerussi, E. Bioeconomy, circular economy and industrial symbiosis: Towards a new concept of productive processes. *Eco-Ind. Parks* **2015**, 43–47.
41. Holmberg, J.; Robèrt, K.H. Backcasting—A framework for strategic planning. International. *J. Sustain. Dev. World Ecol.* **2000**, *7*, 291–308. [CrossRef]
42. Bjørkhaug, H.; Hansen, L.; Zahl-Thanem, A. *Sektorvise Scenarier for Bioøkonomien. (Sectoral Scenarios for the Bioeconomy)*; Ruralis Rapport; RURALIS: Trondheim, Norway, 2018; p. 4. Available online: https://ruralis.no/wp-content/uploads/2018/08/r4_18-sektorvise-scenarier-for-biokonomien-h-bjrkhaug-l--hansen-og-a--zahl-thanem.pdf (accessed on 29 June 2020).
43. Marchetti, M.; Vizzarri, M.; Lasserre, B.; Sallustio, L.; Tavone, A. Natural capital and bioeconomy: Challenges and opportunities for forestry. *Ann. Silvic. Res.* **2014**, *38*, 62–73.
44. Burton, R.J.F. The potential impact of synthetic animal protein on livestock production: The new "war against agriculture"? *J. Rural Stud.* **2019**, *68*, 33–45. [CrossRef]
45. Smetana, S.; Mathys, A.; Knoch, A.; Heinz, V. Meat alternatives: Life cycle analysis of most known meat alternatives. *Int. J. Life Cycle Assess.* **2015**, *20*, 1254–1267. [CrossRef]
46. Hermans, F. The potential contribution of transition theory to the analysis of bioclusters and their role in the transition to a bioeconomy. *Biofuels Bioprod. Biorefin.* **2018**, *12*, 265–276. [CrossRef]

sustainability

MDPI

Article

Cleaning Up Our Act: Systems Engineering to Promote Business Model Innovation for the Offshore Exploration and Production Supply Chain Operations

Karen V. Czachorowski

Faculty of Technology, Natural Sciences and Maritime Sciences, University of South-Eastern Norway, 3184 Vestfold, Norway; kcz@usn.no

Abstract: Oil and gas offshore exploration and production (E & P) will remain necessary to meet increasing global energy demands. However, appraising and exploring these resources has a major impact on sustainability and faces many challenges. Improving the supply chain operations that support E&P activities presents opportunities to contribute to the United Nations (UN) Sustainable Development Goals (SDGs), but relies on organizations being able to adopt new strategies and technology and, innovate their current business models. Business model innovation (BMI) has not been actively pursued in this industry, partially due to the traditional operation management and due to the complexity in changing established models or adopting full-fledged archetypes. Thus, the present study proposes a more flexible and granular approach to BMI by defining elements to be adopted rather than proposing business models archetypes. To define the elements, an application of systems engineering (SE) is adopted through a morphological analysis (MA). They are presented in morphological boxes in three dimensions—technology, organization, and the human element—inspired by sustainable business model (SBM) literature. The elements are proposed as "bricks" for BMI where they can be adopted and re-arranged as necessary, providing granularity and flexibility to facilitate BMI for organizations of varying sizes.

Keywords: offshore exploration and production; offshore supply chain operations; business model innovation; sustainable development goals; morphological analysis; systems engineering

Citation: Czachorowski, K.V. Cleaning Up Our Act: Systems Engineering to Promote Business Model Innovation for the Offshore Exploration and Production Supply Chain Operations. *Sustainability* **2021**, *13*, 2113. https://doi.org/10.3390/su13042113

Academic Editor: Cecilia Haskins
Received: 18 January 2021
Accepted: 9 February 2021
Published: 16 February 2021

Publisher's Note: MDPI stays neutral with regard to jurisdictional claims in published maps and institutional affiliations.

1. Introduction

The offshore exploration and production (E & P) industry will persist as a relevant source of energy for many years as the global energy demand continues to increase [1–3] and the renewable sources of energy are not sufficient to suppress the demand within plausible cost and at large scale [3–5]. This industry contributes with resources that generate electricity, heating, and other sub-products that are used as inputs in the fabrication of plastic, rubber, solvents, and many other items [1,5,6]. Additionally, its petroleum sub-products constitute the main source of fuel for almost all transportation modes utilized in the transport of people and goods worldwide [5–7]. However, emissions from burning fossil fuels impact the environment directly, resulting in global warming and climate change [8–11]; also, minimizing the impact of industrial activity lies within the United Nations (UN) Sustainable Development Goals (SDGs), a set of interconnected directions to guide the international community towards a sustainability agenda [10].

Offshore E & P is an industry that plans, builds, and operates offshore structures in the open sea to extract and retrieve resources through the execution of industrial activities that range from the search for oil and gas and its exploration, to transportation to shore and all steps in between [12]. Its value chain is divided into three major groups: upstream, midstream, and downstream. The upstream consists of exploration and production (E & P) activities that involve field appraisal and development, drilling, operations, maintenance, and decommission activities. Oil refining is handled midstream, while wholesale, distribution, and marketing are part of the downstream. As a general rule, upstream ends at the extraction of crude oil and its transportation to another destination [8,13]. The present study focuses on the E & P activities conducted offshore. These E & P activities are supported by supply chain operations (hereby called SC) that are conducted by a vast network of suppliers, terminals, vessels, and others operating in an intricate web that involves a large amount of money, hazards, and possible environmental impact [14–16]. Figure 1 shows this relationship. Building and delivering platforms and their parts, materials, equipment, and offshore personnel are challenging tasks that depend on numerous stakeholders that are directly and/or indirectly interrelated and inter-dependent to successfully perform the tasks [1,14,15].

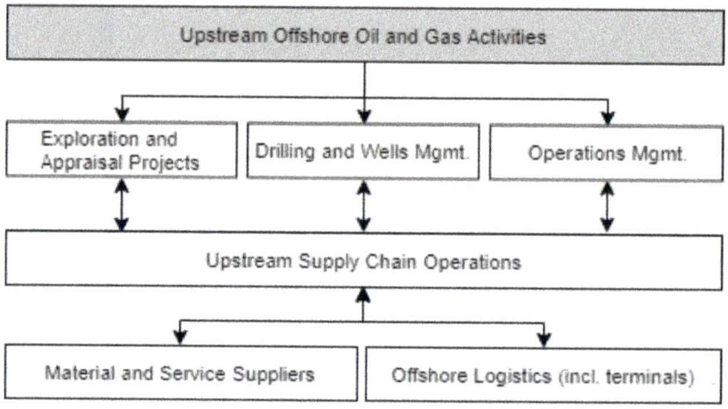

Figure 1. Offshore exploration and production (E & P) activities and its supply chain operations support.

Offshore E & P SC operations increased in complexity as the search for resources moved from shore and shallow waters to reach resources in deeper waters and more remote locations with extreme operating conditions [1,8,13]. In turn, the added complexity contributes to additional safety and environmental risks [1,17]; a higher level of environmental impact, such as carbon dioxide (CO_2) emission [1,18]; and other concerns that have subjected the industry to strict regulation. Therefore, improving these operations can help the industry achieve its SDGs, as suggested in Figure 2.

Figure 2. Supply chain management and Sustainable Development Goals (SDGs) relationships [9].

Despite many initiatives to contribute to SDGs, the industry is failing to meet its goals [4,19], especially CO_2 emission reduction [19]. The existing business models are recognized as the reason because traditional organizational frameworks hinder the industry from succeeding in its improvement efforts [20–22]. Instead, many have suggested that business model innovation (BMI) and the adoption of sustainable business models (SBMs) as a solution [23–25]. However, the literature available on BMI techniques seem to disagree in regard to what is the best approach [24,26,27]. In addition, changing, innovating, and adopting new business models may demand a near-complete organizational restructure [26,28,29], which can be challenging for established organizations and a major task for smaller ones. Therefore, a more flexible and granular approach to BMI has also been recommended [26,28].

The present study addresses this recommendation through the identification and proposal of design elements to be adopted in the BMI process instead of creating another business model archetype. Addressing complex structures, such as the offshore E & P SC operations' ecosystem, can benefit from a systematic approach [30,31]. As the industry is already familiar with systems engineering methods, a morphological analysis [32–34] was conducted to identify and define the elements proposed in this study, and Tukker's [35] product–service system framework is applied to identify the classic and the new business models in the industry. The present study's assumption is that, by adopting the identified elements at different levels in the organization, they work to improve the offshore E & P SC operations and thus contribute to the SDGs. The identified elements are examined in morphological boxes, as presented by Kley, Lerch, and Dallinger [36], and are presented in three main dimensions—technology, organization, and the human element—inspired by the SBM archetypes proposed by Bocken, Short, Rana, and Evans [37]. Figure 3 shows the conceptual framework adopted in this study.

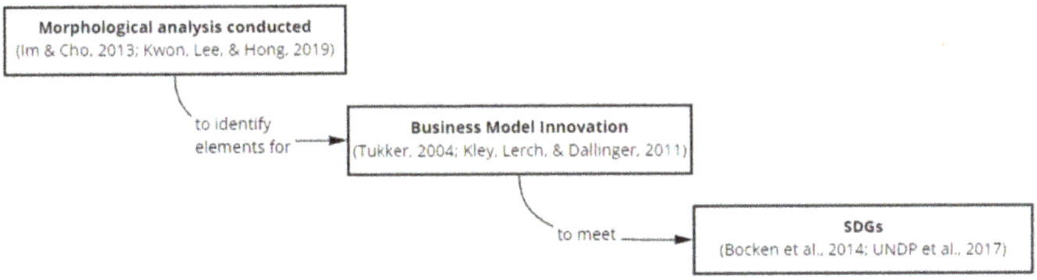

Figure 3. Conceptual framework and study contribution: relationship between morphological analysis (MA) and SDGs.

The study then continues as follows. This section continues to present a literature review on business models and BMI in the context of offshore E & P and its supply chain operations. Section 2 presents the morphological analysis conducted; Section 3 presents the identified elements to be adopted for BMI in morphological boxes and discusses their characteristics. Section 4 discusses how the proposed elements related to "new business models" typology can address SDGs and value proposition, creation, and capture in the context of offshore E & P industry. Finally, Section 5 presents conclusions and future research directions.

Business models and business model innovation. Existing business models have been widely discussed, particularly addressing value realization from technology innovation [20,38,39]. Even though many definitions exist in terms of what a business model is [27], the present study adopts the concept that a business model is a structural tool for companies to operate, manage, assess performance, and innovate their business [37,40,41]. Focusing on technology, Chesbrough [42] argued that there are four functions of a business model: (1) to articulate the value proposition for users, (2) detect a market segment where the technology has a purpose, (3) state the value chain involved in realizing the offering, and (4) evaluate the cost structure and profit potential from the offering(s). Trott [43] complements these definitions by saying that a business model is the framework for an organization to realize profit through the successful creation, marketing, and value delivery to its customer. As such, the business model framework consists of three elements: (1) value proposition, (2) value creation and delivery, and (3) value capture [37,41]. Value proposition concerns the product and/or service offered in order to generate financial return, and which consumer segments are in focus [37,44]. Value creation and delivery concerns the resources utilized to deliver the value proposition, including the performed activities, partners, distribution methods, and technology [37,41]. Finally, value capture concerns how revenue is captured, including cost reductions [37,40,41].

According to Chesbrough [42], beyond describing how an organization works and generates value, these elements present opportunities to capture value from innovation.

Zott et al. [27] affirmed that business models are vehicles for innovation in organizations and subjected to innovation themselves to fulfill that role. Consequently, a business model is not a static framework that an organization must follow, but a transient, dynamic system that must change and adapt so that the organization remains viable and successful in the long term [28,45]. This continuous change process is referred to as business model innovation [28]. Innovating business models go beyond the creation of new products and services. It requires managers to break their cognitive barriers towards going beyond their own or their organization's culture of conservatism and passiveness towards adopting new elements [39,46]. Business model innovation depends on the evaluation of how the business model's components work towards the organization's desired outcome, and a few tools are available to assist this process, such as the business model canvas by Osterwalder and Pigneur [41], the St. Gallen business model navigator [39], and the triple-layered business model canvas [47], which focuses on designing sustainable business models.

The offshore E & P industry has a traditional approach to business models, and disruption of these models requires penetration of a high barrier imposed by conservatism and vested interest [48–50]. Yet, technological advances and regulations are having an effect in fostering the disruption of the traditional business models either by necessity or competitive advantage [50–52]. Many changes in the oil and gas industry occur in response to specific situations, such as low oil prices, and external pressure such as legislation and other regulations [48,50,53]. Thus, in times where challenges are constantly threatening "business as usual", innovation provides opportunities and casts light on possibilities for the industry. Early initiatives proved the possibility of success in changing business models through innovation, such as through coopetition and the creation of joint ventures and alliances [54–56]. A good example comes from the oil and gas industry in Norway, where an alliance allowed Aker BP's, a Norwegian operator, to deliver the Valhall Flank West exploration platform in 14 months and under budget, from the first steel to first oil [57]. This approach to the sharing economy proves that re-thinking how to conduct operations together with suppliers and even competitors can have a long-term positive impact on securing cash flow and stability in low markets, even if profits may be slightly lower. It translates the idea of selling/buying a specific service, rather than managing the whole supply chain. Such innovative initiatives affect business models as they disrupt many parts of the industry. Other innovations include the use of drones for delivery and inspection, robots for performing risky tasks, automation for operational efficiency, 3D printing, etc. [21,22,53]. While not all of these initiatives are fully mature, they show that stakeholders can become cooperative partners with a higher degree of resource sharing to create a service-oriented culture and a communal approach to offshore E & P SC operations. They are also good examples of why and how re-evaluating existing business models is necessary.

2. Materials and Methods

The purpose of this study is to define elements that can be adopted for BMI in the offshore E & P SC, identified through a morphological analysis (MA) [32–34,58]. According to Martin [58], a morphological analysis is conducted to perform a systematic classification and assessment of possible combinations of alternatives that can, together, provide a certain function. To identify the elements to be included in the analysis, business models, BMI, SBMs, offshore E & P, and E & P SC literature was examined (collected from databases such as Scopus and Web of Science, among others); and information available from oil and gas related organizations (such as IPIECA and DNV-GL, among others) was collected through publications and reports made available by these organizations.

In the sequence, the identified elements were classified through the application of Tukker's [35] business model typology, which lead to the identification of the elements that belong to classic business models and those that belong to new business models. Tukker [35] classified a business model by the way it generates value, placing it within a range that starts at value generation mainly from products (tangible) towards value generation from services (intangible). At an operational level, the value generation process moves from being product-oriented towards being use-oriented and result-oriented [35]. Applying this framework (Figure 4) to the offshore E & P industry supply chain results in three main categories: (1) the operator owns and/or manages all parts of its supply chain; (2) the supply chain operations are purchased/offered as a service (defined by its use), and (3) the operations have a communal approach, where assets and others are shared within the supply chain, towards the outcome of the service. The first category represents the classic business models in the industry, whereas the other two range towards new business models.

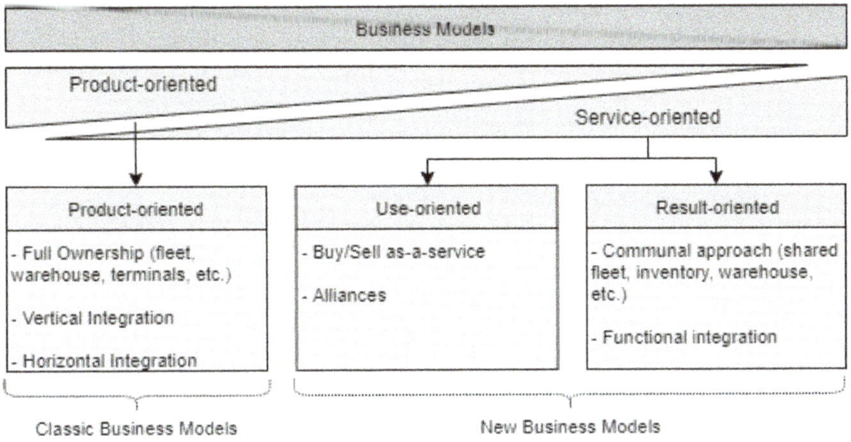

Figure 4. Tukker's [35] typology applied to the offshore E & P supply chain (SC).

To present the elements and their characteristics, the concept of morphological boxes is applied, adapted from Kley et al. [36]. Subsequent to a morphological analysis, morphological boxes "represent a creative way to illustrate all the potential solutions to existing problems in a structured format by defining different features with several configurations with regard to a problem" [36] (p. 3395). Figures 5–7 present three morphological boxes with the identified design elements for business model innovation. Addressing sustainability, these elements are presented based on a list of characteristics related to the three dimensions adapted from Bocken, Short, Rana, and Evans [37]; that is, technology (Figure 5), organization (Figure 6), and the human element (Figure 7). Each box follows a logic in its presentation regarding the infrastructure and organizational changes needed for adoption—the elements more to the left are less complex and require the adoption of less extensive changes. On the other hand, the elements more to the right are more complex and thus require more changes. The proposed elements placed most to the right of the morphological boxes are suggested as an ultimate goal for organizations to adopt, although their complexity means it might not be possible for organizations to adopt them at first. For this reason, the other elements situated to their left in the boxes are proposed to be adopted as interim stages during business model innovation. After presenting the elements to each factor, I discuss how the elements classified in the new business models typology address sustainability and their value proposition, creation, and capture.

3. Results

3.1. Technology-Related Elements: Characteristics and Contributions to SDGs

The morphological box in Figure 5 presents elements and their characteristics related to technological development and deals with different aspects of technology use and application in the offshore E & P industry. From left to right, the presented elements increase system interoperability, proposing a higher level of automation as a value for a business model if the alternative is to be adopted.

Characteristics	Design Elements			
Operations handling & software iteration	Manual Operations	Semi-Autonomous	Automated	Intelligent
Infrastructure	On-Premise	On-Premise/Cloud	Data Warehouse /Cloud	Cloud
Information Flow	Manual / Siloed	Point-to-point integration	Event-based	Orchestrated
Communication Flow	Manual (email, phone, etc.)	Semi-automated	Automated	
Data Collection	Manual input	Hand-held devices	Sensor-based /Beacons	Sensor/beacon + robots
Data Analysis	Manual (excel, etc.)	Historical data + actual	Machine learning and AI	
Cash Flow	Limited cost reduction	Reduced/shared costs	Fee-based (service usage)	

Classic Business Models — New Business Models

Complexity and required level of change

Figure 5. Morphological box for technology-related elements of business models (adapted from [36]).

The first row of design elements depicts technology applications to conduct basic operations and the reliance on manual interaction with systems and software. It ranges from operations being conducted manually, where human resources are needed to input, verify, and exchange information and data throughout the software portfolio in order for operations to be fulfilled. On the other extreme of the array, the presented element is understood as an intelligent element, where the data collected from operations using digital technologies are utilized as input for the system to conduct its own operations, based on algorithms and possibly machine learning and artificial intelligence. Manual interaction can still occur, but it is shifted from data input towards data usage for the supervision of operations, higher-level decision-making and operations' reporting indicators. The next characteristic, infrastructure, evolves from left to right to support the level of automation proposed in line with other elements. The same logic applies to the other characteristics and their elements.

The last element, cash flow, proposes how operational cost is reduced by each element as they help diminish the manual interference to daily activities, manual errors, and other potential issues caused by the lack of integration between activity, process, and software. Its first element relates to the possible level of cost cutting when operations are conducted mostly manually, and technology infrastructure and administration is mostly managed in-house. Once the elements evolve towards interoperability and automation, the cost reduction possibilities increase, as operations are not reliant on manual interactions and cost migrates to payment per service usage. The elements presented for each characteristic introduce stability, efficiency, and resistance to human error due to the increased automation levels they proportionate.

3.2. Organization-Related Elements: Characteristics and Contributions to SDGs

Similar to Figure 5, the presented design options for the organizational characteristics (Figure 6) evolve from left to right with higher levels of interrelationship. In this dimension, the further to the left the element is, the more internal it is to the organization and silo based. The further to the right, the higher the level of partnership and collaboration with stakeholders is necessary and achieved. The presented organizational elements offer value to the business model by providing the opportunity for organizations to change processes that are inefficient and outsource activities that are not core to the organization. This also creates value to the other stakeholders in the value chain, as each stakeholder can become more specialized in its functional area, to innovate and reduce cost by eliminating non-core related activities.

Characteristics	Design Elements			
Structure	Closed/vertically integrated	Horizontally Integrated	Alliances	Communal
Planning Horizon	Short-term	Mid-term	Long-Term	Very Long-term
Operational Planning Strategy	Reactive	Proactive	Strategic	
Soucing Strategy	High number of suppliers	Few strategic suppliers	Pre-certified suppliers + market bidding	
Procuring Invoicing /payment strategy	Manual Purchase order and invoicing	Semi-automated Purchase order and invoicing	Contract/activity triggered – auto Purchase Order and invoicing	
Inventory strategy	Owned	Owned + rented	VMI (Vendor Managed Inventory)	Strategic items/parts at the supplier
Warehouse / terminal strategy	Owned	Leased/rented	At the supplier	
Contract Strategy	Long-term	Spot	Frame-agreement	Strategic / Incentive-based (ex. MLC[1])
Transport Sourcing Strategy	Owned	Long-term rental (FFA)	Spot market and/or Pool	Fully shared (freight-based)
Transport Operations	Manual set-up	Pre-established routes/date plan	Activity and cargo based through Algorithm determination	
Transport Strategy	Exclusive vessel	Shared vessel (intra-company)	Shared vessel (agreed multi-company)	Freight/deck space based (ex. log. hub)
Cash Flow	Limited cost reduction	Reduced/shared costs	Fee-based (service usage)	

Classic Business Models — New Business Models

Complexity and required level of change

Figure 6. Morphological box for organization-related elements of business models (adapted from [36]).

The first characteristic in Figure 6 specifies the organizational structure of the offshore E & P SC, ranging from closer operations handling within the operator (owned vessels, warehouses, etc.) towards partnerships, alliances (strategic suppliers are preferred) and sharing assets and other resources. For example, if the structure consists of alliances involving strategic suppliers, the sourcing strategy will consist of selecting these suppliers that already have certifications and contracts relevant to the operations to be conducted. This logic also applies to the invoicing and payment strategy to be adopted, as well as to the inventory, warehouse/terminal, contract, and transport sourcing ones. Accordingly, the operational planning strategy is replicated through the organizational structure adopted. Following the same example, if the suppliers involved are pre-selected and certified through supply chain processes by the business units responsible for that process, a longer activity span can be planned towards that partnership/alliance, which provides a longer-term planning range. This directly impacts operations, as a longer planning range produces visibility and better resource allocation. Transport, operations, and sourcing strategies are also interrelated and highly dependent on the structure adopted. Continuing to follow the above example, a fewer number of selected suppliers participate in the majority of operations, which makes it possible to share a higher level of information and data within the network that can be used for operations optimization and cargo allocation, reducing the manual input to handle operations. Therefore, data from operations becomes the trigger and input towards reserving space and capacity for sourcing transport (that is, transporting something from A to B for a fee) and handling transportation resource sourcing (that is, the number of vessels needed to deliver items related to an activity). The way in which these elements are combined and selected directly impacts the value capture element proposed—cash flow. If the operational elements adopted relate to a closed structure, the cost reduction opportunities are limited as there is lower flexibility in terms of pooling suppliers and negotiation. On the other side of the array, costs are related to the used services, depending on operational optimization and resource usage.

3.3. Human Element-Related Elements: Characteristics and Contributions to SDGs

In the offshore E & P SC context, the human element is two-fold, consisting of (1) the people working on the offshore platforms, drilling rigs and FPSOs (floating production storage and offloading), and (2) the people working onshore planning and executing commands towards the operations execution. Hence, both aspects need to be considered when evaluating the impact on the human element and its place in the organizational structure and business model. The morphological box in Figure 7 presents business model elements related to the management of the human element.

Characteristics	Design Elements			
Operation Handling	Full Interaction	High interaction	Low interation	No interaction
Work Organization	Business Unit-centric	Asset/Platform centric	Alliance centric	Per contract
Work Specialization	Low (manual input / workers)	Medium (technicians, business analysts)	High (divers, offshore drilling superintendent)	Very High (drilling engineers)
Work Schedule	Fixed hours	Fixed rotation	Flexible/ad-hoc	Activity and/or contract based
Work Location	Onshore office	Offshore	Onshore Control centers/terminals	Flexible
Contract/ Payment strategy (Cash Flow)	Fixed contract	Temporary contract (short and mid-term)	Flexible/ad-hoc (hourly rates)	Activity based (agreed price for an activity)
	Classic Business Models		New Business Models	

Complexity and required level of change

Figure 7. Morphological box for human-related elements of business models (adapted from [36]).

The proposed human-related elements related to operations handling, work organization, specialization, and schedule can nearly or completely eliminate the human involvement in supply chain operations. However, this is not necessarily the intention. The idea is that these elements change how and where humans interact and fulfill activities in the organizations. Hence, when related to humans directly, the presented elements show increased specialization and reduced commitment to one specific contract/organization from left to right in Figure 7. This creates value to organizations by being able to work with highly skilled professionals without worrying about talent retention. However, this does not mean that the link to the organization should be weaker; it simply provides the organization and the professionals with the opportunity to work with multiple parties if they want, and gives flexibility to the professionals to elect the best possible places to work, addressed by the work schedule and location elements. This allows organizations to reduce personnel costs as they will not have to maintain several idle people that do not have constant activity in the company simply to avoid missing the skilled professional to another organization, addressed by the contract/payment strategy element.

4. Discussion

There are many ways in which the offshore E & P industry can contribute to the SDGs [9], although most efforts are directed to minimizing the impact of drilling and platform operations [1,9,19], such as water and waste management [1,59], and managing CO_2 emissions [1,3,60]. However, many opportunities exist in the management of the supply chain operations conducted to support the upstream activities.

In the offshore E & P industry, the supply chain is a complex "ecosystem" that includes many internal and external stakeholders that must comply with many different legislations and overcome many challenges. The SC operations are conducted by specialized suppliers, supply bases, terminals and warehouses (referred to hereafter as support companies) that provide vessels, transportation and storage, manning, and other services, usually managed by the operator hiring them [14,61,62]. The vessels involved in the offshore E & P SC to complete these activities are major contributors to CO_2 emissions in the offshore E & P industry [1] and most sustainable initiatives in this context are linked to the reduction of CO_2 emission from the vessels utilized in transportation, optimal vessel and route allocation, vessel fuel usage, and route optimization for transportation [14,63,64]. Lately, the adoption of technology for this purpose has been widely discussed, such as the implementation of 3D printing of spare parts for local supply to avoid transportation; and the adoption of Blockchain, Internet of Things (IoT), and digital twins for supply chain transparency [19,65,66]. For vessel and deck space optimization, the adoption of big data analytics and machine learning/artificial intelligence has been widely discussed [67–69].

However, there are many other ways in which supply chain operations can improve to contribute to SDGs, such as a more efficient management of inventory to increase inventory usage and reduce new parts purchasing, reducing double-purchases, fostering inventory sharing throughout the supply network, and other initiatives that could be adopted to reduce the production of new items that rely on global natural resources [4,70,71]. A more collaborative approach to supply chain handling can also provide results; instead of each operator managing their own vessels, these can be pooled to promote sharing vessel deck transportation capacity, leading to a better vessel capacity allocation and an overall reduction in the number of vessels and vessel voyages required to transport cargo [4,70,72]. Table 1 summarizes potential areas of change that would effectively contribute to meeting SDGs.

Table 1. Summary of potential contribution to SDGs.

SDG Number	Main Potential Contribution
SDG 14	Minimize the impact of drilling and platform operations; water and waste management in platforms and vessels
SDG 13	Managing CO_2 emission
SDG 12	More efficient management of inventory leads to better purchasing behavior and less consumption
SDG 9	Inventory sharing throughout the supply network reduces the production of new items that rely on global natural resources
SDG 7	Less stranded inventory leads to less scrapping, leading to higher savings and potentially reducing the final cost of energy provided.

Therefore, a more flexible and granular approach to setting and meeting goals toward SDGs is suggested for designing business models that can support innovation, technology adoption, and other changes without having to restructure the whole business model and/or organization. The objective of the present study is to present elements that can be more easily adopted and replaced as wished by organizations to create new business models. Systems engineering methods are ideal to support this objective as they promote a systematic approach to solving an array of complex issues in systems present in several domains. The SE method adopted in this study—morphological analysis—provides this systematic approach and supports non-quantified modeling that allows an in-depth analytical exploration to solve a complex issue [30,31,73]. Specifically, the present study used SE methods to identify and propose design elements for BMI. The resulting elements presented function as "construction bricks" that, once combined, allow the creation/innovation of a business model that fits the organizations' objectives and addresses sustainability at the same time at different intra- and inter-organizational levels. This flexibility enables organizations of different types and sizes to implement design elements that address sustainability without having to replace their entire existing business model or implement traditional sustainable business model archetypes. Therefore, if one desired element cannot be implemented initially, another one can be utilized instead as an interim state, while the desired element works as a final goal. Beyond sustainability, the elements are proposed to foster technology adoption where it is needed, organizational innovation and collaboration as value creation. These elements are presented according to Tukker's [35] typology, categorized as "classic business models" and "new business models". The elements included in the latter are suggested for the offshore E & P industry to adopt to manage its supply chain operations. Once combined, these elements address SDGs in different ways, and their recombination by the organizations can be re-adjusted to meet organization's strategies and purpose. How these elements address business models' value proposition, creation, and capture and contribute to the SDGs is summarized in Figure 7 and discussed in the sequence.

4.1. Value Proposition

The elements value proposition is to offer the integration of offshore E & P SC operations through suitable technology to allow the stakeholders to reach a communal approach to operations handling. By integrating operations, operators and other stakeholders can make better usage of their own operational data, which provides insight regarding the items to be purchased and transported and the services needed. Consolidating this demand through data collection and predictive analysis provides opportunities for the support companies to adjust their fleet according to the demand and forecast with a higher accuracy, which, in turn, allows the support companies to adjust the level of service to the capacity needed. Thus, vessels and other resources can be pooled and shared instead of each operator renting its own. The proposed elements also address the interconnection among vessels, ports/terminals, platforms, other stakeholders and the human element in the industry. This is essential to prepare the industry for implementing innovation and other technological developments such as autonomous vessels, cranes, ports, etc. These are in constant evolution but cannot be adopted if the industry is not ready to adapt. The autonomous assets must be able to intercommunicate to work beyond their technical aspects—an autonomous vessel still needs to dock, just like an autonomous crane still needs to know what it will lift, from where, and where to place the cargo. However, manual interference is not the solution, as machines must be interoperable through machine understandable languages. Hence, the automated and "intelligent" elements in this category define the solutions offered and the stakeholders involved towards operations' completion.

4.2. Value Creation

The elements create value creation through the adoption of automation and intelligent elements to promote a more efficient handling of the supply chain operations and to allow supply chain partners to offer a better service to its customers. Re-thinking how the stakeholders collaborate, communicate, and work is a necessity not only for the stakeholders themselves, but also for the industry to incorporate and benefit from the innovations brought from technological advances being presented to address efficiency, safety and environmental concerns. Even though the elements emphasize integration and automation, people are not eliminated from the system. People have unique characteristics, such as creativity, negotiation, and problem-solving, that extend beyond what was coded into a software for conducting operational tasks. However, people are being used as a cog in the system, whereas they should be placed where interpersonal skills are needed the most, leaving the machines to do what they do best and enabling people to do what they do best: create, develop, and innovate. Hence, technology is part of this change, and technological efforts to digitalize the synergies among the network participants are needed for them to work together. Similarly, it is just as necessary for the offshore E & P domains to extend their willingness for technology to go beyond machines and engineering to reach supply chain operations, so that innovative ways of working can be developed and implemented, thus creating value throughout the organization's value chain.

4.3. Value Capture

The elements capture value by reducing costs and waste through better utilization of resources, such as fleet, vessels, inventory, personnel, etc. This promotes operational efficiency and reduces operational costs, increasing revenue and additional profit opportunities for organizations. Using inventory as an example: better inventory visibility enables a more efficient management of inventory and purchasing, which enables the use of the available items instead of unnecessary double-spending for "emergency" purposes, thus reducing the overall quantity of general and unused inventory. With less stranded inventory, less scrapping is executed, and more money and taxes are saved and made available to be used elsewhere, potentially reducing the final cost of energy provided.

4.4. SDGs

The "new business models'" elements address SDGs by changing how work is conducted, as summarized in Figure 8. This shift from manual work to automation reduces human interactions in operations, resulting in fewer error opportunities that could lead to accidents, thus enhancing safety in offshore logistics operations. Together with automation, increased collaboration allows streamlining planning and provides operational synchrony. This results in better use of resources such as vessels, leading to a higher level of vessel deck capacity usage and a reduced number of voyages from and to the offshore platforms, contributing to safety and addressing environmental concerns due to reduced chances of spillage and reductions of CO_2 levels, in line with the UN's SDGs. Finally, better purchasing and inventory handling reduces the number of items that need to be purchased, which reduces the need for transportation and fabrication. With fewer items to produce, less raw material is needed, reducing the depletion of natural resources.

Value Proposition	Value Creation and Delivery	Value Capture	SDGs
An automation-based, data-based and integrated approach to conducting supply chain operations to upstream oil and gas, deriving a communal approach to asset usage and capacity handling.	Through a communal approach, sharing resources leads to a more efficient resource allocation, decreasing the number of resources needed and their usage to conduct operations.	Cost reduction through better utilization of resources; Increased profit/revenue opportunities from selling stranded capacity (e.g. extra space in vessels)	Focus on SDGs #13 (Take urgent action to combat climate change and its impacts) and #14 (Conserve and sustainably use the oceans, seas and marine resources for sustainable development)

Figure 8. Value proposition, creation, capture and contribution to SDGs from the elements proposed in the "new business models" category (adapted from [37]).

5. Conclusions

Potential new and adaptive business models that can lead to the success of innovations and their diffusion are yet to be widely explored. In the meantime, new products, services, and technology are constantly being released to the market. This gap between technological advances and organizational needs must be addressed if the industry is to succeed in a demanding future, where regulation is increasingly challenging due to environmental worries, technology develops faster than ever, and a globalized world created a complex industry that requires pioneering alternatives to overcome constraints and competition. The offshore E & P industry must research, develop, and adopt not only technical innovation, but also have a holistic approach to business model innovation that will help understand which innovation and technological initiatives will promote sustainability and value creation in the organizations and their industry context. The existing business models that are available no longer meet the industry's needs to overcome its challenges and address sustainability and SDGs. However, finding and adopting business model archetypes that can handle such complexity and allow adopting innovative solutions might not be possible. At the same time, the existing archetypes might be too complex or robust for organizations to adopt as they imply changes necessary throughout the whole organization.

This conceptual study has applied systems engineering methods to explore the business model innovation possibilities in the offshore E & P industry and its support ecosystem. It offers a more flexible and granular approach to business model innovation through the adoption of SE to propose elements to be adopted interchangeably that can be adopted at different organizational levels and timeframes and can function as interim or final stages for the organization. Through examining business models typology and the conduction of a morphological analysis, possible elements for business models' innovation are identified according to classic and new business model typology and presented against three dimensions: technology, organization and the human element. The elements are presented in morphological boxes and can be combined and reorganized to change and build different business models. Given the limitations that more established firms may face in adopting certain elements, this study proposes elements as pieces for the organizations to adopt instead of full-fledged archetypes, providing modularity and granularity to organizations to replace certain parts when necessary without having to change the whole business model many times. The presented elements show different levels of complexity and organizational change for their adoption, and the more complex ones are suggested as an ultimate goal for adoption in business model innovation.

By adopting the elements in the new business model category, the expected end-state for the offshore E & P SC is an ecosystem that includes stakeholders in the network as collaborative partners to deliver higher operational standards. These include taking responsibility not only for operational execution, but also over safety and the environment, thus addressing sustainability and SDGs in offshore E & P operations. How these elements propose, create, and capture value has also been discussed. These elements create value through giving organizations an opportunity to become more strategic as they shift the daily activities from manually conducted to automated, conferring stability and reliability to activity execution and, therefore, generating value from technology application and organizational restructuring. Through more efficient information sharing, information propagates to stakeholders in the value chain more quickly, which can change how the stakeholders conduct their activities as well, taking the opportunity to remove inefficiencies from their part of the operations. Finally, as the different design elements are adopted by offshore E & P companies through business model innovation, a new method of collaboration within the industry can surge and evolve the industry to an ecosystem that addresses sustainability, innovation, the organization, technology application, and focus on their consequences to the human element. The measurement of the extent to which these presented elements address sustainability is a limitation of this study and is suggested as future research.

Funding: This research was funded by the industrial PhD funding by the Norwegian Research Council and Aker BP, grant number 291198 and open access funding was financed by the University of South-Eastern Norway.

Institutional Review Board Statement: Not applicable.

Informed Consent Statement: Not applicable.

Data Availability Statement: Not applicable.

Acknowledgments: The author is grateful to Associate Ziaul Haque Munim and Cecilia Haskins for feedback on earlier drafts of the manuscript.

Conflicts of Interest: The author declares no conflict of interest. The funders had no role in the design of the study; in the collection, analyses, or interpretation of data; in the writing of the manuscript, or in the decision to publish the results.

References

1. Cassagne, T.; International Petroleum Industry Environmental Conservation Association; International Association of Oil & Gas Producers. *Environmental Management in the Upstream Oil and Gas Industry*; IOGP: London, UK, 2020.
2. Alvik, S.; Bakken, B.E.; Onur, O.; Horschig, H.; Koefoed, A.L.; McConnel, E.; Rinaldo, M.; Shafiei, E.; Zwarts, R.J. *Energy Transition Outlook 2020—A Global and Regional Forecast to 2050*; DNV GL: Hovik, Norway, 2020.
3. International Energy Agency. *The Oil and Gas Industry in Energy Transitions: Insights from IEA Analysis*; IEA: Amsterdam, The Netherlands, 2020.
4. Alvik, S.; Bakken, B.E.; Onur, O.; Horschig, H.; Koefoed, A.L.; McConnel, E.; Rinaldo, M.; Shafiei, E.; Zwarts, R.J. *Energy Transition Outlook 2020—Power Supply and Use*; DNV GL: Hovik, Norway, 2020.
5. Rodrigue, J.-P. *The Geography of Transport Systems*, 5th ed.; Routledge: New York, NY, USA, 2020.
6. Demirbas, A.; Bafail, A.; Nizami, A.-S. Heavy oil upgrading: Unlocking the future fuel supply. *Pet. Sci. Technol.* **2016**, *34*, 303–308. [CrossRef]
7. U.S Energy Information Administration. Energy Use for Transportation. Use of Energy Explained. 2020. Available online: https://www.eia.gov/energyexplained/use-of-energy/transportation.php#:~{}:text=Petroleumisthemainsource,innaturalgaspipelinecompressors (accessed on 19 November 2020).
8. Oil Industry International Exploration and Production Forum; United Nations Environment Programme Industry and Environment Centre. *Environmental Management in Oil and Gas Exploration and Production: An Overview of Issues and Management Approaches*; UNEP: Oxford, UK, 1997.
9. United Nations Development Programme; International Finance Corporation; International Petroleum Industry Environmental Conservation Association; Center on Sustainable Investment. Mapping the Oil and Gas Industry to the Sustainable Development Goals: An Atlas. United Nations. 2017. Available online: http://www.ipieca.org/resources/awareness-briefing/mapping-the-oil-and-gas-industry-to-the-sustainable-development-goals-an-atlas/ (accessed on 19 November 2020).
10. United Nations. United Nations Sustainable Development Goals—Climate Change. UN Knowledge Platform. 2015. Available online: https://sustainabledevelopment.un.org/post2015/transformingourworld/publication (accessed on 19 November 2020).
11. Zalasiewicz, J.; Williams, M.; Steffen, W.; Crutzen, P. The new world of the anthropocene. *Environ. Sci. Technol.* **2010**, *44*, 2228–2231. [CrossRef] [PubMed]
12. Van Dokkum, K. *Ship Knowledge—A Modern Encylopedia*; Dokmar: Enkhuzien, The Netherlands, 2003.
13. Inkpen, A.; Moffett, M.H. *The Global Oil & Gas Industry: Management, Strategy & Finance*; PennWell Books: St. Tusla, OK, USA, 2011.
14. Aas, B.; Halskau, Ø., Sr.; Wallace, S.W. The role of supply vessels in offshore logistics. *Marit. Econ. Logist.* **2009**, *11*, 302–325. [CrossRef]
15. Olesen, T.R. *Offshore Supply Industry Dynamics: Business Strategies in the Offshore Supply Industry*; Copenhagen Business School: Copenhagen, Denmark, 2016.
16. Stopford, M. *Maritime Economics*; Routledge: New York, NY, USA, 2013.
17. American Petroleum Institute; International Association of Oil & Gas Producers. Sustainability Reporting Guidance for the Oil and Gas Industry. 2020. Available online: https://www.ipieca.org/media/5115/ipieca_sustainability-guide-2020.pdf (accessed on 19 November 2020).
18. Jære, L.; Norwegian Environment Agency. Environmental Impact of Oil and Gas Activities. *BarentsWatch*, 25 August 2016. Available online: https://www.barentswatch.no/en/articles/Environmental-impact-of-oil-and-gas-activities/ (accessed on 19 November 2020).
19. Bech, C.; Rashidbeigi, S.; Roelofsen, O.; Speelman, E. The future is now: How oil and gas companies can decarbonize. *McKinsey*, 7 January 2020. Available online: https://www.mckinsey.com/industries/oil-and-gas/our-insights/the-future-is-now-how-oil-and-gas-companies-can-decarbonize (accessed on 21 November 2020).
20. Baden-Fuller, C.; Haefliger, S. Business models and technological innovation. *Long Range Plan.* **2013**, *46*, 419–426. [CrossRef]
21. Forbes-Cable, M.; Liu, W. Digital Disruption: Upstream Supply Chain Threats and Opportunities. Available online: https://www.woodmac.com/reports/upstream-oil-and-gas-digital-disruption-upstream-supply-chain-threats-and-opportunities-310260/ (accessed on 19 November 2020).
22. World Economic Forum and Accenture. Digital Transformation Initiative—Oil and Gas Industry. 2017. Available online: http://reports.weforum.org/digital-transformation (accessed on 19 November 2020).
23. Eglash, S.; Fisher, K. *Sustainable Energy for All: Opportunities for the Oil and Gas Industry*; Accenture: New York, NY, USA, 2012.

24. Evans, S.; Vladimirova, D.; Holgado, M.; Van Fossen, K.; Yang, M.; Silva, E.A.; Barlow, C.Y. Business model innovation for sustainability: Towards a unified perspective for creation of sustainable business models. *Bus. Strat. Environ.* **2017**, *26*, 597–608. [CrossRef]
25. Florescu, M.S.; Ceptureanu, E.G.; Cruceru, A.F.; Ceptureanu, S.I. Sustainable supply chain management strategy influence on supply chain management functions in the oil and gas distribution industry. *Energies* **2019**, *12*, 1632. [CrossRef]
26. Al-Debei, M.M.; Avison, D. Developing a unified framework of the business model concept. *Eur. J. Inf. Syst.* **2010**, *19*, 359–376. [CrossRef]
27. Zott, A.; Aamit, R. The business model: Recent developments and future research. *SSRN Electron. J.* **2011**, *37*, 1019–1042. [CrossRef]
28. Bucherer, E.; Eisert, U.; Gassmann, O. Towards systematic business model innovation: Lessons from product innovation management. *Creat. Innov. Manag.* **2012**, *21*, 183–198. [CrossRef]
29. Veit, D.; Clemons, E.; Benlian, A.; Buxmann, P.; Hess, T.; Kundisch, D.; Leimeister, J.M.; Loos, P.; Spann, M. Business models: An information systems research agenda. *Bus. Inf. Syst. Eng.* **2014**, *6*, 45–53.
30. Arciszewski, T. Morphological analysis in inventive engineering. *Technol. Forecast. Soc. Chang.* **2018**, *126*, 92–101. [CrossRef]
31. Buzuku, S.; Farfan, J.; Harmaa, K.; Kraslawski, A.; Kässi, T. A case study of complex policy design: The systems engineering approach. *Complexity* **2019**. [CrossRef]
32. Im, K.; Cho, H. A systematic approach for developing a new business model using morphological analysis and integrated fuzzy approach. *Expert Syst. Appl.* **2013**, *40*, 4463–4477. [CrossRef]
33. Kwon, M.; Lee, J.; Hong, Y.S. Product-service system business modelling methodology using morphological analysis. *Sustainability* **2019**, *11*, 1376. [CrossRef]
34. Zwicky, F. *Discovery, Invention, Research through the Morphological Approach*; MacMillan: New York, NY, USA, 1969.
35. Tukker, A. Eight types of product-service system: Eight ways to sutainability? Experiences from suspronet. *Bus. Strateg. Environ.* **2004**, *260*, 246–260. [CrossRef]
36. Kley, F.; Lerch, C.; Dallinger, D. New business models for electric cars—A holistic approach. *Energy Policy* **2011**, *39*, 3392–3403. [CrossRef]
37. Bocken, N.M.P.; Short, S.W.; Rana, P.; Evans, S. A literature and practice review to develop sustainable business model archetypes. *J. Clean. Prod.* **2014**, *65*, 42–56. [CrossRef]
38. Ahmad, N.K.W.; De Brito, M.P.; Rezaei, J.; Tavasszy, L.A. An integrative framework for sustainable supply chain management practices in the oil and gas industry. *J. Environ. Plan. Manag.* **2016**, *60*, 577–601. [CrossRef]
39. Gassmann, O.; Frankenberger, K.; Csik, M. The St. Gallen business model navigator. *Int. J. Prod. Dev.* **2013**, *18*, 249–273.
40. Bouncken, R.B.; Kraus, S.; Roig-Tierno, N. Knowledge- and innovation-based business models for future growth: Digitalized business models and portfolio considerations. *Rev. Manag. Sci.* **2019**, *20*, 1–14. [CrossRef]
41. Osterwalder, A.; Pigneur, Y. *Business Model Generation: A Handbook for Visionaries, Game Changers, and Challengers*; John Wiley & Sons: Hoboken, NJ, USA, 2010.
42. Chesbrough, H. The role of the business model in capturing value from innovation: Evidence from Xerox Corporation's technology spin-off companies. *Ind. Corp Chang.* **2002**, *11*, 529–555. [CrossRef]
43. Trott, P. Business Models. In *Innovation Management and New Product Development*, 6th ed.; Pearson: New York, NY, USA, 2017; pp. 412–444.
44. Chesbrough, H. Business model innovation: Opportunities and barriers. *Long Range Plan.* **2010**, *43*, 354–363. [CrossRef]
45. Morris, M.; Schindehutte, M.; Allen, J. The entrepreneur's business model: Toward a unified perspective. *J. Bus. Res.* **2005**, *58*, 726–735. [CrossRef]
46. Markides, C. Disruptive innovation: In need of better theory. *J. Prod. Innov. Manag.* **2006**, *23*, 19–25. [CrossRef]
47. Joyce, A.; Paquin, R.L. The triple layered business model canvas: A tool to design more sustainable business models. *J. Clean. Prod.* **2016**, *135*, 1474–1486. [CrossRef]
48. Ebneyamini, S.; Bandarian, R. Explaining the role of technology in the dynamics of the players business models in the global oil playground. *Int. J. Energy Sect. Manag.* **2019**, *13*, 556–572. [CrossRef]
49. Raut, R.D.; Narkhede, B.; Gardas, B.B. To identify the critical success factors of sustainable supply chain management practices in the context of oil and gas industries: ISM approach. *Renew. Sustain. Energy Rev.* **2017**, *68*, 33–47. [CrossRef]
50. Stevens, P. *International Oil Companies*; Chatham House: London, UK, 2016.
51. Gardas, B.B.; Raut, R.D.; Narkhede, B. Determinants of sustainable supply chain management: A case study from the oil and gas supply chain. *Sustain. Prod. Consum.* **2019**, *17*, 241–253. [CrossRef]
52. Wendel, D. *Maritime Forecast to 2050*; DNV-GL: Hovik, Norway, 2017; pp. 1–17.
53. Hassani, H.; Silva, E.S.; Al Kaabi, A.M. The role of innovation and technology in sustaining the petroleum and petrochemical industry. *Technol. Forecast. Soc. Chang.* **1027**, *119*, 1–17. [CrossRef]
54. Bengtsson, M.; Kock, S. "Coopetition" in Business Networks—To cooperate and compete simultaneously. *Ind. Mark. Manag.* **2000**, *29*, 411–426.
55. Cooper, M.C.; Lambert, D.M.; Pagh, J.D. Supply chain management: More than a new name for logistics. *Int. J. Logist. Manag.* **1997**, *8*, 1–14. [CrossRef]

56. Squire, B.; Cousins, P.D.; Brown, S. Cooperation and knowledge transfer within buyer-supplier relationships: The moderating properties of trust, relationship duration and supplier performance. *Br. J. Manag.* **2009**, *20*, 461–477. [CrossRef]
57. Mackenzie, W. *Upstream Supply Chain in Brief*; Wood Mackenzie: Edinburgh, UK, 2019.
58. Martin, M.J.C. *Managing Innovation and Entrepreneurship in Technology Based Firms*; John Wiley and Sons Inc.: New York, NY, USA, 1994.
59. Bergman, L. Ways the Oil & Gas Industry Is Trying to Become More Sustainable and Green. *Biofriendly Planet*, 7 March 2019. Available online: https://biofriendlyplanet.com/eco-awareness/air-quality/emissions/ways-the-oil-gas-industry-is-trying-to-become-more-sustainable-and-green/ (accessed on 19 November 2020).
60. International Petroleum Industry Environmental Conservation Association. Exploring Low-Emissions Pathways for Transport. 2019. Available online: www.ipieca.org (accessed on 19 November 2020).
61. Albjerk, N.B.; Danielsen, T.K.; Krey, S. Operational Planning and Disruption Management in Offshore Logistics. Master's Thesis, Norwegian University of Science and Technology, Trondheim, Norway, 2015.
62. Borch, O.J.; Batalden, B.-M. Business-process management in high-turbulence environments: The case of the offshore service vessel industry. *Marit. Policy Manag.* **2014**, *42*, 481–498. [CrossRef]
63. Elhedhli, S.; Merrick, R. Green supply chain network design to reduce carbon emissions. *Transp. Res. Part D Transp. Environ.* **2012**, *17*, 370–379. [CrossRef]
64. Seuring, S.; Müller, M. From a literature review to a conceptual framework for sustainable supply chain management. *J. Clean. Prod.* **2008**, *16*, 1699–1710. [CrossRef]
65. Wanasinghe, T.R.; Gosine, R.G.; James, L.A.; Mann, G.K.; De Silva, O.; Warrian, P.J. The Internet of things in the oil and gas industry: A systematic review. *IEEE Internet Things J.* **2020**, *7*, 8654–8673. [CrossRef]
66. Wanasinghe, T.R.; Wroblewski, L.; Petersen, B.K.; Gosine, R.G.; James, L.A.; De Silva, O.; Mann, G.K.; Warrian, P.J. Digital twin for the oil and gas industry: Overview, research trends, opportunities, and challenges. *IEEE Access* **2020**, *8*, 104175–104197. [CrossRef]
67. Ehret, M.; Wirtz, J. Unlocking value from machines: Business models and the industrial internet of things. *J. Mark. Manag.* **2017**, *33*, 111–130. [CrossRef]
68. Nguyen, T.; Gosine, R.G.; Warrian, P. A systematic review of big data analytics for oil and gas industry 4.0. *IEEE Access* **2020**, *8*, 61183–61201. [CrossRef]
69. West, B.; Lafferty, T. Synchronizing the offshore supply chain creates new value. *Offshore* **2008**, *68*, 78–81.
70. KonKraft. *Project Competitiveness—Changing Tide on the Norwegian Continental Shelf*; KonKraft: Oslo, Norway, 2018.
71. McKinsey. Supply Chain 4.0—The Next-Generation Digital Supply Chain. 2016. Available online: https://www.mckinsey.com/business-functions/operations/our-insights/supply-chain-40--the-next-generation-digital-supply-chain (accessed on 19 November 2020).
72. KonKraft. *The Energy Industry of Tomorrow on the Norwegian Continental Shelf Climate Strategy towards and 2030–2050*; KonKraft: Oslo, Norway, 2020.
73. Hall, A.D. Three-Dimensional Morphology of Systems Engineering. *IEEE Trans. Syst. Sci. Cybern.* **1969**, *5*, 156–160. [CrossRef]

sustainability MDPI

Article

Systemic Modeling of the Peace–Development Nexus

Bernard Amadei

Department of Civil Engineering, University of Colorado, Boulder, CO 80309-0428, USA; amadei@colorado.edu; Tel.: +1-303-929-8167

Abstract: As we enter the third decade of the 21st century, the value proposition of promoting sustainability and peace in the world has become more imperative than ever. It is an appropriate time to pause and reflect on what a post-pandemic COVID-19 world will look like and what constitutes a new mindset toward a more sustainable, stable, peaceful, and equitable world where all humans live with dignity and at peace. As emphasized in this paper, the new mindset must acknowledge that sustainability and peace are two entangled states of dynamic equilibrium. It is hard to envision a sustainable world that is not peaceful and a peaceful world that has not endorsed sustainable practices. This paper looks more specifically at the value proposition of adopting a systems approach to capture the linkages between selected development sectors (e.g., SDGs) and peace sectors (e.g., positive, negative, and cultural). Basic system dynamics (SD) models are presented to illustrate the peace–development nexus dynamics. The models are general enough to be used for different contexts and scales.

Keywords: peace–development nexus; SDGs; system dynamics; mindset; COVID-19 pandemic

Citation: Amadei, B. Systemic Modeling of the Peace–Development Nexus. *Sustainability* **2021**, *13*, 2522. https://doi.org/10.3390/su13052522

Academic Editor: Cecilia Haskins

Received: 16 February 2021
Accepted: 22 February 2021
Published: 26 February 2021

Publisher's Note: MDPI stays neutral with regard to jurisdictional claims in published maps and institutional affiliations.

1. Introduction

The year 2020 is likely to be remembered as a pivotal moment in socio-economic development worldwide. Starting with an outbreak believed to have started at the end of 2019, COVID-19 has evolved into a "democratic super disease" [1] that "respects no borders" [2]. The pandemic has revealed existing forms of vulnerability, injustice, and inequality; pushed more people into poverty [3]; and impacted peace negatively worldwide [4].

The pandemic has shown the fragility, interconnectedness, and diversity of human life and the close linkages between humans and their environment. It has negatively affected all the systems involved in socio-economic development and negatively disrupted public life, predominately for humanity's most vulnerable sections. As summarized in the 2020 sustainable development report [5], "the world is facing the worst public health and economic crisis in a century". The World Bank predicts a "lost decade" for the world economy [6]. On the flip side, the pandemic "has brought out some of the best human characteristics; self-sacrifice in helping others; empathy and solidarity despite the need for social distancing" [7]. It is also a "wake-up call and a training ground to enhance our joint and resilient response to future pandemics and other external disturbances" [7].

The pandemic has also stressed the challenges in planning, designing, and implementing humanitarian and development interventions and programs. According to the Alliance for Peace Building [8], "In 2020, nearly 168 million people needed humanitarian assistance and protection—about 1-in-45 people in the world—the highest figure in decades". Combined with the many planetary challenges already existing before 2020 (e.g., population growth and migration, urbanization, climate change, and environmental protection, among many others), the pandemic has added confusion and uncertainty about intervening appropriately in the systems (social, economic, cultural, ecological, and technical) deemed responsible for these challenges. There is a realization that the uncertainty and predictability of these systems can only be handled using systems thinking and tools of complexity science. In the context of a post-2020 world, systems thinking can be used to (i) capture various socio-economic development dynamics, (ii) explore different intervention scenarios, and (iii) develop integrated and collaborative solutions that transcend local and national boundaries [9].

As we enter the third decade of the 21st century and navigate through what Tooley [10] referred to as "VUCA (volatile, uncertain, complex and ambiguous) times", it is appropriate to pause and reflect on (i) what a post-pandemic world would look like for a growing and more urbanized world's population [11]; (ii) how to build back better (rather than return to the old normal) more resilient and equitable inclusive societies that put all people back at the center of human and economic development [12]; and (iii) how to build capacity and increase resilience at different scales (household, community, national, and regional) to cope and adapt to future adverse events and challenging conditions (e.g., climate change and natural disasters). Together, answering these questions will hopefully contribute to creating a much anticipated sustainable and peaceful world for all.

Since 1990, many authors and organizations have advocated the need for a paradigm shift in the way humans interact with each other and the environment on which they depend. The overarching goal of building a more sustainable, stable, peaceful, and equitable world where all humans live with dignity and peace is not new. It has been in the mind of many constituencies during the 20th century, starting with the League of Nations (1920–1946), followed by the United Nations ever since. It was the underlying thrust behind the publication of *Common Future* [13] and *Agenda 21* [14], and in establishing the Millennium Development Goals (1990–2015) and, more recently, the Sustainable Development Goals (SDGs, 2015–2030) agenda, also known as Agenda 2030. This agenda represents a comprehensive plan of action that involves five critical aspects of sustainability: people, planet, prosperity, partnerships, and peace [15].

The noble vision of a more sustainable, peaceful, equitable, and stable world emphasized by the SDGs since 2015 is worthy of consideration. Paraphrasing Albert Einstein, "Problems cannot be solved with the same level of thinking [mindset] that created them"; a question arises as to whether the original SDGs' vision and its associated mindset and precepts are still relevant after 2020 or need to be updated. There seems to be a consensus on the need to readjust socio-economic development to reflect the new normal of a post-2020 world [16,17]. As noted by Sachs et al. [5], "COVID-19 will have several negative impacts on most SDGs". Implementing the SDGs agenda and mitigating these impacts may take more years to implement than planned in 2015 or even early 2020.

As noted by UNDP [3], the value proposition of sustainability and peace in the world of tomorrow has become more imperative than ever. Simply put, it is hard to envision a peaceful and sustainable world in the foreseeable future with yesterday's normal(s). It is time to acknowledge that "a world divided by wealth and poverty, health and sickness, food and hunger, cannot long remain a stable [and peaceful] place for civilization to thrive" [18]. More than ever, science, technology, and innovation (STI) have a critical role to play in creating a peaceful and sustainable world for all [7].

This paper presents first some insights into the new development mindset for a post-2020 world. It builds on the joint dynamic of peace and development discussed in the article by Amadei [19]. The paper looks more specifically at using system dynamics to model the linkages and dynamic interactions between some development and peace sectors of interest rather than modeling how to achieve a specific or a combination of several SDGs. The generic models presented herein can explore the peace–development nexus in a high abstraction and strategic level of decision-making and implementation and different contexts and scales. However, the models require policymakers and practitioners to develop a systems thinking mindset first and be willing to explore the various interdependencies between peace and development.

2. A New Post-2020 Mindset

A new mindset is urgently needed to implement humanitarian and development interventions and programs in a post-2020 world. This recommendation is based on Meadows' [20] observation that changing the mindset in any system represents the place (i.e., leverage) to intervene with the highest return on the investment. However, such intervention is not as easy as it sounds, since the new mindset components must first be clearly outlined, understood, and adopted by multiple stakeholders (e.g., insiders and outsiders in development projects and programs) before they are implemented and assessed over time. In general, adopting a new mindset is initially challenging since it implies behavior change, which takes time and can be difficult for certain groups of stakeholders, policymakers, and practitioners involved in humanitarian and development aid.

It is interesting to note that the recommendation for a new mindset is already part of the post-2020 vernacular literature discussion. It is common to read, for instance, that changing the mindset for many constituencies is about "adopting a new normal" and "pivoting". This naïve concept ignores the fact that there was never a one-size-fits-all normal to start with before the pandemic but rather multiple normal(s), some better than others. In addition, pivoting is fine if there is a vision attached to it. Without it, there is a danger of ending up pointing in the original direction. With that in mind, creating a more sustainable, stable, and equitable world where all humans live with dignity and peace must be done in an intelligent, systemic, fair, and compassionate manner where normal is seen "as a plural". This approach departs from the dominant neoliberal capitalistic one that was the de facto mindset of the 20th century and the first two decades of the 21st century, characterized by determinism, compartmentalization, fear, greed, and benefit a few [21].

Let us explore some of the many characteristics of the new mindset that humanity must embrace to address the challenges mentioned above and adequately handle potential future crises (natural or human-made). First, the mindset must acknowledge that the challenges facing humanity cannot be tackled in isolation. They are complex and involve multiple interconnected components (social, economic, cultural, and technical) specific to the context and scale of the landscape in which they unfold. At times, these components may even transcend national boundaries requiring regional and international collaboration. The uncertainty, ambiguity, and unpredictable nature of the systems involved in the various challenges facing humanity imply that a systems/integrated approach to sustainability and peace is better suited in the new mindset than a deterministic one to capture their dynamics and linkages, explore different intervention scenarios, and develop integrated solutions. There is enough evidence that repeating the past business-as-usual mindset of ignoring the complexity of the systems at play in socio-economic development and treating them in isolation has the potential to do more harm than good [22].

A second aspect of the new mindset is that it requires humanity to reconsider its values and socio-economic priorities (i.e., how it sees reality) and put them into practice. A more mature level of consciousness in the day-to-day management and operation of our institutions and our occupations [23–25] is needed. As noted by Tanabe [2], "a critical question is emerging that faces humanity as a whole: what should come in the first place—society or economy, strong public health or profit, citizen's physiological, psychological, intellectual and spiritual well-being or plutocracy". Another remark is how to incentivize "each human citizen as a critical and transformative agent to contribute building sustainable global peace". These two recommendations represent a departure from the traditional human development dynamics where citizens are passive actors subject to policies that "maximize the economic advantage [for a strong capitalist class] while directing little energy to humanity's social, cultural, and even spiritual self-improvement or maturity" [2].

Third, the COVID-19 pandemic has demonstrated "the power of [scientific research] collaboration to create solutions quickly" [26]. The new mindset must build on such success stories and emphasize the importance of innovation in human and economic development at different scales (individual, household, community, country, regional, and planetary). As suggested by TWI2050 [7], "new thoughts, frameworks, and methods for the STI [Science, Technology, and Innovation] ecosystem to promote innovation, efficiency, and sufficiency for the achievement of the SDGs" are needed. Innovation must lead to solutions that embody the five aspects of Agenda 2030, i.e., they must be good for people and the environment, be profitable, promote human security and social justice, and create meaningful and just partnerships [27].

Fourth, collective activities at different scales from local to international are urgently needed to prevent further decline in human development in a post-COVID-19 world [3,5]. According to Moritz [28], to avoid further instability worldwide, socio-economic partnerships and collaboration must simultaneously address short-term improvements to the current situation and long-term sustainability planning along five tracks: (1) repair what is currently most damaged; (2) rethink change without going back to how things were, i.e., without rebuilding the vulnerability of business-as-usual; (3) reconfigure change so that it can happen; (4) restart change with the recommendations mentioned above; (5) and report how change progresses with the ability for course correction through monitoring and evaluation. Simply put, yesterday's socio-economic development tools have a limited range of applications in developing the world of tomorrow. The metaphor of not placing new wine into old wineskins but using new wineskins [29] comes to mind. Innovative development tools and priorities are needed to operate in a new socio-economic structure.

An open-ended question arises as to how Moritz's five changing tracks affect how the SDGs and their respective targets and indicators must be addressed now and in the future. Another aspect of these five tracks is that they emphasize the importance of capacity building and resilience at different scales (individual, household, community, country, region, and global) in the overall discussion on sustainable development for the 21st century. These two concepts are not new and have been part of the development vocabulary for a long time. A traditional approach to building capacity and resilience is to identify and address in a fragmented manner specific issues at play in the systems (institutional, economic, social, environmental, and infrastructure) that may prevent the delivery of services and meeting particular SDGs. This compartmentalized approach, driven by a need to reach some form of satisfactory equilibrium, fails to account for possible states of synergy and trade-offs at play between these systems and the changing, adaptive, and dynamic nature of social networks [30].

Since the 2020 pandemic is a "wake-up call and a training ground to enhance our joint and resilient response to future pandemics and other external disturbances" [7], the concepts of capacity and resilience must be reimagined, redefined, and strengthened to handle future crises (e.g., health). As discussed further in a paper by Amadei [31], a systemic approach to capacity and resilience is needed to explore possible synergies and trade-offs at play in humanitarian and development interventions in a specific context and scale. At the community scale, capacity and resilience cut horizontally across multiple vertical silos of community development. One type of capacity building to achieve a specific goal, such as providing a reliable service (e.g., water, energy, food, transportation, etc.), could affect achieving another goal associated with a different service type over time. Likewise, the interaction between various systems at play (e.g., institutional, sociocultural, infrastructure, environmental, economic, etc.) at some scale may contribute to resilience when exposed to small and large adverse events over time.

Fifth, the dynamic between insiders and outsiders involved in development and humanitarian projects and programs needs to be participatory. Until about 30 years ago, the Western world's traditional approach to humanitarian and development work was top-down contractual and consultative, with limited input from the bottom-up beneficiaries [32,33]. More recently, there has been more emphasis on promoting collaborative and collegial approaches and transformation through empowerment [32,33]. Yet, development and humanitarian work shaped by external actors remains a dominant way of doing things today.

Sixth, the new mindset must recognize the linkages between sustainability and peace. Simply put, it is hard to envision a sustainable world that is not peaceful and a peaceful world that has not endorsed sustainable practices. Peace is fully integrated into the sustainable development agenda with SDG 16 (Peace, Justice, and Strong Institutions), which is to "Promote peaceful and inclusive societies for sustainable development, provide access to justice for all and build effective, accountable and inclusive institutions at all levels" [5].

Meeting SDG 16 is about the right relationships, which is contained in principle 16f of the Earth Charter [34], defining peace as " ... the wholeness created by right relationships with oneself, other persons, other cultures, other life, Earth, and the larger whole of which all are a part". According to Gittins and Velasquez-Castellanos [35], such relationships' core characteristics include freedom and "empathy, acceptance, and honesty". The National Peace Academy [36] sees the right relationships as part of a peace system that encompasses five interactive spheres of peace at the personal, social, political, institutional, and ecological levels.

Finally, the new mindset must also include the inner dimension of human development, i.e., what Maslow [37] referred to as esteem and self-actualization. It is interesting to note that such needs are not explicitly emphasized in the SDGs agenda, which prioritizes addressing outwardly the bottom tiers of Maslow's pyramid (i.e., meeting physiological and safety needs). This inner dimension is critical if the SDGs need to be met and the world does not revert to past practices. A quote of Meister Eckhart [38] to support the inner dimension of human development is pertinent here: "the outer work can never be minor when the inner work is a major one, and the outer work can never be major or good when the inner work is a minor one and without value". With that in mind, one can question the quality of our western society's inner work, institutions, and decision-makers based on how such groups (including ourselves) have managed planetary challenges over the past 200 years. Lessons can be learned from less well-known non-Western forms of human development that emphasize strengthening the whole person's inner dimension concurrently with socio-economic development.

3. An Integrated Approach to the SDGs

The 17 SDGs and associated targets and indicators were introduced in 2015 as a new 15-year long road map for worldwide sustainable development at the country level. In launching the so-called Agenda 2030, the General Assembly of the United Nations "recognize[d] that eradicating poverty in all its forms and dimensions, including extreme poverty, is the greatest global challenge and an indispensable requirement for sustainable development" [15]. Compared to the preceding Millennium Development Goals (1990–2015), the SDGs apply to all countries regardless of their development level.

Fulfilling the SDGs by 2030 is indeed a daunting task, a truly formidable undertaking, unparalleled in human history. Since 2015, the comprehensive SDGs agenda has been a work in progress, and the SDGs' targets and indicators have been refined further. Significant additions include introducing the SDG index and dashboards to quantify the progress of different countries on the SDGs [39] and introducing six SDG transformations to operationalize the SDGs' implementation at the country level [7,40]. They include (i) education, gender, and inequality; (ii) health, well-being, and demography; (iii) energy decarbonization and sustainable industry; (iv) sustainable food, land, water, oceans; (v) sustainable cities and communities; and (vi) digital revolution for sustainable development. The 2020 sustainable report [5] proposes short-term and long-term guidelines to address these six transformations considering the COVID-19 pandemic. Despite such efforts, the jury is still out on how the SDGs' targets and indicators need to be updated individually and together to match the post-COVID-19 world's new reality.

Another question that has always been pertinent to the SDGs agenda since its inception is how to meet the goals across different physical scales (country, cities, communities, households, and individuals) and temporal scales (short-, medium- and long-term). Scharlemann et al. [41] remarked that progress toward sustainable development might have synergistic benefits at one physical or temporal scale but create negative impacts, requiring trade-offs, at other scales.

Since 2015, there has been an increasing interest in understanding and quantifying how the SDGs interact with each other, since sustainable development is more than meeting a series of independent goals [42]. Zelinka and Amadei [43] noted that addressing the connections among the SDGs in a multi-sectoral integrated approach is crucial to ensure the coherence of Agenda 2030. Although the SDGs represent an "indivisible whole" [44], some goals are likely to affect others positively (i.e., creating synergies) or negatively (i.e., requiring trade-offs). In contrast, others may only have indirect interactions or no interaction at all. Furthermore, as noted by Scharlemann et al. [41], the nature of the linkages across the SDGs being considered depends on (i) the context and "groups of actors" at the country level; and (ii) the perspective (socio-economic, geopolitical, geographic) used to explore the interactions (e.g., the environment-human linkage perspective).

The socio-economic development literature is rich in contributions that emphasize, mostly qualitatively, the value proposition of using an integrated approach to Agenda 2030. Landmark papers, among many others, include those of Griggs et al. [44]; Nilsson et al. [45]; Griggs et al. [46]; Waage et al. [47]; Coopman et al. [48]; Vladimirova and Le Blanc [49]; Barbier and Burgess [50]; Morton et al. [51]; Lim et al. [52]; and TWI2050 [53]. Qualitative and semi-quantitative tools have been used to quantify the SDGs' interactions [54–58]. A noteworthy contribution to understanding the SDGs' interdependence is the report entitled *A Guide to SDGs' Interactions* [59]. The report examined the interdependence at the target level between SDGs # 2 (zero hunger), 3 (good health and well-being), 7 (energy), and 14 (life below water) with the other goals, using a semi-quantitative impact factor ranging over a seven-point scale: neutral impact (0); different levels of positive impact (+1 to +3); and different levels of adverse impact (−1 to −3). More recently, Scharlemann et al. [41] provided an extensive review of the different formulations proposed in the literature since 2015 that capture the interaction between the SDGs, emphasizing other interaction types between humans and their environment. Many of these formulations use a double-causality analysis, which, in the case of the 17 SDGs, consists of creating a 17 × 17 matrix describing the direct influence and dependence of each SDG on the other goals. The analysis would become more complicated if one were to analyze the SDGs' interaction at the target level.

Quantitative tools borrowed from systems science have also been proposed to model the SDGs' interactions. They include, for instance, neural network analysis [60,61], cross-impact analysis [43], and system dynamics [62]. As summarized by Zelinka and Amadei [43], using such tools to address the SDGs has a strong value proposition when exploring how complexity and uncertainty in the country's systems affect the decision-making process of policymakers and practitioners. More specifically, using the habits of systems thinking developed by the Waters Foundation as a guide [63], systems science tools can be used to:

- Make meaningful connections across the overall Agenda 2030 and within and between the various systems that affect the SDGs.
- Understand the big picture of sustainable development at the country level while simultaneously paying attention to specific details.
- Explore and evaluate different perspectives in the eyes of various stakeholders involved in decision-making from the private and public sectors, civil societies, and others.
- Appreciate how different mental models of development shape views and actions when addressing the SDGs and selecting strategies.
- Recognize that sustainable development is a dynamic process that requires flexible and adaptive decision-making while recognizing patterns and trends.
- Explore the role that assumptions in decision-making play in shaping outcomes.
- Realize that the structure of the systems involved in sustainable development influences its dynamic.
- Account for time-delays (short and long-term) between making country-level decisions and observing the associated outcomes and how such delays require monitoring and evaluation.
- Consider possible intended and unintended implications of decision-making and policies.
- Explore the role of some structural variables, archetypes, reinforcing and balancing feedback loops of cause and effect, and patterns of sustainable development play in shaping emerging behavior and identifying possible leverage points in meeting the SDGs with a higher return on actions taken.
- Identify how accumulations and rates of change control the behavior of multiple systems.

- Realize that sustainable development requires a flexible and adaptive approach to decision-making, leading to good enough solutions.
- Recognize that multiple strategies, approximations, parametric and sensitivity studies need to be considered before coherent solutions are outlined, evaluated, and an implementation plan can be selected and implemented.
- Better understand how meeting the SDGs (and their targets) depends on the initial country capacity and resilience and the potential for capacity building over time.
- Predict how countries may respond to different strategies of capacity development under constraints and disturbances.

In summary, using a systems approach to address the interactions among the SDGs (and their targets) represents an alternative to the traditional deterministic and rigid decision process used in development worldwide over the past 50 years. As discussed in the previous section, a prerequisite for using such an approach is that decision-makers and practitioners involved in development interventions and programs must be willing to adopt a new mindset of systems thinking when working in partnership with other stakeholders. Unfortunately, the history of human and economic development over the past 50 years shows that a lack of will often hinders any change and progress [7].

4. The Peace–Development Nexus

Since its inception, Agenda 2030 has acknowledged the coherence between peace and development and that "there can be no sustainable development without peace and no peace without sustainable development" [15]. The two are entangled. Understanding and modeling the peace–development nexus requires some preliminary discussion about peace. This section summarizes key concepts about peace necessary to understand the narratives behind the system dynamics simulations presented below.

The peace studies and conflict management literature is rich in contributions exploring the different aspects of peace. Peace can mean different things to different people and cultures [64–67]. As discussed in a paper by Amadei [19], there is no such thing as a one-size-fits-all unified and optimized static state of peace, the same way as there is no individual united and optimized stationary state of sustainability. It is more realistic to talk about "many [dynamic] peaces" [68] and, along the same line, "many dynamic sustainabilities". What works in a specific context and scale does not necessarily work and translate somewhere else.

This paper focuses on outer peace, which seems to be of higher priority in Western cultures and religions. Another dimension of peace, inner peace at the individual level, is also "an essential component and precondition for a peaceful world". Inner peace is more in line with the world's rich spiritual-religious traditions, such as Hinduism and Buddhism [64]. Regardless of the culture, inner peace is positively correlated with outer peace and is often seen as the place where peace builds outward at the individual and institutional levels [65].

Diamond and McDonald [69] suggested that peace is more than the absence of hostility and violence, and should be considered as "a potential, a possibility, an ever-changing condition [state] ... a direction in which to head, one step at a time". Peace can be understood as a state or a process. As a state, peace emerges from the interaction of multiple socio-economic, infrastructure, and environmental systems operating in a constrained landscape of specific context and scale. These systems constitute what is referred to in the peace studies and conflict management literature as peace infrastructure [70,71].

In that infrastructure, peace can be defined as "an organizing principle and an enabling violent-free state of dynamic equilibrium emerging from the right relationships among different populations and their interaction with the various systems in the landscape upon which they depend" [19]. This systems-based definition of peace builds on that of sustainability proposed by Ben-Eli [72] as: "an organizing principle and a dynamic [symbiotic] equilibrium in the processes of interaction between a population and the carrying capacity of an environment such that the population develops to express its full potential without adversely and irreversibly affecting the carrying capacity of the environment upon which it depends". Both aforementioned definitions lend themselves well to using systems modeling tools to capture how the two states of peace and sustainability interact.

As a process, peace unfolds over time through peacebuilding (building conditions for peace), peacemaking (getting parties to find common ground), and/or peacekeeping (supporting sustainable peace) efforts. Another way of looking at peace is to see it as a noun (outcome) or a verb (process), depending on how peace is being addressed. It should be noted that this last remark can also be made about health, sustainability, and resilience.

As noted by Gittins and Velasquez-Castellanos [35], "there are about 35 theories of peace, at least at the university level". The conflict and peace studies literature frequently refers to Johan Galtung's work, who pioneered the concepts of negative and positive peace in the early 1960s [73,74]. In short, negative peace relates to the absence of war and direct or organized violence. Undesirable violence and fear of violence cease to exist due to activities such as "ceasefires, disarmament, prevention of terrorism and state terrorism, nonviolence" [75].

On the other hand, positive peace relates to the presence and prevalence of positive attributes, conditions, and priorities that promote "social and economic justice, environmental integrity, human rights, and development" and contribute to the structural "integration of human society" [73]. As remarked by Fischer [75], positive peace activities may range from "building a life-sustaining economy at the local, national and global level in which everyone's basic needs are met" to "good governance and participation, self-determination, human rights".

Galtung [74] added another dimension of peace (cultural peace) to positive and negative peace. It refers to the "aspects of a culture that serve to justify and legitimize direct [negative] peace and structural [positive] peace". Cultural peace activities may include: "promotion of a culture of peace and mutual learning; global communication and dialogues; development of peaceful deep cultures and deep structures; peace education; peace journalism" [75].

Since peace is not a direct "measurable commodity" [69] and is difficult to conceptualize at different contexts and scales, questions arise as to how to (i) measure it indirectly through indicators and proxies and (ii) monitor and evaluate peace over time. The challenge here is how to measure a state that is the outcome of many interacting systems and sub-systems (e.g., social, economic, environmental, infrastructure) with different levels of complexity, uncertainty, and adaptability, and subject to multiple constraints (geopolitical, environmental, cultural, etc.).

At the national level, a measure of peace was suggested by the Institute for Economics and Peace (IEP) in Sydney, Australia [76–78]. Two indices were proposed to semi-quantify peace: the Positive Peace Index (PPI) and the Global Peace Index (GPI). The IEP considers positive peace as being founded on eight interdependent pillars or domains, each containing three weighted indicators (Figure 1). The domains contributing to positive peace include (i) a well-functioning government; (ii) a sound business environment; (iii) equitable distribution of resources; (iv) acceptance of the rights of others; (v) good relations with neighbors; (vi) a free flow of information; (vii) high levels of human capital; and (viii) low levels of corruption. Mathematically, the PPI is the weighted average of 24 indicators.

A second index called the Global Peace Index (GPI; values of country GPIs can be found at http://visionofhumanity.org/indexes/global-peace-index/, accessed on 10 Janaury 2021) was also proposed by the IEP to measure negative peace (i.e., the level of country peacefulness) for the same countries as the PPI [77]. It consists of 23 indicators (10 external and 13 internal) distributed over three domains: ongoing domestic and international conflict, societal safety and security, and militarization. Analysis of the PPI and GPI overall scores at the country level indicates that both indices are highly positively correlated with a coefficient of correlation of 0.75 for data obtained from 2008–2017 [78].

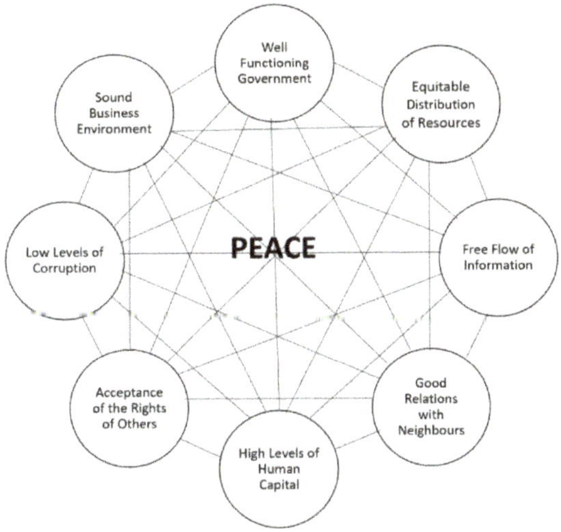

Figure 1. The eight pillars (domains) that create positive peace according to the IEP [76–78].

Based on the values of their GPI and PPI ranging between 1 (most positive peace) and 5 (least positive peace), the IEP [78] ranks countries into four potential peacefulness states (i) countries with sustainable peace (high positive and negative peace); (ii) countries with a positive peace deficit (low positive peace and high negative peace) that are likely to experience violence in the future; (ii) countries with a positive peace surplus (high positive peace and low negative peace) with potential to become more peaceful over time; and (iv) countries trapped into violence (low negative and positive peace).

Although not initially proposed by the IEP, a third index, the Cultural Peace Index (CPI), can be introduced to measure cultural peace [79]. Even though its indicators are yet to be determined, the CPI is assumed to range between 1 and 5 for consistency with the PPI and GPI.

The interdependence of the three PPI, GPI, and CPI indices is illustrated in Figure 2 using the peace triangle representation of Galtung [74]. In this diagram, each type of peace influences and depends on the other two. One can interpret the area of that triangle as representing the extent of the enabling environment in which peace unfolds over time. Outside the triangle is the external environment of a specific context that influences the three interacting components of peace over time.

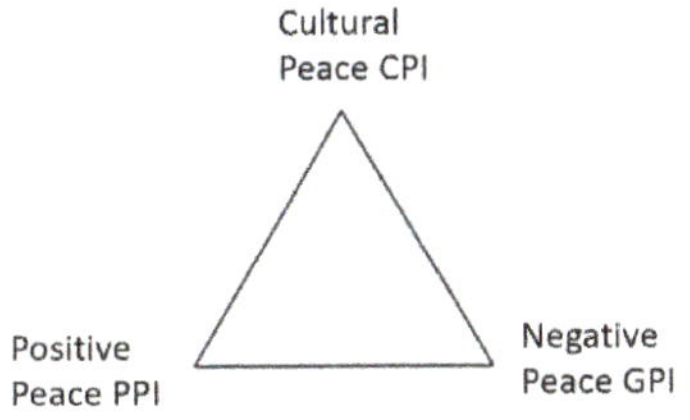

Figure 2. The peace triangle. The state of peace unfolds in the triangle (enabling environment), where the three aspects of peace overlap.

Although peace is intimately linked to socio economic development [80,81], not all forms of peace contribute positively to development and vice-versa. Whether peace is beneficial to development depends mostly on how peace resolution and transformation are designed, implemented, and evaluated. For instance, not addressing the root causes of conflict may result in resuming the conflict and negatively affecting the development in the foreseeable future [82].

Likewise, development can positively or negatively impact peace, depending on the type of development being implemented. It can have intended consequences or create unintended issues [83]. For instance, inappropriate decision-making and trade-offs in activities such as the supply and demand of water, energy, and food resources, inadequate associated infrastructure planning and design, poor decisions in resource management and allocation, and poor governance may result in divisions, unrest, conflict, violence, and insecurity.

In the SDGs sustainable agenda, peace is introduced through SDG 16 (Peace, Justice, and Strong Institutions) and contains 12 targets [5]. Peace also appears in the SDG cross-cutting issues of gender equality, governance, health, inequalities, security, support of vulnerable states, and sustainable cities [84].

5. System Dynamics Modeling

5.1. Background

Since the 1940s, modeling tools have been proposed in various complexity science disciplines to address ill-defined problems (see the map by Castellani [85]). One of the challenges when modeling complex systems is to select the most appropriate tools to model their dynamics. As noted by Rahmandad and Sterman [86], modeling the dynamics of a given problem depends on "the purpose of the model and the level of aggregation appropriate for that purpose". In short, the selected level of aggregation must match the level of details in the available data sources and provide a balance between "simplicity and realistic depiction of the underlying mechanisms" expected to be at play in the problem of interest. A discussion of the pros and cons of three commonly used modeling methods (system dynamics, discrete event modeling, and agent-based modeling) can be found in Borshchev and Filippov [87].

System dynamics is a branch of systems science originating from Dr. Forrester's work in the 1960s and 1970s [88]. The technique and its multiple applications are well documented in the landmark books by Richmond [89], Sterman [90], and Ford [91], among many others. The value proposition of the SD method in modeling the dynamics of complex systems includes being able to (i) capture both qualitatively and quantitatively how systems continuously change over time due to changes in their components and their interactions; (ii) account for non-linearities, delays, and feedback mechanisms; and (iii) illustrate that as the structure of a system changes, so does its behavior and vice-versa. System dynamics models are causally closed and require selecting closed boundaries. Only endogenous components and factors (those originating from within) are assumed to form the system structure and predominantly dictate the systems' behavior. Compared to other systems modeling tools, the SD method is top-down and can be applied to systems with high aggregation levels (i.e., high abstraction levels). It is appropriate at the strategic level of decision-making [87].

In general, SD models use two types of graphical representations of systems dynamics. Causal loop diagrams (CLDs), not used in this paper, show qualitatively how elements of feedback mechanisms interact causally. The other graphical representation, stock-and-flow diagrams, consists of combining several building blocks (Figure 3) to visualize qualitatively and quantitatively accumulation, flows, delay, and dissipation.

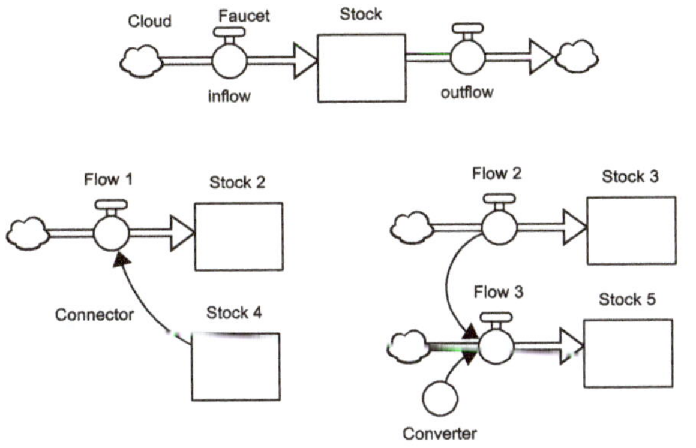

Figure 3. Basic building blocks of system dynamics models.

- Stocks correspond to accumulations of something that can be measured at one point in time. They are state variables [88] that define the current state of a system (e.g., peace and development).
- Flow (inflow, outflow) is represented in the form of pipelines (with a faucet controlling the flow). Flow (i.e., flux or rate) results in changes (dynamic behavior) in the stock accumulations and in the entire system. Flows are control variables [88] that create changes in the state (e.g., peace and development) of a system.
- Clouds indicate infinite sources or sinks, somewhere outside of the system boundaries.
- Converters are used to convert or transform information from one stock-and-flow path to another, or to feed information into an existing flow. A converter can also represent a stock if there is no flow in and out of the stock. They are converting variables. Converters can change over time and be described in a functional form.
- Connectors indicate transmission or links of actions and information (i.e., causal connections) between variables such as stock-to-flow, flow-to-flow, or between converters. One or several variables can provide input to and have some influence on another variable through connectors.

In general, system dynamics models (SD) consist of combinations of these five building blocks. However, it must be kept in mind that there is no such thing as a one-size-fits-all SD model that would capture all the possible dynamics in the interaction between peace and development in multiple contexts and scales. We present two examples of SD models to illustrate how to capture the general dynamics at play in the interactions between peace and development. Both were developed using the STELLA Architect software (Version 2.0) by isee systems (www.iseesystems.com, accessed on 15 January 2021). The two models come with two interactive interfaces available online that can be used by the readers to explore different scenarios.

5.2. A Simple Peace–Development Nexus Model

Figure 4 shows a simple goal-seeking model of nonlinear interaction between development and peace. The two main dynamic variables (stocks) are the current states of development (D) and peace (P). Both are assumed to involve several endogenous factors that are context-specific and interact within a specified boundary (e.g., country). An interactive user interface for this model can be found on the web (https://exchange.iseesystems.com/public/bernardamadei/peace-development-example-1, accessed on 15 February 2021).

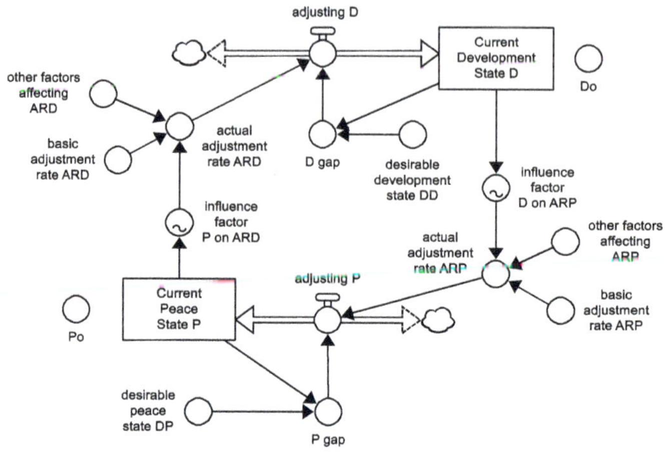

Figure 4. Systems dynamics (SD) model of the dynamic between states of development (D) and peace (P).

Starting with an initial baseline (D_o, P_o), both D(t) and P(t) adjust over time toward their respective desirable states of development (DD) and peace (DP) at different adjustment rates, ARD (per year) and ARP (per year), respectively. The actual ARD (P) rate is assumed to be the product of an estimated basic rate and a factor that depends on the current state of peace (~influence factor P on ARD). In Figure 4, the change in the stock D over time ("adjusting D") is equal to the product between ARD(P) and the gap between the desired development state DD and the actual state D as follows:

$$\frac{dD}{dt} = ARD(P) \times (DD - D)$$

Likewise, the actual ARP(D) rate is assumed to be the product of an estimated basic rate and a factor that depends on the state of development (~influence factor D on ARP). In Figure 4, the change in the stock P over time ("adjusting P") is equal to the product between ARP(D) and the gap between the desired peace state DP and the actual state P as follows:

$$\frac{dP}{dt} = ARP(D) \times (DP - P)$$

Note that both ARP(D) and ARD(P) may also depend on other factors (e.g., socio-economic, political, cultural, etc.) that may influence peace and development.

Solving these two nonlinear first-order differential equations with (D_o, P_o) as initial conditions would give an expression for D(t) and P(t) if we knew the functional forms of ARD(P) and ARP(D) and associated parameters and variables (dependent and independent). It is noteworthy that Figure 4 contains two bi-flows instead of two uni-flows feeding the current D and P stocks. The bi-flows model a possible increase or decrease in the two stocks. The current states (D, P) may decrease, for instance, if the 'adjusting D' and 'adjusting P' flow rates become negative. In that case, the development and peace states would degrade over time.

Let us consider that (D, DD), and (P, DP) can be expressed respectively in generic development units (du) and peace units (pu) ranging over two 0–100 scales. Both units are arbitrary and are introduced here as semi-quantitative measures of development and peace. Examples of such measures are discussed in Section 5.4.

The development and peace scales can be broken down into several achievement level groups on an as-needed basis. Each group is specific to the context in which the development–peace nexus analysis is carried out. As an example, the state of development is divided into five levels of development achievement: very low development level (1–20);

low development level (21–40); medium development level with unlikely sustainability (41–60); sustainability possible (61–80); and sustainability likely (81–100). The same approach is used for the state of peace by introducing five levels of peace achievement: very low (1–20), low (21–40), medium (41–60), peace possible (61–80), and peace likely (81–100).

The approach of using a semi-quantitative rating scale to describe the qualitative state of a variable has been used by many authors. For instance, the Institute for Sustainable Infrastructure [92] proposed a framework called *Envision*™ to evaluate and rate the sustainability of infrastructure projects over their life cycle. The rating system consists of five categories with credits. They include (i) quality of life (well-being, mobility, and community); (ii) leadership (collaboration, planning, and the economy); (iii) resource allocation (materials, energy, and water); (iv) the natural world (siting, conservation, and ecology); and (v) climate and risks (emissions and resilience). Points are assigned to each credit for different project sustainability achievement levels: improved, enhanced, superior, conserving, or restorative. Each achievement level has specific characteristics.

Similarly, Schweitzer and Mihelcic [93] proposed an assessment tool to score the sustainability of rural water systems in the developing world. It is based on eight indicators: the activity level, participation, governance, tariff payment, accounting transparency, financial durability, repair service, and system function. Based on the indicators' values, rural water systems are scored into three groups: sustainability likely, sustainability possible, and sustainability unlikely.

Finally, Bouabid and Louis [94] considered eight categories of capacity involved in delivering municipal sanitation services: service level, institutional, human resources, technical resources, economic and finances, energy, environmental, and social and cultural. Each category consists of several requirements. Each requirement is rated with a score ranging between 0 and 100, broken down into five rating groups with 20 units each. For each category of capacity, a capacity factor is calculated as the weighted sum of its requirement scores. The lowest capacity factor is understood as the most vulnerable place in the community where intervention to improve a specific service is first needed. According to Bouabid and Louis [94], it can be interpreted as a semi-quantitative measure of the stage of development of a community and its readiness to provide the service. Based on the value of the lowest capacity factor, the stage of development varies between 1 (no capacity) and 5 (capacity to manage centralized systems).

An SD analysis was carried out assuming that (i) the development and peace levels are initially low with D_o = 20 development units (du) and P_o = 10 peace units (pu); (ii) DD = 100 du and DP = 100 pu; (iii) the basic ARD = 0.02/year, the basic ARP = 0.01/year; (iv) P does not impact ARD; and (v) D does not impact ARP. The other factors affecting ARP and ARD remain constant and equal to 1. Figure 5 shows the corresponding asymptotic increase in D and P toward their desired values. Peace increases from being initially low to reach a low to medium achievement level. Likewise, development starts at a low achievement level and reaches a "sustainability possible" level after 50 years.

As a second numerical example, let us consider the case where the "influence factor of P on ARD" and the "influence factor of D on ARP" have functional forms. They are both assumed to vary linearly between −0.2 and 1 as P and D vary between 0 and 100 peace and development units, respectively. In this example, the initial development and peace levels are very low with D_o = 20 du and P_o = 5 pu. Figure 6 shows the variation of D and P over 50 years. Both development and peace decrease over time to very low values.

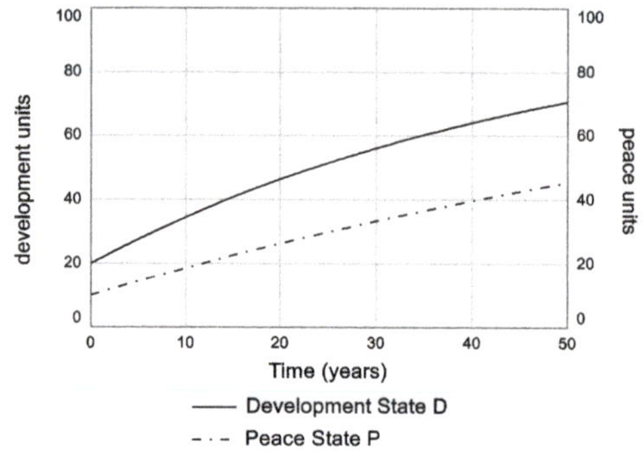

Figure 5. A numerical example showing the variation of development and peace states with time. The two influence factors are equal to 1.

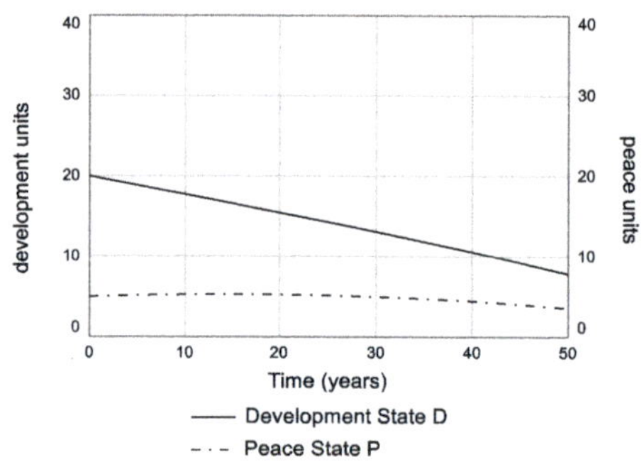

Figure 6. A numerical example showing the variation of development and peace states with time. The two influence factors in Figure 3 are linear functions of P and D.

Note that other numerical examples can be conducted with the STELLA-Architect software. The software includes various functionalities that allow users to carry out sensitivity and parametric studies and optimization.

5.3. A More Complex Peace–Development Nexus Model

The SD model in Figure 3 represents a simplified picture of the more complex dynamic usually at play between different aspects of the peace and development states at some specific scale and context. The second model shown in Figure 7 generalizes the dynamics of Figure 3 when several interacting development and peace sectors are considered.

To illustrate some possible interactions, we will consider three peace sectors P_i (i = 1–3) (positive, negative, and cultural peace) interacting with three development sectors D_i (i = 1–3) (food security, energy security, and water security). In Figure 6, layered stocks represent the current development and peace states as (3×1) arrays.

Compared to Figure 3, the converter "~influence factor P on ARD" is now a 3×3 array with nine components. Each represents how each peace sector affects the adjustment rate of a development sector in a functional form. Since three peace sectors influence each development sector, a weighted average of that influence is determined in the 3×1 array converter "weighted influence P on ARD". The user selects the weights. For instance, positive, negative, and cultural peace may affect the adjustment rate in water security differently. The weighted average determines how peace, in general, involves a change in water security. The same can be done for energy and food security. The actual adjustment rate ARD_i (i = 1–3) for each development sector D_i is calculated as the product between the "weighted influence P on ARD" for that sector and the basic adjustment rate and other factors affecting that sector.

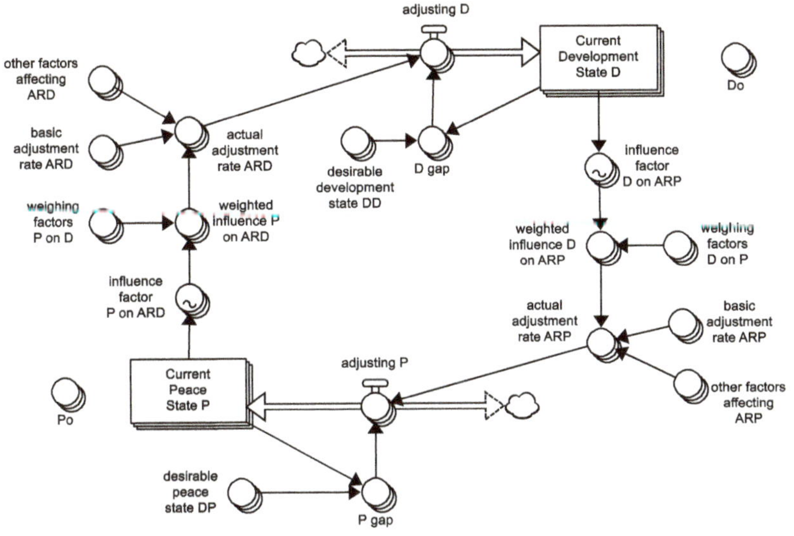

Figure 7. SD model of the dynamic between states of development (D) and peace (P). Three sectors of development and peace are considered.

The same approach is used for the "~influence factor D on ARP". The user selects nine functions to describe how each development sector affects the peace sectors. In our case, how water, energy, and food security affect positive, negative, and cultural peace. Since three development sectors influence each peace sector, a weighted average of that influence is determined in the (3 × 1) array converter "weighted influence D on ARP". For instance, water, energy, and food security may affect the adjustment rates of the three peace sectors differently. The actual adjustment rates ARP_i (i = 1–3) for each peace sector P_i are calculated as the product between the "weighted influence D on ARP" for that sector and the basic adjustment rates and other factors affecting that sector.

A numerical example is shown in Figure 8. In this example, the three peace sectors and the three development sectors' initial values are equal to 5 pu (very low peace) and 20 (low development) du, respectively. All weighing factors P on D and D on P are equal to 1/3. The desirable development and peace sectors values are equal to 100 du and 100 pu, respectively. The basic development and peace adjustment rates are constant and equal to 0.02/year and 0.01/year, respectively. Linear functions were selected to capture the positive influence of peace on development and development on peace. An interactive user interface for this model can be found on the web (https://exchange.iseesystems.com/public/bernardamadei/peace-development-example-2, accessed on 22 February 2021).

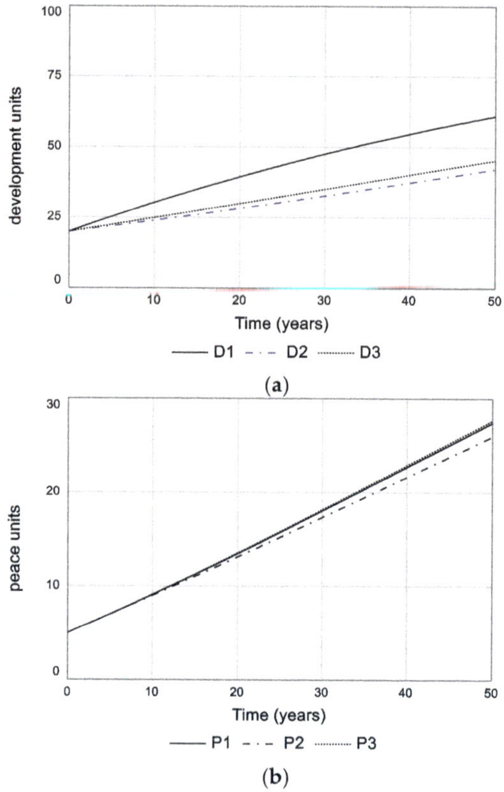

Figure 8. Variation of the three sectors of development (**a**) and peace (**b**) with time.

5.4. Discussion

The models considered above are two of many possible SD models that could be developed to explore the linkages and feedback mechanisms between development and peace at the country level. Both models are general enough to allow users to consider specific development and peace sectors deemed necessary for the case study.

The arbitrary development units (du) and peace units (pu) used in the two models can be related to existing measures of peace and sustainability. For instance, in the first model, the overall state of peace P could be related to SDG 16, the positive peace index (PPI), the global peace index (GPI), or a combination of PPI and GPI. Likewise, the D variable could be some measure of development. As an example, if food security (SDG 2), energy security (SDG 7), and water resources security (SDG 6) at the country level are assumed to be indicators of development, D could be related to the FEW Security Index proposed by Willis et al. [95]. This index is calculated as the geometric mean of three sub-indices related to the food, energy, and water sectors. Finally, a third option to quantify D is to relate it to the SDG index, which is a recent index introduced to monitor and evaluate in an integrated way the evolution of the 17 SDGs at the country level [40].

In the second model, the development and peace states are represented by multiple sectors. More development sectors can be included in the model, such as various SDGs, the six SDG transformations [7,40], or the five sectors of sustainability [15]. One sector (SDG 16) or multiple sectors (positive, negative, and cultural) can represent the state of peace.

In Figures 3 and 7, the basic adjustment rates of peace and development (ARD and ARP) and the other factors affecting peace and development, can themselves be time-dependent functions if necessary. Whether they are positive or negative will dictate how one or several development sectors and peace sectors may increase or decrease in value.

It should be noted that the second model does not account for the interaction between the sectors that define the state of development and those that define the state of peace. Accounting for such interactions would require developing more complicated SD models [19].

Finally, it is the experience of the author that one of the challenges of developing system dynamics models of the peace–development nexus is finding a balance between simplicity and a realistic depiction of the structural mechanisms underlying the problem being analyzed. SD models can quickly become overwhelming, as it is easy to fall into selecting too many details that lead to "model paralysis in analysis". Another challenge related to the previous one is finding realistic data sources to estimate the model parameters and their linkages. A third challenge is to recognize that system dynamics modeling is not a random process. As shown in Figure 9, the comprehensive modeling of complex peace–development nexus problems requires the following stages: (i) identification of development and peace problems; (ii) definition of these problems; (iii) formulation of possible SD models using causing loop diagrams and/or stock-and-flow diagrams; (iv) selection of input parameters; (vi) model calibration to reproduce problem current and past dynamics; and (vii) conducting parametric and sensitivity analyses toward (viii) policies and decision making. These eight stages involve multiple feedback loops.

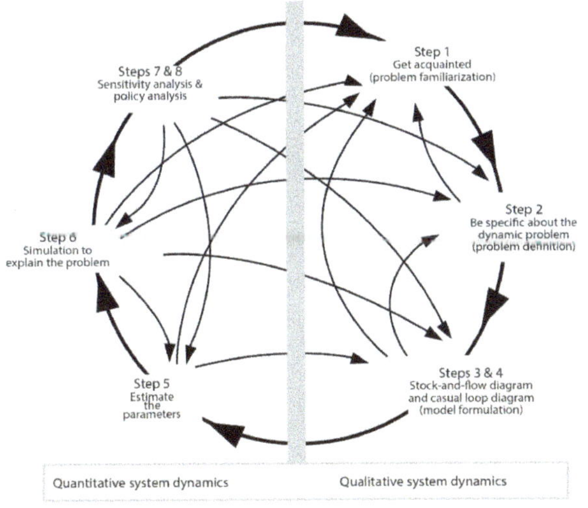

Figure 9. Suggested stages in system dynamics modeling (from Ford 2010). The steps consist of those involved in qualitative modeling (right-hand side) and those involved in quantitative modeling (left-hand side).

6. Conclusions

Extraordinary times can create unique opportunities. As we enter the third decade of the 21st century, there is a unique opportunity to build back better and develop "dynamic new normals" (if we want to call it that way) that do not bring back the different forms of injustice and inequality (ecological, racial, social, economic, and gender) from the past. Addressing the consequences of the COVID-19 pandemic while dealing with the global challenges facing humanity today and in the near future requires adopting a new mindset. The details of that mindset are yet to be agreed upon and implemented by the international community. As discussed in this paper, some of the characteristics of the mindset include: (i) using an integrated approach to socio-economic development (and humanitarian aid) based on principles of complexity and systems science; (ii) adopting a more mature level of consciousness in the management and operation of our institutions and occupations; (iii) investing in scientific and technical innovation that embody the five aspects of sustainability (people, planet, profit, partnership, and peace); (iv) developing socio-economic partnerships and collaborations that respect participation and empowerment; and (iv) account for the inner and outer dimensions of human development.

There is a need to reconsider Agenda 2030's priorities for the next ten years and beyond with two goals in mind. An immediate goal is to prevent further decline in socio-economic development that would affect society's poorest sections the most. Another goal is to plan for medium- to long-term sustainability. Meeting both goals requires working on multiple tracks of change simultaneously. As suggested by Moritz [28], they include: repairing what is currently damaged; rethinking change without building back past and current vulnerabilities; reconfiguring development without re-adopting the business-as-usual mindset; and reporting how change takes place through monitoring and evaluation and proposing course correction. To that list, one can add reconnecting with the inner dimension of human development.

Addressing these priorities and developing an action plan that guarantees a certain level of success is not easy. As noted in the TWI2050 [7] report, "success is a matter of choice. Choice requires the deployment of economic, political, and social instruments, technological and cultural innovations, and changes in lifestyles to bring about the needed transformational changes at every scale". Unfortunately, the past 30 years have shown that the lack of will to change from policymakers and practitioners and other geopolitical issues often hinder socio-economic development progress. Inherent to the entire history of socio-economic development are multiple intended and unintended roadblocks that limit progress.

The value proposition of sustainability and peace in the world of tomorrow has become more imperative than ever. Both are entangled. Not effectively addressing SDG 16 may jeopardize all other SDGs [96]. However, addressing how peace interacts with SDGs using a systems approach is not straightforward. As shown in the simple SD models herein, understanding the peace–development nexus requires selecting the sectors that define peace and development. Another challenge is quantifying how development changes the state of peace and how peace changes the state of development. Data and case studies are needed to quantify such complex interactions to be able to consider trade-offs and synergies.

Sustainable development is more than meeting independent goals [97,98]. There is no one-size-fits-all approach to addressing the linkages between the SDGs systematically. All models are context- and scale-specific and are based on an interpretation of reality, but not the reality itself. What works at one scale may not work on another scale.

Finally, a limitation of the SDGs is that they are defined at the country level but cannot be scaled down to other levels. A question remains as to how relevant the SDGs are at the local level when outside experts define them with limited or no input from those who face the actual problems [35].

Funding: This research received no external funding.

Institutional Review Board Statement: Not applicable.

Informed Consent Statement: Not applicable.

Data Availability Statement: Not applicable.

Conflicts of Interest: The authors declare no conflict of interest.

References

1. Cherkaoui, M. The Shifting Geopolitics of Coronavirus and the Demise of Neoliberalism-(Part 1). 2020. Available online: https://studies.aljazeera.net/en/reports/shifting-geopolitics-coronavirus-and-demise-neoliberalism-%E2%80%93-part-1 (accessed on 1 February 2021).
2. Tanabe, J. Exploring a post-covid-19 sustainable peace model. *Soc. Ethics Soc. J. Appl. Philos.* **2020**, *6*, 73–103.

3. United Nations Development Programme (UNDP). *COVID-19 and Human Development: Assessing the Crisis, Envisioning the Recovery*; UNDP: New York, NY, USA, 2020.
4. IEP (Institute for Economics & Peace). Global Peace Index 2020: Measuring Peace in A Complex World. Available online: http://visionofhumanity.org/reports (accessed on 1 January 2021).
5. Sachs, J.; Schmidt-Traub, G.; Kroll, C.; LaFortune, G.; Fuller, G.; Woelm, F. The Sustainable Development Goals and COVID-19. In *Sustainable Development Report 2020*; Cambridge University Press: Cambridge, UK, 2020.
6. Hayashi, Y. *Covid-19 Aftermath Could Spell a 'Lost Decade' for Global Economy, World Bank Says*; American Bankruptcy Institute: Alexandria, VA, USA, 2021.
7. The World in 2050 (TWI2050). Innovations for Sustainability. In *Pathways to an Efficient and Post-Pandemic Future*; Report Prepared by The World in 2050 Initiative; International Institute for Applied Systems Analysis (IIASA): Laxenburg, Austria, 2020.
8. Alliance for Peacebuilding (AfP). Reimagining The Path To Peace-Priorities For The New Presidential Agenda. 2020. Available online: https://www.allianceforpeacebuilding.org/afp-publications/afp-transition-memo-2020 (accessed on 23 January 2021).
9. Organization for Economic Co-operation and Development. Water and Violent Conflict. Issues Brief. Available online: http://www.eda.admin.ch/dam/deza/en/documents/themen/fragile-kontexte/92767-water-violent-conflict_EN.pdf (accessed on 10 June 2019).
10. Tooley, C. *What Systems Thinking Actually Means—and Why It Matters for Innovation Today*; The World Economic Forum: Geneva, Switzerland, 2021.
11. IEP (Institute for Economics & Peace). COVID-19 and Peace. Available online: http://visionofhumanity.org/reports (accessed on 14 January 2021).
12. United Nations (UN). Shared Responsibility, Global Security: Responding to the Socio-Economic Impacts of COVID-19. Available online: https://unsdg.un.org/resources/shared-responsibility-global-solidarity-responding-socio-economic-impacts-covid-19 (accessed on 2 June 2020).
13. World Commission on Environment and Development (WCED). *Our Common Future*; Oxford University Press: Oxford, UK, 1987.
14. UNCED (United Nations Conference on Environment and Development). Available online: http://ww.un.org/esa/sustdev/documents/agenda21/english/Agenda21.pdf (accessed on 15 January 2019).
15. United Nations (UN) A/RES/70/1. Transforming Our World: The 2030 Agenda for Sustainable Development. Available online: https://www.un.org/ga/search/view_doc.asp?symbol=A/RES/70/1&Lang=E (accessed on 7 June 2020).
16. Lu, J. What Will COVID-19 Do to the Sustainable Development Goals? Available online: https://www.undispatch.com/what-will-covid-19-do-to-the-sustainable-development-goals/ (accessed on 1 June 2020).
17. Sumner, A.; Hoy, C.; Ortiz-Juarez, E. *Estimates of the Impact of COVID-19 on Global Poverty*; United Nations University: Tokyo, Japan, 2020.
18. National Academy of Engineering (NAE). NAE Grand Challenges for Engineering. Available online: www.engineeringchallenges.org (accessed on 20 September 2018).
19. Amadei, B. A systems approach to the sustainability-peace nexus. *Sustain. Sci.* **2021**. [CrossRef]
20. Meadows, D.H. Places to intervene in a system in increasing order of effectiveness. *Whole Earth Winter* **1997**, *91*, 78–84.
21. Cherkaoui, M. The Shifting Geopolitics of Coronavirus and the Demise of Neoliberalism-(Part 2). 2020. Available online: https://studies.aljazeera.net/en/reports/shifting-geopolitics-coronavirus-and-demise-neoliberalism-%E2%80%93-part-2 (accessed on 1 February 2021).
22. Dörner, D. *The Logic of Failure: Recognizing and Avoiding Error in Complex Situations*; Perseus Books: Cambridge, MA, USA, 1997.
23. Huesemann, M.; Huesemann, J. *Techno-Fix: Why Technology Won't Save Us and the Environment*; New Society Publishers: Gabriola Island, BC, Canada, 2011.
24. Korten, D. *The Great Turning: From Empire to Earth Community*; Berrett-Koehler: San Francisco, CA, USA, 2006.
25. Jaworski, J. *Source: The Inner Path of Knowledge Creation*; Berrett-Koehler Publ.: San Francisco, CA, USA, 2012.
26. Gil, D. COVID-19 a Year Later: What Have We Learned? World Economic Forum: Geneva, Switzerland, 2021.
27. United Nations Economic and Social Commission for Western Asia (UNESCWA). Available online: https://www.unescwa.org/sites/www.unescwa.org/files/u593/the_5ps_of_the_sustainable_development_goals.pdf (accessed on 10 January 2021).
28. Moritz, R.E. To Reinvent the Future, We Must All Work Together. Available online: https://www.weforum.org (accessed on 17 July 2020).
29. Mark 2:18-22. Available online: http://www.biblehub.com (accessed on 9 February 2021).
30. Zolli, A.; Healy, A.M. *Resilience—Why Things Bounce Back*; Free Press: New York, NY, USA, 2012.
31. Amadei, B. A systems approach to community capacity and resilience. *Challenges* **2020**, *11*, 28. [CrossRef]
32. Cornwall, A.; Jewkes, R. What is participatory research? *Soc. Sci. Med.* **1995**, *41*, 1667–1676. [CrossRef]
33. Checkland, P.; Poulter, J. *Learning for Action: Soft Systems Methodology and Its Use for Practitioners, Teacher, and Students*; John Wiley & Sons: Chichester, UK, 2006.
34. Earth Charter. Principle 16f: Promote a Culture of Tolerance, Nonviolence, and Peace. Available online: https://earthcharter.org/library/the-earth-charter-text/ (accessed on 25 January 2021).
35. Gittins, P.; Velasquez-Castellanos, I.O. *Peace and Conflict in Bolivia*; Konrad Adenauer Stiftung: La Paz, Bolivia, 2016.
36. National Peace Academy. A Conceptual Framework for Peace Education and Peacebuilding Programs. Available online: https://nationalpeaceacademy.us/images/files/ProgramFramework1.pdf (accessed on 21 January 2021).

37. Maslow, A.H. A theory of human motivation. *Psychol. Rev.* **1943**, *50*, 370–396. [CrossRef]
38. Meister Eckhart and Parke, S. *Conversation with Meister Eckhart*; White Crow Books: Esher, UK, 2010.
39. LaFortune, G.; Fuller, G.; Moreno, J.; Schmidt-Traub, G.; Kroll, C. *SDG Index and Dashboards Detailed Methodological Paper*; Bertelsmann Stiftung and Sustainable Development Solutions Network (SDSN): New York, NY, USA, 2018.
40. Sachs, J.; Schmidt-Traub, G.; Kroll, C.; LaFortune, G.; Fuller, G. *Sustainable Development Report 2019*; Bertelsmann Stiftung and Sustainable Development Solutions Network (SDSN): New York, NY, USA, 2019.
41. Scharlemann, J.P.W.; Brock, R.C.; Nicholas, B.; Brown, C.; Burgess, N.D.; Guth, M.K.; Ingram, D.J.; Lane, R.; Martin, J.G.C.; Wicander, S.; et al. Towards understanding interactions between Sustainable Development Goals: The role of environment–human linkages. *Sustain. Sci.* **2020**, *15*, 1573–1584. [CrossRef]
42. Costanza, B.R.; Daly, L.; Fioramonti, L.; Giovannini, E.; Kubiszewski, I.; Fogh, L.; Pickett, K.; Ragnarsdóttir, K.V.; Vogli RDe Wilkinson, R. The UN Sustainable Development Goals and the Dynamics of Well-being. *Solutions* **2016**, *7*, 20–22. [CrossRef]
43. Zelinka, D.; Amadei, B. A systems approach for modeling interactions among the Sustainable Development Goals Part1: Cross-impact network analysis. *Int. J. Syst. Dyn. Appl.* **2019**, *8*, 23–40.
44. Griggs, D. A Systems Approach: Imperative to Achieve the Sustainable Development Goals. Available online: https://futureearth.org/publications/explainers/a-systems-approach/ (accessed on 19 January 2021).
45. Nilsson, M.; Lucas, P.; Yoshida, T. Towards an integrated framework for the SDGs: Ultimate and enabling goals for the case of energy. *Sustainability* **2013**, *5*, 4124–4151. [CrossRef]
46. Griggs, D.; Smith, M.S.; Gaffney, O.; Noble, I. Sustainable development goals for people and planet. *Nature* **2013**, *495*, 305–307. [CrossRef]
47. Waage, J.; Yap, C.; Bell, S. Governing the sustainable development goals: Interactions, infrastructures, and institutions. *Lancet Glob. Health* **2015**, *3*, e251–e252. [CrossRef]
48. Coopman, A.; Osborn, D.; Ullah, F.; Auckland, E.; Long, G. *Seeing the Whole: Implementing the SDGs in an Integrated and Coherent Way*; The Stakeholder Forum: Herne Bay, UK, 2016.
49. Vladimirova, K.; Le Blanc, D. Exploring Links between Education and Sustainable Development Goals through the Lens of UN Flagship Reports. *Sustain. Dev.* **2016**, *24*, 254–271. [CrossRef]
50. Barbier, E.B.; Burgess, J.C. The Sustainable Development Goals and the systems approach to sustainability. *Econ. E-J.* **2017**, *11*, 1–23. [CrossRef]
51. Morton, S.; Pencheon, D.; Aquires, N. Sustainable Development Goals (SDGs), and their implementation: A national global framework for health, development, and equity needs a systems approach at every level. *Br. Med. Bull.* **2017**, *124*, 81–90. [CrossRef]
52. Lim, M.M.L.; Jørgensen, P.S.; Wyborn, C.A. Reframing the sustainable development goals to achieve sustainable development in the Anthropocene—A systems approach. *Ecol. Soc.* **2018**, *23*, 22. [CrossRef]
53. The world in 2050 (TWI2050). The digital revolution and sustainable development: Opportunities and challenges. In *Report Prepared by The World in 2050 Initiative*; International Institute for Applied Systems Analysis (IIASA): Laxenburg, Austria, 2019.
54. International Council for Science (ICSU). *Review of the Sustainable Development Goals: The Science Perspective*; International Council for Science: Paris, France, 2015.
55. Schmidt-Traub, G. Indicators and a Monitoring Framework for the SDGs: Launching a Revolution for the SDGs; Sustainable Development Solutions. Available online: https://indicators/report (accessed on 15 August 2020).
56. Nilsson, M.; Griggs, D.; Visback, M. Map the interactions between Sustainable Development Goals. *Nature* **2016**, *534*, 320–322. [CrossRef]
57. Nilsson, M.; Griggs, D.; Visbeck, M.; Ringler, C.; McCollum, D. A framework for understanding sustainable development goal interaction. In *A Guide to SDG Interactions: From Science to Implementation*; International Council for Science: Paris, France, 2017.
58. Zhang, Q.; Prouty, C.; Zimmerman, J.B.; Mihelcic, J.R. More than target 6.3: A systems approach to rethinking sustainable development goals in a resource-scarce world. *Engineering* **2016**, *2*, 481–489. [CrossRef]
59. International Council for Science (ICSU). *A Guide to SDG Interactions: From Science to Implementation*; Griggs, D.J., Nilsson, M., Stevance, A., McCollum, D., Eds.; International Council for Science: Paris, France, 2017.
60. Zhou, X.; Moinuddin, M. *Sustainable Development Goals Interlinkages and Network Analysis: A Practical Tool for SDG Integration and Policy Coherence*; IGES Research Report No. RR1602; Institute for Global Environmental Strategies: Kanagawa, Japan, 2017.
61. Gue, I.H.V.; Ubando, A.T.; Tseng, M.L.; Tan, R.R. Artificial neural networks for sustainable development: A critical review. *Clean Tech. Environ. Policy* **2020**, *22*, 1449–1465. [CrossRef]
62. Zelinka, D.; Amadei, B. A systems approach for modeling interactions among the Sustainable Development Goals Part 2: System dynamics. *Int. J. Syst. Dyn. Appl.* **2019**, *8*, 41–59. [CrossRef]
63. Benson, T.; Marlin, S. *The Habit-Forming Guide to Becoming a Systems Thinker*; The Waters Foundation Systems Thinking Group Publ.: Pittsburgh, PA, USA, 2017.
64. Groff, L. Contributions of different cultural-religious traditions to different aspects of peace—Leading to a holistic, integrative view of peace for a 21st century independent world. *FUTUREtakes* **2008**, *7*, 1–14.
65. Cortright, D. *Peace: A History of Movements and Ideas*; Cambridge University Press: Cambridge, UK, 2008.
66. Dietrich, W. *Interpretations of Peace in History and Culture*; Palgrave Macmillan: New York, NY, USA, 2012.
67. Stearns, P.N. *Peace in World History*; Routledge: New York, NY, USA, 2014.

68. Dietrich, W.; Pearce, J. Many violences, many peaces. *Peacebuilding* **2019**, *7*, 1–15.
69. Diamond, L.; McDonald, J. *Multi-Track Diplomacy: A Systems Approach to Peace*, 3rd ed.; Kumarian Press: Boulder, CO, USA, 1996.
70. Lederach, J.-P. *Sustainable Reconciliation in Divided Societies*; United States Institute for Peace: Washington, DC, USA, 1999.
71. Davis, Q. *Building Infrastructures for Peace: The Role of Liaison Offices in Myanmar's Peace Process*; Center for Peace and Conflict Studies, Australian Government Department of Foreign Affairs and Trade: Canberra, Australia, 2016.
72. Ben-Eli, M. Sustainability: Definition and five core principles: A systems perspective. *Sustain. Sci.* **2018**, *13*, 1337–1343. [CrossRef]
73. Galtung, J. An editorial. *J. Peace Res.* **1964**, *1*, 1–4.
74. Galtung, J. Cultural violence. *J. Peace Res.* **1990**, *27*, 291–305. [CrossRef]
75. Fischer, D. Peace as a self-regulating process. In *Handbook of Peace and Conflict Studies*; Webel, C., Galtung, J., Eds.; Chapter 13; Routledge: New York, NY, USA, 2007.
76. IEP (Institute for Economics & Peace). *Positive Peace Report 2017: Tracking Peace Transitions Through a Systems Thinking Approach*; Report Number 54; IEP: Sydney, Australia, 2017.
77. IEP (Institute for Economics & Peace). *Global Peace Index 2018: Measuring Peace in a Complex World*; IEP: Sydney, Australia, 2018; Available online: http://visionofhumanity.org/reports (accessed on 1 October 2018).
78. IEP (Institute for Economics & Peace). *Global Peace Index 2019: Measuring Peace in a Complex World*; IEP: Sydney, Australia, 2019; Available online: http://visionofhumanity.org/reports (accessed on 1 October 2019).
79. Amadei, B. Revisiting positive peace using systems tools. *J. Tech. Forecast. Soc. Chang.* **2020**, *158*, 120149. [CrossRef]
80. Milante, G.; Oxhorn, P. No Development without Peace; World Bank Open Knowledge Repository. Available online: https://openknowledge.worldbank.org/handle/10986/4582 (accessed on 25 January 2021).
81. Dews, F. UN Deputy Secretary-General Jan Eliasson: No Peace without Development, no Development without Peace; Brookings. Available online: http://www.brookings.edu/blog/brookings-now/2013/10/17/un-deputy-secretary-general-jan-eliasson-no-peace-without-development-no-development-without-peace (accessed on 15 March 2019).
82. Ricigliano, R. *Making Peace Last: A Toolbox for Sustainable Peacebuilding*; Paradigm Publishers: Boulder, CO, USA, 2012.
83. Bush, K. *A Measure of Peace: Peace and Conflict Impact Assessment (PCIA) of Development Projects in Conflict Zones*; Peacebuilding and Reconstruction Program Initiative, International Development Research Center—Canada: Ottawa, ON, Canada; Johannesburg, South Africa, 1998.
84. United Nations (UN). *Global Sustainable Development Report 2019: The Future is Now—Science for Achieving Sustainable Development*; United Nations: New York, NY, USA, 2019; Available online: https://sustainabledevelopment.un.org/globalsdreport/2019 (accessed on 8 June 2020).
85. Castellani, B. Map of the Complexity Sciences. In *Art and Science Factory*; The Austrian Institute of Technology Press: Vienna, Austria; Available online: http://scimaps.org/mapdetail/map_of_complexity_sc_154 (accessed on 1 June 2020).
86. Rahmandad, H.; Sterman, J. System Dynamics or Agent-Based Models? Wrong Question! Seek the Right Level of Aggregation. Available online: https://www.systemdynamics.org/assets/docs/sdorabm.pdf (accessed on 20 April 2020).
87. Borshchev, A.; Filippov, A. From system dynamics and discrete event to practical agent-based modeling: Reasons, techniques, tools. In Proceedings of the 22nd International Conference of the System Dynamics Society, Oxford, UK, 25–29 July 2004.
88. Forrester, J.W. *World Dynamics*; Productivity Press: Portland, OR, USA, 1971.
89. Richmond, B. *An Introduction to Systems Thinking, STELLA Software*; ISEE Systems, Inc.: Lebanon, NH, Middle East, 2004.
90. Sterman, J. *Business Dynamics: Systems Thinking and Modeling for a Complex World*; Irwin, McGraw Hill: New York City, NY, USA, 2000.
91. Ford, A. *Modeling the Environment*; Island Press: Washington, DC, USA, 2010.
92. ISI (Institute for Sustainable Infrastructure). Envision. Sustainable Infrastructure Report; Version 3.0. Available online: https://v3.sustainableinfrastructure.org/uploads/user-materials/6e23716858c46844adfc57f13026a826.pdf (accessed on 1 January 2021).
93. Schweitzer, R.W.; Mihelcic, J.R. Assessing sustainability of community engagement of rural water systems in the developing world. *J. Water Sanit. Hyg. Dev.* **2012**, *2*, 20–30. [CrossRef]
94. Bouabid, M.; Louis, G. Capacity factors for evaluating water and sanitation infrastructure choices for developing communities. *J. Environ. Manag.* **2015**, *161*, 335–343. [CrossRef] [PubMed]
95. Willis, H.H.; Groves, D.G.; Ringel, J.S.; Mao, Z.; Efron, S.; Abbott, M. *Developing the Pardee RAND Food-Energy-Water Security Index: Toward a Global Standardized, Quantitative, and Transparent Resource Assessment*; RAND Corporation: Santa Monica, CA, USA, 2016; Available online: https://www.rand.org/pubs/tools/TL165.html (accessed on 15 January 2021).
96. Virji, H.; Sharifi, A.; Kaneko, S.; Simangan, D. The sustainability–peace nexus in the context of global change. *Sustain. Sci.* **2019**, *14*, 1467–1468. [CrossRef]
97. Rippin, N. How to avoid the silo structure of the Millennium Development Goals (MDGs)? In *The United Nations Post-2015 Agenda for Global Development: Perspectives from China and Europe*; Fues, T., Ye, J., Eds.; German Development Institute: Bonn, Germany, 2014.
98. Collste, D.; Pedercini, M.; Cornell, S.E. Policy coherence to achieve the SDGs: Using integrated simulation models to assess effective policies. *Sustain. Sci.* **2017**, *12*, 921–931. [CrossRef]

sustainability MDPI

Article

The Crossovers and Connectivity between Systems Engineering and the Sustainable Development Goals: A Scoping Study

Lan Yang [1,*] and Kathryn Cormican [1,2]

1 School of Engineering, College of Science and Engineering, National University of Ireland, Galway, H91 TK33 Galway, Ireland; kathryn.cormican@nuigalway.ie

2 Lero—The Irish Software Research Centre, School of Engineering, National University of Ireland, Galway, H91 TK33 Galway, Ireland

* Correspondence: lan.yang@nuigalway.ie

Abstract: The United Nation's sustainable development goals (SDGs) are interconnected and indivisible and need to be addressed in a systematic and holistic way. However, a lack of stakeholder perspective, fragmented responses, and a dearth of integration across sectors have long been perceived as the SDGs' main pitfalls. In recent years, scholars are calling to address these issues by adopting a systems engineering perspective, as this approach espouses a stakeholder-focused position, embraces a holistic and dynamic mindset, and provides a variety of technical and managerial toolkits, which can help to untangle the complexity and interactions inherent in global sustainability. Nevertheless, little has been done to map the existing literature, comprehensively review, and synthesize research evidence in this field. Therefore, this paper aims to conduct a scoping study that analyzes the extant evidence to uncover the contributions of systems engineering in advancing the SDGs. A three-phase methodology integrating natural language processing and systematic literature review is used to investigate this space. We conclude that systems engineering has been an active catalyst promoting the SDGs, and that systems engineering has the potential to support more transdisciplinary research to achieve long-term transformational and sustainable change across sectors and disciplines.

Keywords: sustainable development goals; systems engineering; systems thinking; complex adaptive systems; socio-technical systems; system of systems; cyber physical systems; natural language processing

Citation: Yang, L.; Cormican, K. The Crossovers and Connectivity between Systems Engineering and the Sustainable Development Goals: A Scoping Study. *Sustainability* **2021**, *13*, 3176. https://doi.org/10.3390/su13063176

Academic Editor: Cecilia Haskins

Received: 26 January 2021
Accepted: 11 March 2021
Published: 14 March 2021

1. Introduction

Sustainable development has become an underlying strategy to guide global transformation [1]. The goal of sustainable development has evolved from pursuing the single goal of sustainable use of natural resources to resolving the complexities of economic development, environmental protection, and societal polarization [2]. In September 2000, the United Nations (UN) Millennium Summit published the Millennium Development Goals (MDGs) comprising 8 key areas and 21 operational goals [3]. These goals guided national development and international cooperation for the first 15 years in the new century. In 2012, the UN Conference on Sustainable Development in Rio de Janeiro called for a revision of the MDGs [4]. On 25 September 2015, the General Assembly of the UN adopted a resolution "Transforming our world: the 2030 Agenda for Sustainable Development" [5]. The agenda announced 17 sustainable development goals (SDGs) and 169 specific targets. They replaced the MDGs and strive to shift the world onto a more sustainable path.

Since the SDGs were announced, scholars have invested considerable effort in identifying new sustainable solutions to face the complex and interwoven challenges. The key is to take an integrated and balanced consideration of the three dimensions relating to sustainable development: economic, social, and environmental. Barbier and Burgess [6] were among the first to advocate using a systems approach to characterize sustainability in terms of the optimization of goals across social, economic, and environmental systems through an adaptive process of trade-offs.

According to the agenda, the 17 SDGs should be viewed in an integrated manner and cannot be separated from each other [5]. However, in reality, the SDGs are very broad and diverse. They are overly fragmented in their formulation and largely sectoral [7]. In other words, there is a lack of integration across sectors in terms of strategies, policies and implementation. Little guidance has been given on how to trade off and balance the goals [6]. In addition, the SDGs have been criticized for lacking stakeholder perspectives and being ambiguous about their target audience [8]. Furthermore, Laurent et al. [9] advocated that more attention should focus on the method and approach to achieve these goals. Therefore, the achievement of the SDGs needs a holistic and systematic approach. The dynamics and complexity of the SDGs should be considered when prioritizing and optimizing the goals [10]. It is clear that there is an urgent need for a holistic and systematic perspective when seeking a feasible solution to harmonizing the SDGs [11,12].

However, addressing sustainability in such a wide range of contexts can be highly complex. This diversity suggests a need for a systematic and multidisciplinary approach to help tackle these problems. Systems engineering is a suitable methodology to help with this. Systems engineering [13] is an integrative and transdisciplinary approach that focuses on establishing, balancing and integrating stakeholders' requirements. It also focuses on the entire life cycle process to achieve goals, considering the levels of uncertainty, variety, change, and complexity. Systems thinking [14] is at the heart of systems engineering, as it considers the interconnections, dynamics and emergent behavior of system components. In other words, it can focus on the patterns of change rather than a static 'snapshot' of the current problem. In addition, the systems approach advocates and follows a system life cycle process, which can identify potential risks at the earliest stakeholder requirement definition and system conceptualization stage. This can help recognize ecologically or socially problematic decisions and manage the sustainability impacts throughout the entire life cycle of a process, product, or service.

Although systems engineering has many functions and benefits, there is a dearth of research that analyzes the connectivity between systems engineering and the SDGs. For example, little is known about where systems engineering has been used to advance the SDGs, and what systems engineering knowledge has been used in previous studies.

In order to address these deficits, our paper presents a scoping study that identifies existing synergies between systems engineering and the SDGs and helps build a foundation for future research directions. Whereas there are various methodological approaches for scoping studies [15], we focused on the most cited ones [16–18] and tailored them to our needs. Furthermore, we integrated natural language processing (NLP) [19] and systematic literature review (SLR) [20] in the process to pursue and ensure a rigorous and transparent method for synthesis and analysis. We adopted a three-phase analytical approach, namely (1) data collection and pre-processing, (2) data visualization and analysis, and (3) results interpretation. NLP is used in the first two phases and the principles of the SLR guides us throughout the study.

This research identifies the relevant literature and analyzes and interprets the crossovers and synergies between systems engineering and the SDGs. The results identify specific areas in which sustainable development can apply systems engineering knowledge to obtain sustained competitive advantage while attempting to address the sustainability expectations of the SDGs.

2. Materials and Methods

The research methodology involves three phases. As shown by the schematic figure (Figure 1), they are: Phase I, data collection and pre-processing, Phase II, data visualization and analysis, and Phase III, results interpretation. The following subsections further explain each phase.

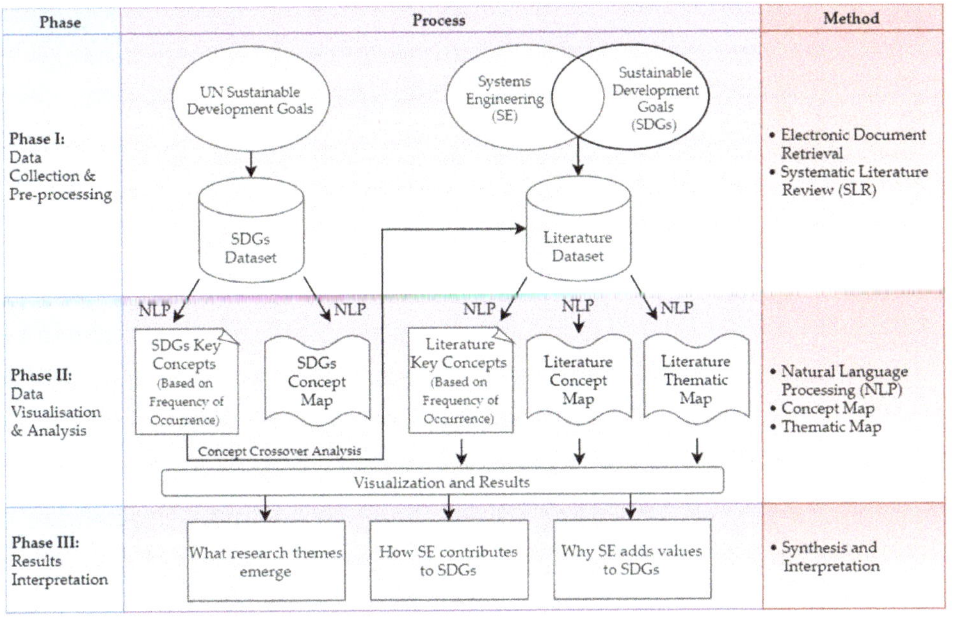

Figure 1. Research process.

2.1. Phase I: Data Collection and Pre-Processing

The first phase of the study involves assembling an SDGs dataset and a literature dataset. The SDGs dataset is made up of the 17 goals and their 169 targets and was obtained from the UN Sustainable Development Goals website. The literature dataset is composed of a group of scientific papers that uses systems engineering to advance the SDGs. We used SLR techniques to attain this scientific paper dataset. More specifically, we designed our search strings to retrieve papers that apply systems engineering core concepts to the context of sustainable development. The search query ensures that it returns any document whose title, abstract or keyword contains at least one term related to systems engineering and at least one term related to sustainable development. The search terms are listed in Table 1. We repeated the same query in many leading databases to ensure an adequate coverage, including Scopus, Web of Science, ProQuest, Science Direct, and IGI Global, and then we removed any duplicates.

Table 1. Search terms.

Search Terms Related to Systems Engineering	Search Terms Related to SDGs
systems engineering, systems thinking, systems theory, systems science, systems approach, systems modelling, systems life cycle, systems analysis, system of systems, complex systems, complex adaptive systems, cyber-physical systems, sociotechnical systems, systems engineer	sustainable development, sustain, sustainment, sustainability, SDGs

The review was completed progressively over several weeks. To ensure their relevance and to confirm that each paper aligned well with our study, papers were carefully reviewed.

The results were filtered to include only peer-reviewed, English-language academic articles for which full-text papers were available. A dataset of 45 unique papers resulted in this.

2.2. Phase II: Data Visulization and Analysis

The second phase involved an application of an emerging method in artificial intelligent called natural language processing (NLP). This method is proven to be useful in concept extraction, topic modeling, and content categorization to explore large datasets and has been employed by a diverse range of users. We used it to iteratively build up a map of associated concepts to analyze the concept crossovers and themes within the topic area. To apply this method, we used a text processing tool called Natural Language Toolkit (NLTK). NLTK is one of the most powerful NLP libraries which contains packages to parse text datasets. The approach taken is as follows.

First, we defined the key concepts. Here, a concept comprises a collection of terms that describe the same thing and often travel together throughout the text. To determine whether a term is related to a concept, it was weighted according to how frequently it occurs in a sentence that contains the concept, compared to how frequently it occurs elsewhere. The occurrence of each word in a phrase makes an adequate contribution to the cumulative evidence for the presence of a concept, since terms are weighted. To eliminate duplicates, group similar terms, remove incorrectly defined terms, and maintain a stable collection of outcomes, the resulting concepts were manually vetted. Next, we created themes which comprise a cluster of interconnected concepts. Though clustering, a thematic view of relationships between concepts was generated. In other words, we used NLTK to interpret document sets and built maps of key concepts, where their relationships are suggested by their connectedness or distance on the map. Qualitative analyses can be carried out on the basis of quantitative, algorithmic analysis by analyzing the resulting concept map, frequency distributions, and relationships between both concepts and themes. NLP was used to detect the main concepts present in the datasets (according to their frequency of occurrence), cluster these concepts, and consequently uncover the relationships between them.

There are two datasets involved in the study, i.e., the SDGs dataset and the literature dataset. As shown by Figure 1, they are both analyzed using NLP.

The NLTK processed a digital copy of the SDGs and associated targets. To obtain a frequency distribution of the concepts, the full text of the 17 objectives and the corresponding 169 target specifications were processed over several iterations. For clarification, terms inappropriately defined as concepts (e.g., 'including', 'particular', 'inclusive') were discarded manually from the study. As a result, we generated a list of key concepts that contain the core information with their frequency of occurrence in the SDGs text. We then developed a concept map that demonstrate how these concepts are thematically linked.

Regarding the literature dataset, the full texts of the 45 papers were analyzed from four dimensions. First, we used the key concepts generated from the SDGs dataset as seed concepts and applied them to the literature dataset to examine how these concepts are addressed and covered by the literature. This approach allowed us to evaluate the concept crossovers and connectivity between the SDGs and the literature. Next, the literature dataset was analyzed through NLP and resulted in a frequency distribution of the key concepts within the literature and a concept map. We then compared the outcomes with the concept map of the SDGs dataset to distinguish the similarity and differences. Finally, to gain a better understanding of the themes covered by the literature, we developed a thematic map. Together with the analysis above, the thematic map was used to identify research trends, as well as to interpret research gaps.

This phase was not intended to investigate specific cases of how systems engineering is applied to advance SDGs, but to establish the concept crossovers between the SDGs and the extant research performed via systems engineering perspectives.

2.3. Phase III: Results Interpretation

The third phase produced a detailed and extensive interpretation of the integrated results from the NLP analysis as well as from the literature review. Our goal was to provide insights on the following three aspects, i.e., what research themes emerged from the literature, how systems engineering supports the implementation of the SDGs, and why systems engineering can add value to the SDGs. By analyzing the above questions, we investigated where systems engineering and the SDGs overlap, and consequently determine the crossovers and connectivity between them.

3. Results

3.1. Results from Phase I

A total of 45 academic articles that focus on systems engineering and the SDGs were identified through the search and initial review, including 35 journal articles, 6 conference papers, and 4 book chapters. Publication per year generally followed an upward trend (as shown in Figure 2), with only 1 paper published in 2009, 2012, 2015, and 2016, respectively, 5 papers in 2017, 13 papers in 2018, and 15 papers in 2019. Since this time, the numbers have decreased marginally, with 8 published in 2020.

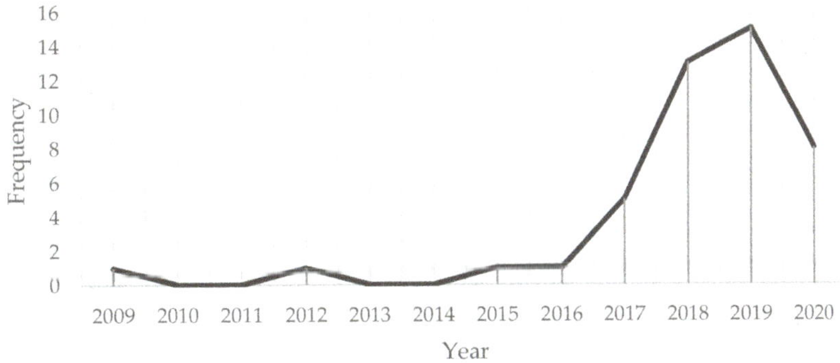

Figure 2. Trend of the publications over the timeframe 2009–2020.

The journals in this area are written in the fields of environmental science, social science, energy, engineering, public health, earth and planetary science, chemistry, sociology and political science, education, and ecology. The literature focuses on the use of systems thinking, systems approach, and integrated tools to implement the SDGs, spanning the three pillars (environmental, social and economic) of sustainable development. Major application areas include information, energy, production, health, technology, climate, and policy.

3.2. Results from Phase II

3.2.1. NLP Analysis for the SDGs

Figure 3 shows the stable concept map that depicts the relationships between the concepts generated from the SDGs dataset through NLP. The results provide 38 key concepts that characterize the SDGs and later become the seed concepts that are used to analyze the crossovers with the literature dataset.

In Figure 3, the size of the dots represents the frequency of occurrence (the larger the dots, the higher the frequency). The length of the lines linking the dots represent the co-occurrence of two concepts in the corpus (the shorter the distance, the more often they co-occur in a sentence). As the generation of the concept map is based on how often two concepts travel together in the SDGs text and how closely they are to each other, we can use the concept map to present whether two concepts are mentioned together repeatedly in the same context. As a result, several clusters were constructed represented by the same color in Figure 3, where the concepts gather and become a theme.

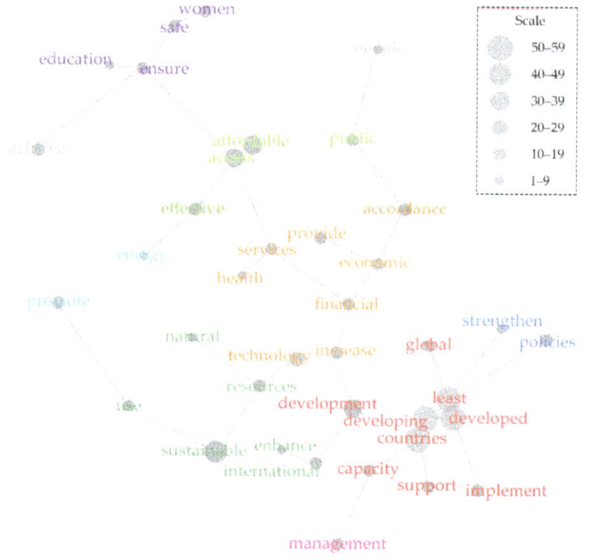

Figure 3. Concept map with clusters derived from natural language processing (NLP) analysis of the sustainable development goals (SDGs).

Major collocations that emerge from the NLP analysis are 'least developed' and 'development countries', which is unsurprising given the context in which they were released. Both are at the center of the red cluster. We found 'capacity' and 'support' are close to 'developing countries', while 'development' is at the boundary of the red cluster, linking the yellow cluster as well. The second largest dot is 'sustainable', which appears in the green cluster. This cluster is related to the sustainable use of ecosystems and resources (key concepts including 'energy', 'resources', 'sustainable', 'natural', and 'technology'). The green and the red clusters are located closely next to each other, indicating that they have concepts that overlap between the two, such as 'resources', 'enhance' and 'international'. The green and the yellow clusters share a fuzzy boundary, where 'resources' and 'technology' links with each other. The rest of the clusters are relatively far away from the main body, but they form several distinct themes. For example, there is a separate theme that stands out. This is colored in purple. It emphasizes the importance of the safety of women and the assurance of education and access to needs for women ('women', 'safe', 'education', 'access' and 'affordable'). We found 'affordable access' is frequently mentioned. It is in the middle of the purple cluster and the yellow cluster, linked with the purple cluster by 'ensure' and with the yellow cluster by 'services'. In addition, 'services' is connected with 'health' closely. Based on these observations, we infer that the SDGs emphasize an affordable access to health-related services.

However, in the stable concept map, 'management' is often located on the outside (in pink), with few connections, suggesting that this idea is not addressed to any great degree in the text of SDGs.

3.2.2. NLP Analysis for the Literature

We conducted two types of NLP analysis on the literature dataset. The first analyzed the concept crossovers and connections between the literature and the SDGs. In other words, we used the 38 concepts as seed concepts and analyzed how frequently they are discussed in the 45 papers. The second analysis was conducted only on the literature dataset itself and was done to identify the key concepts and themes of the 45 papers. From the first analysis, the concepts identified from the SDGs (as seed concepts) and also shown in the literature dataset, along with their frequency counts and relevance scores are shown in Table 2. Figures 4 and 5 are the concept map and thematic map developed from the literature dataset, respectively.

Table 2. Concept crossovers and connectivity between the SDGs and the literature.

Concept	Count	Relevance %	Concept	Count	Relevance %
energy	1128	100	natural	289	26
development	1095	97	services	283	25
sustainable	1071	95	developing	246	22
use	939	83	capacity	241	21
management	625	55	policies	217	19
health	581	52	achieve	200	18
economic	578	51	international	176	16
global	568	50	effective	165	15
countries	438	39	women	150	13
access	424	38	ensure	140	12
people	407	36	financial	127	11
technology	404	36	promote	117	10
resources	401	36	least	117	10
provide	359	32	safe	103	9
education	352	31	enhance	82	7
public	345	31	affordable	80	7
support	341	30	implement	61	5
developed	308	27	accordance	16	1
increase	293	26	strengthen	14	1

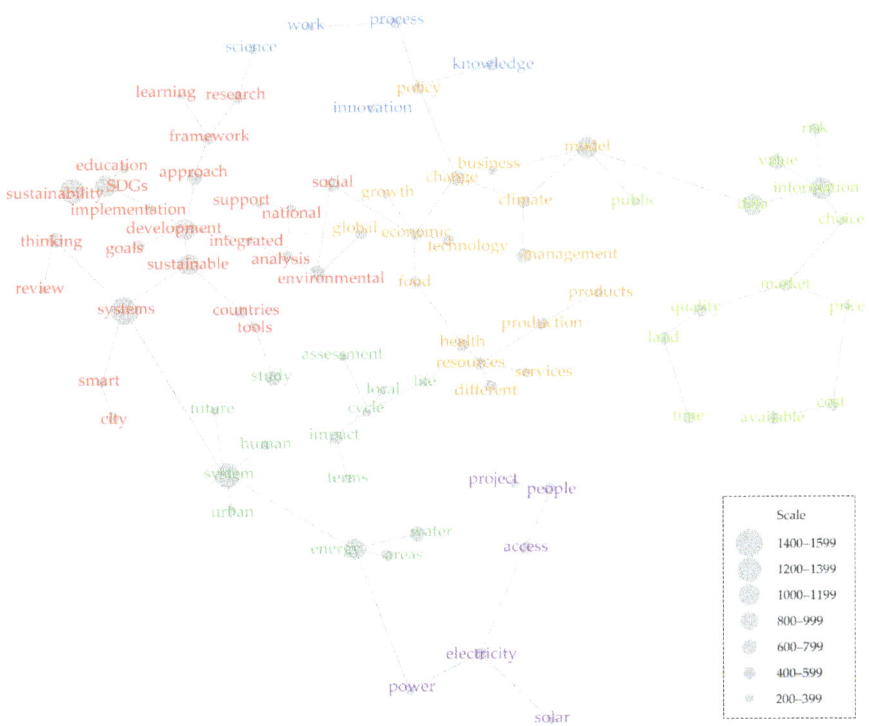

Figure 4. Concept map with clusters derived from NLP analysis of the literature.

From Table 2, we found that of the 38 concepts identified as key in the analysis of the SDGs (concepts in Figure 3), all 38 are directly identified as having some degree of relevance in the literature dataset. This relevance varies from 'energy' at 100% and 1128 mentions in the literature dataset, to 'strengthen', which receives only 14 mentions and has a relevance score of 1%. The crossovers between the SDGs and the literature in the field of systems engineering and sustainable development are representative of the fact that all concepts appear in the literature dataset.

Figure 4 shows the concept map of the 45 papers, with the key concepts highlighted in large dots and the clusters forming several different topic groups. The literature focuses on key concepts including 'systems', 'sustainability', 'model', 'data', 'energy', 'SDGs', 'information', and their interaction with 'sustainable development', along with 'social', 'economic' and 'environmental' factors. These concepts are further grouped into themes based on their relationships in a concept map, as shown by Figure 5.

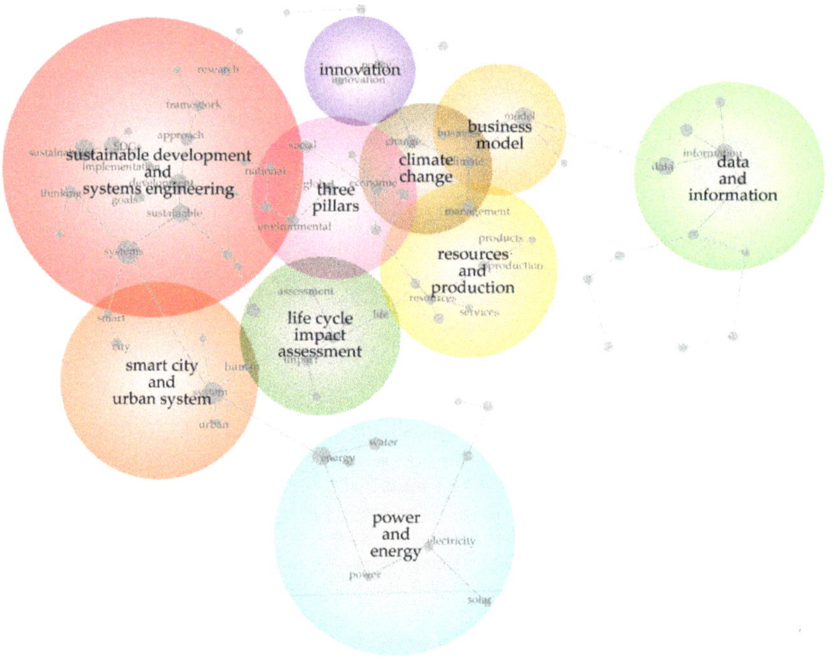

Figure 5. Thematic map derived from NLP analysis of the literature.

Comparing Figure 4 with Figure 3, we found that Figure 4 contains many identical concepts that appeared in the SDGs but also include many other concepts that are relative to systems engineering. Moreover, in Figure 4, the clustering is more obvious. The core concepts of the red cluster and the yellow cluster are relevant to systems engineering and sustainability. The three pillars of the SDGs (social, environmental and economic) are on the boundary. What is interesting is that the concepts 'environmental' and 'social' are closer to the red cluster, whereas 'economic' is close to 'food', 'growth' and 'change'. The blue cluster is more about 'innovation', 'process' and 'knowledge', with 'policy' acting as a glue to hold them together. There is a special focus on 'information' and 'data' in the literature, which stands out alone (the lime green cluster). To better illustrate the themes, we developed a thematic map shown by Figure 5.

In Figure 5, we found that the literature contains several themes, such as sustainable development and systems engineering (including systems thinking, approach, framework, etc.), smart city and urban system, life cycle impact assessment, climate change, business model, research that is relevant to the three pillars, innovation, resources and production, power and energy, as well as data and information. The results obtained from the NLP analysis of the literature help us to understand the general research foci and trends. They offer a bird-eye view of the extant literature that focuses on using systems engineering to advance sustainable development.

3.3. Results from Phase III

In the third phase of the analysis, we conducted an in-depth analysis of the 45 papers. They are categorized by their theoretical foundations in relation to systems engineering, such as systems thinking, socio-technical systems, complex adaptive systems, system of systems, etc. For each domain, we analyzed the definitions of the key concept related to systems engineering especially for the context of sustainable development research. We are particularly interested in what research themes have emerged from the literature, how systems engineering supports the implementation of the SDGs, and how systems engineering can add value to the SDGs.

3.3.1. Systems Engineering

By definition, systems engineering is a transdisciplinary and integrative approach to enable the successful realization, use, and retirement of engineered systems, using systems principles and concepts, and scientific, technological, and management methods [13]. In the overall life cycle context, systems engineering deals with the analysis and design, operation and maintenance of large interconnected systems and helps to maintain the scale and constraints of a multidisciplinary research problem [21]. In the study of sustainable development, systems engineering is seen as a provider of knowledge principles and practices that better facilitate the integration of economic, environmental and social efficiency and effectiveness in achieving sustainable development [22]. It provides the holistic perspective needed to guide the analysis, design, implementation, integration and maintenance of complex systems composed of different interrelated components interacting together to reach a specific goal [23].

System engineering is important to sustainable development mainly due to the following three aspects.

First, model-based systems engineering enables the design of complex systems, especially for large systems that support sustainability [24–26]. For example, building the architecture of large earth observation systems heavily relies on successful systems modelling [27]. These systems are essential to analyze natural environment changes, predict the impact of the changes on humans, and offer a decision-making support in response to the environmental changes. Model-based systems engineering plays a role in ensuring the design and implementation of such systems.

Second, systems engineering has the potential to significantly contribute to the evaluation of socio-technical systems [28–30]. For instance, Palmer et al. [31] used systems engineering to develop a framework for a systematic and systemic mapping of bioeconomic transitions within and between socio-technical systems. The conceptualization of bioeconomy transitions naturally leads itself towards systems engineering, as systems engineering offers a socio-technical perspective.

Third, systems engineering can assist decision-makers to implement the SDGs in an improved, consistent, traceable, and transparent manner. Deshpande and Aspen [21] provide a systems engineering framework to make SDGs more practical. This stepwise framework, embedded in system engineering, uses models to help create clear, traceable, and transparent connections between SDGs, targets, strategies for change and decisions.

These three points prove the positive impact that system engineering has on the SDGs. Next, we analyze in detail how the core concepts of systems engineering are applied in sustainable development research.

3.3.2. Systems Thinking

Systems thinking has been interpreted in numerous ways. The core features include a recognition of interrelationships and the use of holistic approaches [32,33], an investigation of dynamic and emergent behavior arising from feedback [34], and a focus on patterns of change rather than static 'snapshots' [35]. Systems thinking is founded on the conviction that parts of the system will behave differently when viewed in isolation [36].

According to our analysis, systems thinking contributes to the realization of the SDGs in two ways. First, understanding the critical linkages among the SDGs requires systems thinking [37,38]. Second, systems thinking is applied to solve specific problems that contribute to the implementation of one or some SDGs [39–41].

Regarding the first point, many researchers have discussed the importance of systems thinking in analyzing SDGs. Allen et al. [38] endorsed the use of systems thinking to define and evaluate interlinkages among the SDGs and examine synergies and trade-offs to promote the formulation of visions, plans and strategies. As the SDGs framework gains momentum, all 17 goals and their interrelationships should be taken into consideration [42]. Additionally, as the SDGs cannot be approached in a linear and deterministic way, it is important to consider the dynamic condition that is inherent in the SDGs. Systems thinking provides a greater understanding of how the dynamics exist by exploring the relations and changes within the system [43]. In addition, the application of the SDGs explicitly calls for an integrated, comprehensive, multi-stakeholder solution to be created [44]. This signifies the need for systems thinking in practice to enhance better conversations and cooperation globally.

Regarding the second point, many studies have shown that pivoting from unsustainable to sustainable in a particular space requires systems thinking. While these studies come from several application areas, special attention has been paid to integrating systems thinking into education. For example, many researchers have agreed and proved that incorporating systems thinking into the chemistry curriculum can equip the next generation with the skills required to produce more sustainable and cleaner technologies, bringing the planet closer towards a sustainable society [34,40,45]. A systems thinking mindset is also crucial to engineers because a sustainable engineering system will not only require engineers have analytical skills, creativity, and a wide breadth of education but also take into account the multiple impacts on the natural, economic and social environment [46,47]. Therefore, the promotion of systems thinking in engineering education is crucial for equipping engineers with a sense of sustainability. Eustachio et al. [48] believed the key is to relate the complexity of sustainability to educational learning outcomes. It is system thinking that offers tools for complexity management by moving problem frameworks from linear explanations of cause-effects to an understanding of the broader context in which interventions may occur. Reynolds et al. [44] discussed the role of systems thinking in practice for supporting SDGs implementation capability. Kioupi and Voulvoulis [49] developed a framework that enables stakeholders in education to create a shared vision of sustainability, with the SDGs as endpoints. Systems thinking is used as a methodology to look at the overall picture of the importance of education in facilitating such transformation. A systems thinking approach has also been used in other studies to optimize the sustainable use of the system, such as scaling-up local measures for better medical care [32], scaling-up food production [50], growing sustainable aquaculture [36], understand the impact of land management [51], smart and sustainable city [43]. Although these endeavors happened in different domains, systems thinking offers opportunities for improving knowledge exchange through wider engagement.

3.3.3. System Theory, Systems Analysis, System Dynamics, and Systems Modeling

Systems theory has a long history and is inseparable from systems science and systems thinking. General system theory [52] discusses the holism and wholeness of a system, the need for integration of knowledge, open and closed systems, feedback loops, interactions between elements of the system, and emergence. Systems theory is closely related to cybernetics. Now, system theory has spread and evolved to other fields such as business, management and biodiversity [53]. System theory has also been used to better understand social systems and predict behavior [54]. It is beneficial to the SDGs because it helps to develop a holistic viewpoint when developing a new framework for sustainable development, enabling traceability of stakeholder requirements and equality of stakeholders from all over the world [55]. Consequently, the SDGs can be viewed in a systemic way, reducing complexity for governments, scholars and civil society engaged in the process of expanding awareness to take steps towards a more sustainable system.

Systems science is broken down into systems theory, systems analysis and cybernetics [56]. The perspectives gained from systems theory and cybernetics contributes to various interdisciplinary approaches. Especially notable is the approach called system dynamics, which is used to help explain nonlinear behavior in complex systems. It is a systems analysis approach that is used to research behavioral patterns of systems. Obersteiner et al. [57] proposed a design methodology focused on the dynamics of systems to determine the effects of earth observation systems, which have tremendous potential to help ensure the planet's sustainable future. Systems analysis is an evolving tool for the evaluation of interlinkages between SDGs that can be used to help the prioritization of targets [58]. The application of systems analysis leads to more equitable, more resilient and more sustainable assistive technology [35].

Ideally, a sustainability evaluation should incorporate the assessment of a consequence caused by a certain mechanism or transition in our environment. To this end, the integrated system modeling of complex systems will form the basis of a comprehensive assessment of sustainability and can provide a clearer vision of the dynamics of the globe [59].

3.3.4. Socio-Technical Systems

A socio-technical system is a collection of stakeholders, their networks, practices, and knowledge; the technologies they use; their collective representations; and the standards and rules they adopt [30]. The broad range of sustainability transition literature addresses the long-term transformation of socio-technical systems into sustainable systems in various fields, such as renewable power [29,60,61], transportation [62], and agricultural and food systems [30,39]. In the SDGs, for instance, the energy system is represented mainly as a technological system rather than a complicated socio-technical one. Nevertheless, for the latter, improvements in consumer behavior or other behavioral changes tend to balance several of the modern technology's beneficial effects. In addition, socio-technical systems also have low- and no-tech responses to challenges [29].

3.3.5. Complex Adaptive Systems

Complex adaptive systems are systems whose behaviors emerge on different spatial and temporal scales as a result of non-linear interrelationships, among a large number of elements without central control [50]. Within a system and between a system and its environment, they contain several complex, interconnected components. They are open systems, which means that elements and systems have an effect on the system beyond the boundaries of the observed system and its influence, and vice versa. Rusoja et al. [33] conceptualized healthcare systems as a complex adaptive system which is a set of interacting entities that change quickly in relation to each other and their shared environment. The value of this approach is that complex adaptive systems necessarily reflect the evolving environment, its key actors and their connections in medical comprehension over time, helping planners to consider and optimize health outcomes more effectively. Jagustović et al. [50] discussed the contribution of systems thinking as a conceptual method and the characteristics of complex adaptive systems as a basis for supporting the scaling-up of sustainable agriculture.

3.3.6. System of Systems

A complex system of systems, ranging from individually unique non-physical influencers to a wider collection of social and environmental influencers who have a common effect on the larger society-environment-economy system, strongly influences human, ecological and economic outcomes. Via several layers of systems that interact in many ways, sustainability plays a major role [63]. At the micro-level, there are individual layer systems. At meso-level, there are various forms of infrastructure, such as production and transportation, power generation, wastewater treatment and control, etc. At macro-level, the infrastructure consists of interconnected networks, including the power grid and the oil and gas storage and networking. System of systems engineering is explicitly meant to address problems of this essence. It allows decision-makers to solve the challenges of creating a robust system of systems that tackles the measurement of sustainability performance [64].

System of systems engineering involves identifying interconnections between technology, management, environmental and social issues, finances, and corporate strategies. The system of systems architecture and the system life cycle approach are essential to addressing various requirements and planning for long-term sustainable operation and maintenance [65]. System of systems engineering promises a toolset that have been adapted to modelling, simulation, optimization, and decision analysis of complex systems with both quantitative and qualitative characteristics. To resolve the need for a straightforward description of the process to develop the required technical, financial, and managerial support modules, a system of systems hierarchy spanning social, economic, and operational levels can allow sustainable development and strategic planning [23].

3.3.7. Cyber Physical Systems

Cyber physical systems can be characterized as systems that interconnect physical and software components to generate large-scale effects, each functioning on different temporal and spatial scales, demonstrating multiple and distinct behavioral mechanisms, and communicating with each other in a variety of context-changing ways [66]. It is a revolutionary technology when combined in manufacturing, wherein knowledge is strictly managed and coordinated from all relevant perspectives between the physical assembly line and the cyber computational space [67]. A cyber physical system connects a layer of physical components, such as workstations, components, and equipment, with a cyber layer of information systems, through sensors and devices. In other words, the virtualization of the process is achieved by cyber-physical systems. To attain a sustainable community on a worldwide scale as soon as practicable, it will be vital to allow communication and cooperation, such as knowledge sharing, triggering actions, and autonomous control between humans, machines, and materials.

The Japanese Cabinet's 'Society 5.0' initiative aims to establish a sustainable society through a cyber-physical system for human security and well-being [68]. Via cyber-physical systems, it is possible to analyze and visualize various big data artifacts obtained from low-power intelligent sensor networks and kept in information cloud storage using high-computing analytical tools in cyberspace. Such useful data, often difficult to notice for humans alone, can inform the actions taken by decision-makers to resolve social challenges and economic development in the physical realm.

4. Discussion

This paper presents the first scoping study on the crossovers and connectivity between systems engineering and the SDGs.

To do this, we adopted an innovative approach and integrated the emerging technique of NLP to the established SLR process. Consequently, we present a comprehensive analysis of the connectivity between systems engineering and the SDGs.

Our paper contributes a comprehensive and theoretically grounded picture of the crossovers and connectivity between systems engineering and the SDGs. This research provides a useful lens for analyzing the existing resolutions in moving towards more holistic, dynamic and integrated perspectives for sustainable development. Systems engineering practitioners and the sustainability research community can refer to this body of work when designing and improving the current implementation framework according to their value creation goals.

As with other studies, our review has the potential for bias. For instance, if the scoping review does not consider all available data, selection bias can occur. Our search query (as shown in Table 1) contains the key concepts within systems engineering domain. However, it is inevitable that papers that contain other systems engineering concepts (that are not considered in Table 1) cannot be retrieved from the search. As our study aims to ensure a broad coverage of literature related to systems engineering and sustainable development, we believe the design of the search terms is appropriate for this particular study.

Due to the large coverage of the search inherent in the method, scoping reviews take a considerable amount of time to finish. The number of studies included in the review process can be sizable by design. Thus, a large research team could enhance the screening of a greater number of studies and consider other references for potential inclusion in the scoping analysis.

This scoping study is not an end but a beginning which is intended to lead into further work. It has looked at a broad range of systems engineering core concepts for helping to achieve the SDGs but has not been able to consider each of the goals in details. This again highlights the gaps in current practices for using systems engineering core concepts to understand and explain the complexity behind sustainability. The need for examining existing frameworks and methods that can be used to assess this type of work is therefore urgent.

5. Conclusions

The present study was designed to analyze the existing literature that uses a systems engineering perspective to advance the SDGs. One of the significant findings to emerge from this study is that systems engineering has great potential to contribute to global sustainability. It offers various approaches and toolsets to enable the success of complex systems, drives the evolution of socio-technical systems, and assists decision-makers to implement the SDGs in an improved and traceable manner. Our investigation also determines the relationship between systems engineering and sustainable development and proposes interactions that may contribute to the achievement of the SDGs. It confirms that systems engineering has been recognized as an effective way to reframe the SDGs.

The insights gained from this study may be of assistance in creating a better understanding of the concept crossovers between systems engineering and sustainable development. It lays the groundwork for future research into the holism and dynamics of the SDGs and adds to the growing body of research that advocates using systems engineering methods to integrate sectoral efforts towards global sustainability. Systems engineering can act as a shared foundation for future research in harmonizing, optimizing and advancing the SDGs.

We hope that our findings, analysis, and discussion can pave the way for an appreciation of the value of systems engineering for the SDGs. We expect that future studies can take advantage of the crossovers between systems engineering and the SDGs to discover specific ways that systems engineering can contribute to a sustainable future.

Author Contributions: Data curation, L.Y.; Formal analysis, L.Y.; Investigation, L.Y.; Methodology, L.Y.; Supervision, K.C.; Visualization, L.Y.; Writing—original draft, L.Y. and K.C.; Writing—review and editing, L.Y. and K.C. All authors have read and agreed to the published version of the manuscript.

Funding: This research received no external funding.

Institutional Review Board Statement: Not applicable.

Informed Consent Statement: Not applicable.

Data Availability Statement: The data presented in this study are available on request from the corresponding author.

Conflicts of Interest: The authors declare no conflict of interest.

References

1. Sachs, J.D.; Schmidt-Traub, G.; Mazzucato, M.; Messner, D.; Nakicenovic, N.; Rockström, J. Six Transformations to achieve the Sustainable Development Goals. *Nat. Sustain.* **2019**, *2*, 805–814. [CrossRef]
2. Jabareen, Y. A New Conceptual Framework for Sustainable Development. *Environ. Dev. Sustain.* **2008**, *10*, 179–192. [CrossRef]
3. United Nations. United Nations Millennium Declaration. In Proceedings of the Millennium Development Summit of the United Nations, New York, NY, USA, 6–8 September 2000.
4. United Nations. The future we want. In Proceedings of the United Nations Conference on Sustainable Development, Rio de Janeiro, Brazil, 20–22 June 2012.
5. United Nations. Transforming our World: The 2030 Agenda for Sustainable Development. In Proceedings of the United Nations Summit, New York, NY, USA, 25–27 September 2015.
6. Barbier, E.B.; Burgess, J.C. The Sustainable Development Goals and the systems approach to sustainability. *Econ. Open Access Open Assess. E J.* **2017**. [CrossRef]
7. Le Blanc, D. Towards Integration at Last? The Sustainable Development Goals as a Network of Targets. *Sustain. Dev.* **2015**, *23*, 176–187. [CrossRef]
8. Scharlemann, J.P.W.; Brock, R.C.; Balfour, N.; Brown, C.; Burgess, N.D.; Guth, M.K.; Ingram, D.J.; Lane, R.; Martin, J.G.C.; Wicander, S.; et al. Towards understanding interactions between Sustainable Development Goals: The role of environment–human linkages. *Sustain. Sci.* **2020**, *15*, 1573–1584. [CrossRef]
9. Laurent, A.; Molin, C.; Owsianiak, M.; Fantke, P.; Dewulf, W.; Herrmann, C.; Kara, S.; Hauschild, M. The role of life cycle engineering (LCE) in meeting the sustainable development goals—Report from a consultation of LCE experts. *J. Clean. Prod.* **2019**, *230*, 378–382. [CrossRef]
10. Selomane, O.; Reyers, B.; Biggs, R.; Hamann, M. Harnessing Insights from Social-Ecological Systems Research for Monitoring Sustainable Development. *Sustainability* **2019**, *11*, 1190. [CrossRef]
11. Iandolo, F.; Barile, S.; Armenia, S.; Carrubbo, L. A system dynamics perspective on a viable systems approach definition for sustainable value. *Sustain. Sci.* **2018**, *13*, 1245–1263. [CrossRef]
12. Diwekar, U. Perspective on pursuit of sustainability: Challenges for engineering community. *Clean Technol. Environ. Policy* **2015**, *17*, 1729–1741. [CrossRef]
13. Hillary, S.; James, M.; Dorothy, M.; Regina, G.; Dov, D.; Daniel, K.; Patrick, G.; Eileen, A.; Scott, J. Systems Engineering and System Definitions. Available online: https://www.incose.org/docs/default-source/default-document-library/final_-se-definition.pdf?sfvrsn=340b9fc6_0 (accessed on 19 February 2021).
14. Pearce, O.J.D.; Murry, N.J.A.; Broyd, T.W. Halstar: Systems engineering for sustainable development. *Proc. Inst. Civ. Eng. Eng. Sustain.* **2012**, *165*, 129–140. [CrossRef]
15. Colquhoun, H.L.; Levac, D.; O'Brien, K.K.; Straus, S.; Tricco, A.C.; Perrier, L.; Kastner, M.; Moher, D. Scoping reviews: Time for clarity in definition, methods, and reporting. *J. Clin. Epidemiol.* **2014**, *67*, 1291–1294. [CrossRef] [PubMed]
16. Peters, M.D.J.; Godfrey, C.M.; Khalil, H.; McInerney, P.; Parker, D.; Soares, C.B. Guidance for conducting systematic scoping reviews. *Int. J. Evid. Based. Healthc.* **2015**, *13*, 141–146. [CrossRef]
17. Levac, D.; Colquhoun, H.; O'Brien, K.K. Scoping studies: Advancing the methodology. *Implement. Sci.* **2010**, *5*, 69. [CrossRef] [PubMed]
18. Arksey, H.; O'Malley, L. Scoping studies: Towards a methodological framework. *Int. J. Soc. Res. Methodol.* **2005**, *8*, 19–32. [CrossRef]
19. Bird, S.; Klein, E.; Loper, E. *Natural Language Processing with Python: Analyzing Text with the Natural Language Toolkit*; O'Reilly Media, Inc.: Sebastopol, CA, USA, 2009; ISBN 0596555717.
20. Pickering, C.; Byrne, J. The benefits of publishing systematic quantitative literature reviews for PhD candidates and other early-career researchers. *High. Educ. Res. Dev.* **2014**, *33*, 534–548. [CrossRef]
21. Deshpande, P.C.; Aspen, D.M. A Framework to Conceptualize Sustainable Development Goals for Fishing Gear Resource Management. In *Handbook of Sustainability Science and Research*; Leal Filho, W., Ed.; Springer: Cham, Switzerland, 2018; pp. 727–744.
22. Sage, A.P. Systems engineering and management for industrial ecology and sustainable development. In Proceedings of the 1997 IEEE International Conference on Systems, Man, and Cybernetics. Computational Cybernetics and Simulation, Orlando, FL, USA, 12–15 October 1997; Volume 1, pp. 784–790.
23. Anderson, A.; Suryanarayanan, S. An Enterprise Systems Engineering Approach to Electrification: Looking at the Bigger Picture Through Life-Cycle Analysis of Community Microgrids: A Case Study in Papua New Guinea. *IEEE Electrif. Mag.* **2018**, *6*, 18–31. [CrossRef]
24. Bertheau, P.; Hoffmann, M.M.; Eras-Almeida, A.; Blechinger, P. Assessment of Microgrid Potential in Southeast Asia Based on the Application of Geospatial and Microgrid Simulation and Planning Tools. In *Sustainable Energy Solutions for Remote Areas in the Tropics*; Gandhi, O., Srinivasan, D., Eds.; Springer: Cham, Switzerland, 2020; pp. 149–178.
25. Machado, P.G.; Mouette, D.; Rathmann, R.; dos Santos, E.; Peyerl, D. Is Energy Planning Moving Towards Sustainable Development? A Review of Energy Systems Modeling and Their Focus on Sustainability. In *International Business, Trade and Institutional Sustainability*; Springer International Publishing: New Year, NY, USA, 2020; pp. 629–644.

26. Rauner, S.; Budzinski, M. Holistic energy system modeling combining multi-objective optimization and life cycle assessment. *Environ. Res. Lett.* **2017**, *12*, 124005. [CrossRef]
27. Reid, J.; Zeng, C.; Wood, D. Combining Social, Environmental and Design Models to Support the Sustainable Development Goals. In Proceedings of the 2019 IEEE Aerospace Conference, Big Sky, MT, USA, 2–9 March 2019; pp. 1–13.
28. Schot, J.; Steinmueller, W.E. Three frames for innovation policy: R&D, systems of innovation and transformative change. *Res. Policy* **2018**, *47*, 1554–1567. [CrossRef]
29. Hillerbrand, R. Why Affordable Clean Energy Is Not Enough. A Capability Perspective on the Sustainable Development Goals. *Sustainability* **2018**, *10*, 2485. [CrossRef]
30. Serhan, H.; Yannou-Lebris, G. The engineering of food with sustainable development goals:policies, curriculums, business models, and practices. *Int. J. Sustain. Eng.* **2020**, 1–14. [CrossRef]
31. Palmer, E.; Burton, R.; Haskins, C. A Systems Engineering Framework for Bioeconomic Transitions in a Sustainable Development Goal Context. *Sustainability* **2020**, *12*, 6650. [CrossRef]
32. Tan, D.T.; Siri, J.G.; Gong, Y.; Ong, B.; Lim, S.C.; MacGillivray, B.H.; Marsden, T. Systems approaches for localising the SDGs: Co-production of place-based case studies. *Global. Health* **2019**, *15*, 85. [CrossRef] [PubMed]
33. Rusoja, E.; Haynie, D.; Sievers, J.; Mustafee, N.; Nelson, F.; Reynolds, M.; Sarriot, E.; Swanson, R.C.; Williams, B. Thinking about complexity in health: A systematic review of the key systems thinking and complexity ideas in health. *J. Eval. Clin. Pract.* **2018**, *24*, 600–606. [CrossRef] [PubMed]
34. Michalopoulou, E.; Shallcross, D.E.; Atkins, E.; Tierney, A.; Norman, N.C.; Preist, C.; O'Doherty, S.; Saunders, R.; Birkett, A.; Willmore, C.; et al. The End of Simple Problems: Repositioning Chemistry in Higher Education and Society Using a Systems Thinking Approach and the United Nations' Sustainable Development Goals as a Framework. *J. Chem. Educ.* **2019**, *96*, 2825–2835. [CrossRef]
35. MacLachlan, M.; McVeigh, J.; Cooke, M.; Ferri, D.; Holloway, C.; Austin, V.; Javadi, D. Intersections Between Systems Thinking and Market Shaping for Assistive Technology: The SMART (Systems-Market for Assistive and Related Technologies) Thinking Matrix. *Int. J. Environ. Res. Public Health* **2018**, *15*, 2627. [CrossRef]
36. Stead, S.M. Using systems thinking and open innovation to strengthen aquaculture policy for the United Nations Sustainable Development Goals. *J. Fish Biol.* **2019**. [CrossRef] [PubMed]
37. Zhang, Q.; Prouty, C.; Zimmerman, J.B.; Mihelcic, J.R. More than Target 6.3: A Systems Approach to Rethinking Sustainable Development Goals in a Resource-Scarce World. *Engineering* **2016**, *2*, 481–489. [CrossRef]
38. Allen, C.; Metternicht, G.; Wiedmann, T. Initial progress in implementing the Sustainable Development Goals (SDGs): A review of evidence from countries. *Sustain. Sci.* **2018**, *13*, 1453–1467. [CrossRef]
39. Whitfield, S.; Challinor, A.J.; Rees, R.M. Frontiers in Climate Smart Food Systems: Outlining the Research Space. *Front. Sustain. Food Syst.* **2018**, *2*. [CrossRef]
40. Hurst, G.A. Systems thinking approaches for international green chemistry education. *Curr. Opin. Green Sustain. Chem.* **2020**, *21*, 93–97. [CrossRef]
41. Pant, A.; Kumar, S. Climate Change, Public Health and Implementation of Sustainable Development Goals in India-Issues and Challenges. *Indian J. Public Heal. Res. Dev.* **2019**, *10*, 60. [CrossRef]
42. Morton, S.; Pencheon, D.; Bickler, G. The sustainable development goals provide an important framework for addressing dangerous climate change and achieving wider public health benefits. *Public Health* **2019**, *174*, 65–68. [CrossRef]
43. Kutty, A.A.; Abdella, G.M.; Kucukvar, M.; Onat, N.C.; Bulu, M. A system thinking approach for harmonizing smart and sustainable city initiatives with United Nations sustainable development goals. *Sustain. Dev.* **2020**, *28*, 1347–1365. [CrossRef]
44. Reynolds, M.; Blackmore, C.; Ison, R.; Shah, R.; Wedlock, E. The Role of Systems Thinking in the Practice of Implementing Sustainable Development Goals. In *Handbook of Sustainability Science and Research*; Leal Filho, W., Ed.; Springer: Cham, Switzerland, 2018; pp. 677–698.
45. Petillion, R.J.; Freeman, T.K.; McNeil, W.S. United Nations Sustainable Development Goals as a Thematic Framework for an Introductory Chemistry Curriculum. *J. Chem. Educ.* **2019**, *96*, 2845–2851. [CrossRef]
46. Duboz, R.; Echaubard, P.; Promburom, P.; Kilvington, M.; Ross, H.; Allen, W.; Ward, J.; Deffuant, G.; de Garine-Wichatitsky, M.; Binot, A. Systems Thinking in Practice: Participatory Modeling as a Foundation for Integrated Approaches to Health. *Front. Vet. Sci.* **2018**, *5*. [CrossRef] [PubMed]
47. Zelinka, D.; Amadei, B. A methodology to model the integrated nature of the Sustainable development goals: Importance for engineering education. In Proceedings of the ASEE Annual Conference and Exposition, Columbus, OH, USA, 25–28 June 2017.
48. Eustachio, J.H.P.P.; Caldana, A.C.F.; Liboni, L.B.; Martinelli, D.P. Systemic indicator of sustainable development: Proposal and application of a framework. *J. Clean. Prod.* **2019**, *241*, 118383. [CrossRef]
49. Kioupi, V.; Voulvoulis, N. Education for Sustainable Development: A Systemic Framework for Connecting the SDGs to Educational Outcomes. *Sustainability* **2019**, *11*, 6104. [CrossRef]
50. Jagustović, R.; Zougmoré, R.B.; Kessler, A.; Ritsema, C.J.; Keesstra, S.; Reynolds, M. Contribution of systems thinking and complex adaptive system attributes to sustainable food production: Example from a climate-smart village. *Agric. Syst.* **2019**, *171*, 65–75. [CrossRef]
51. Keesstra, S.; Mol, G.; de Leeuw, J.; Okx, J.; Molenaar, C.; de Cleen, M.; Visser, S. Soil-Related Sustainable Development Goals: Four Concepts to Make Land Degradation Neutrality and Restoration Work. *Land* **2018**, *7*, 133. [CrossRef]

52. Von Bertalanffy, L. General theory of systems: Application to psychology. *Soc. Sci. Inf.* **1967**, *6*, 125–136. [CrossRef]
53. Fucà, R.; Cubico, S.; Favretto, G.; Leitão, J. The "Local Town Market Area" in Enna, Sicily: Using the Psychology of Sustainability to Propose Sustainable and Developmental Policies. *Sustainability* **2019**, *11*, 486. [CrossRef]
54. Wilcox, B.A.; Aguirre, A.A.; De Paula, N.; Siriaroonrat, B.; Echaubard, P. Operationalizing One Health Employing Social-Ecological Systems Theory: Lessons From the Greater Mekong Sub-region. *Front. Public Health* **2019**, *7*. [CrossRef]
55. Donaires, O.S.; Cezarino, L.O.; Caldana, A.C.F.; Liboni, L. Sustainable development goals—An analysis of outcomes. *Kybernetes* **2019**, *48*, 183–207. [CrossRef]
56. Wieser, A.A.; Scherz, M.; Maier, S.; Passer, A.; Kreiner, H. Implementation of Sustainable Development Goals in construction industry—A systemic consideration of synergies and trade-offs. *IOP Conf. Ser. Earth Environ. Sci.* **2019**, *323*, 012177. [CrossRef]
57. Obersteiner, M.; Rydzak, F.; Fritz, S.; McCallum, I. Valuing the Potential Impacts of GEOSS: A Systems Dynamics Approach. In *The Value of Information*; Laxminarayan, R., Macauley, M., Eds.; Springer: Dordrecht, The Netherlands, 2012; pp. 67–90.
58. Allen, C.; Metternicht, G.; Wiedmann, T. Prioritising SDG targets: Assessing baselines, gaps and interlinkages. *Sustain. Sci.* **2019**, *14*, 421–438. [CrossRef]
59. Schaubroeck, T.; Rugani, B. A Revision of What Life Cycle Sustainability Assessment Should Entail: Towards Modeling the Net Impact on Human Well-Being. *J. Ind. Ecol.* **2017**, *21*, 1464–1477. [CrossRef]
60. Renouf, M.A.; Kenway, S.J. Evaluation Approaches for Advancing Urban Water Goals. *J. Ind. Ecol.* **2017**, *21*, 995–1009. [CrossRef]
61. Ahmad, N.; Derrible, S. Evolution of Public Supply Water Withdrawal in the USA: A Network Approach. *J. Ind. Ecol.* **2015**, *19*, 321–330. [CrossRef]
62. Liu, G.; Zhai, R. Application and Development of Systems Engineering in Road Traffic Management. In Proceedings of the International Conference on Transportation and Development 2018; Wang, Y., McNerney, M.T., Eds.; American Society of Civil Engineers: Reston, VA, USA, 2018; pp. 306–313.
63. Muller, G.; Elvebakk, L.; van der Velde, J.; Lean, F. Mac Roadmapping for Sustainability; How to Navigate a Social, Political, and Many Systems-of-Systems Playing Field? A Local Initiative. In Proceedings of the 2019 14th Annual Conference System of Systems Engineering (SoSE), Anchorage, AK, USA, 19–22 May 2019; pp. 317–322.
64. Searcy, C. Corporate sustainability performance measurement: Lessons from system of systems engineering. In Proceedings of the 2009 IEEE International Conference on Systems, Man and Cybernetics, San Antonio, TX, USA, 11–14 October 2009; pp. 1057–1060.
65. Nativi, S.; Santoro, M.; Giuliani, G.; Mazzetti, P. Towards a knowledge base to support global change policy goals. *Int. J. Digit. Earth* **2020**, *13*, 188–216. [CrossRef]
66. Oláh, J.; Aburumman, N.; Popp, J.; Khan, M.A.; Haddad, H.; Kitukutha, N. Impact of Industry 4.0 on Environmental Sustainability. *Sustainability* **2020**, *12*, 4674. [CrossRef]
67. Bonilla, S.; Silva, H.; Terra da Silva, M.; Franco Gonçalves, R.; Sacomano, J. Industry 4.0 and Sustainability Implications: A Scenario-Based Analysis of the Impacts and Challenges. *Sustainability* **2018**, *10*, 3740. [CrossRef]
68. Shiroishi, Y.; Uchiyama, K.; Suzuki, N. Society 5.0: For Human Security and Well-Being. *Computer* **2018**, *51*, 91–95. [CrossRef]

sustainability

MDPI

Erratum

Erratum: Yang, L.; Cormican, K. The Crossovers and Connectivity between Systems Engineering and the Sustainable Development Goals: A Scoping Study. *Sustainability* 2021, *13*, 3176

Lan Yang [1,*] and Kathryn Cormican [1,2]

1 School of Engineering, College of Science and Engineering, National University of Ireland, Galway, H91 TK33 Galway, Ireland; kathryn.cormican@nuigalway.ie

2 Lero—The Irish Software Research Centre, School of Engineering, National University of Ireland, Galway, H91 TK33 Galway, Ireland

* Correspondence: lan.yang@nuigalway.ie

Citation: Yang, L.; Cormican, K. Erratum: Yang, L.; Cormican, K. The Crossovers and Connectivity between Systems Engineering and the Sustainable Development Goals: A Scoping Study. *Sustainability* 2021, *13*, 3176. *Sustainability* **2021**, *13*, 8930. https://doi.org/10.3390/su13168930

Received: 30 July 2021
Accepted: 2 August 2021
Published: 10 August 2021

The authors would like to make the following corrections to the published paper [1]. The changes are as follows:

(1) Replacement of the sentence and a reference number in "Section 3.3.2. Systems Thinking" on page 10:

Kioupi and Voulvoulis [44] developed a framework that enables stakeholders in education to create a shared vision of sustainability, with the SDGs as endpoints.

with

Kioupi and Voulvoulis [49] developed a framework that enables stakeholders in education to create a shared vision of sustainability, with the SDGs as endpoints.

(2) The addition of a reference to the citation list:

49. Kioupi, V.; Voulvoulis, N. Education for Sustainable Development: A Systemic Framework for Connecting the SDGs to Educational Outcomes. *Sustainability* **2019**, *11*, 6104, doi:10.3390/su11216104.

Reference

1. Yang, L.; Cormican, K. The Crossovers and Connectivity between Systems Engineering and the Sustainable Development Goals: A Scoping Study. *Sustainability* **2021**, *13*, 3176. [CrossRef]

sustainability

MDPI

Article

Applying Roadmapping and Conceptual Modeling to the Energy Transition: A Local Case Study

Gerrit Muller

Faculty of Technology, Natural Sciences and Maritime Sciences, Campus Kongsberg, University of South-Eastern Norway, Kongsberg 3616, Norway; Gerrit.muller@usn.no

Abstract: The climate crisis requires a global transition toward sustainable practices. In this transition, policy makers face the challenge to take along a wide variety of stakeholders with own interests, needs, and concerns. This research explores the combined use of conceptual models and roadmapping to facilitate understanding, communication, reasoning, and decision-making between a large heterogeneous set of stakeholders. We apply these methods, in the form of action research, in several smaller research projects at a small town in the Netherlands. We find that the combination of conceptual modeling and roadmapping facilitates discussions between heterogeneous stakeholders on complex transition problems, such as the energy transition, at a local scale. However, we see a significant gap in the way of thinking and communicating between experts and decision-makers, which requires additional means to connect them.

Keywords: roadmapping; energy transition; sustainability; conceptual modeling

Citation: Muller, G. Applying Roadmapping and Conceptual Modeling to the Energy Transition: A Local Case Study. *Sustainability* **2021**, *13*, 3683. https://doi.org/10.3390/su13073683

Academic Editor: Cecilia Haskins

Received: 9 March 2021
Accepted: 23 March 2021
Published: 26 March 2021

1. Introduction

In this paper, we study methods and techniques from the systems engineering domain to cope with the tremendous complexity of the global challenge of the energy transition. We build upon [1], that proposes the use of roadmapping as a method for analyzing the sustainability transition at the local level and using conceptual modeling to support shared understanding, communication, and reasoning about options.

Phaal et al. [2] propose roadmapping to provide a temporal overview, e.g., from the past to the long-term future. Conceptual models are models that are simple enough to promote shared understanding and communication and, at the same time, substantial enough to create meaning and value. Conceptual modeling builds upon a way of working from various disciplines, such as simulation [3,4], soft systems methodology [5], collaborative working [6], and systems engineering [7].

The sustainability transition is a global scale challenge that relies upon every available human expertise. The United Nations have defined 17 Sustainability Development Goals (SDGs) [8] as an organizing framework. A systems engineering perspective suggests many factors that make the SDGs and the energy transition so challenging:

- Earth as natural system
- The Systems of Systems (SoS) characteristics
- Many solutions need inventing, e.g., they are unknown or uncertain
- The scope is broader than sociotechnical
- The global variety of stakeholders.

The International Council of Systems Engineering (INCOSE) [9] states,

> Systems Engineering is *a* ***transdisciplinary*** *and* ***integrative*** *approach to enable the successful realization, use, and retirement of* ***engineered systems****, using* ***systems principles and concepts****, and scientific, technological, and management methods.*
>
> "We use the terms "engineering" and "engineered" in their **widest sense**: "the action of working artfully to bring something about." "**Engineered sys-**

tems" may be composed of any or all of people, products, services, information, processes, and natural elements.

This paper explores and illustrates systems engineering methods that support building (shared) understanding, communication, and reasoning for complex problems, using roadmapping and conceptual modeling to address the challenges of the energy transition.

The goal of this research is to find ways to support society in dealing with these transition challenges and to accelerate the sluggish transition. The research questions that addressed in this paper are as follows:

- RQ1. How can we help a heterogeneous set of stakeholders to formulate a shared understanding of the current situation and of the options and their consequences to resolve them?
- RQ2. How can a shared understanding help in reasoning about the solutions and, in that way, help in decision-making and preparing governance?
- RQ3. What is a manageable set of conceptual models to achieve understanding, reasoning, communicating, and decision-making?

The research design is that we apply these methods at a local level, e.g., a town of 30 thousand inhabitants, rather than at national level (e.g., tens of millions of inhabitants) or global level (billions of humans). We hope to gain insights from using the methods on a smaller scale that may help to address the even more complex transition at national and global levels and apply the insights to the SDG framework.

The contribution of this paper is the combined use of roadmapping and conceptual modeling and their application in a societal context toward the energy transition, as part of the sustainability transition.

The paper continues with background on the sustainability and energy transition in general and the challenges in addressing these transitions, followed by a brief description of the town that we used in the case. The next section provides the theory behind roadmapping and conceptual modeling. The materials and methods section discusses the research design. In the results section, we show various results achieved so far. Finally, we discuss the progress of the municipality toward the sustainability transition and the findings related to the research questions.

1.1. The Global Sustainability Challenge

The United Nations provides a scientific framework through the Intergovernmental Panel on Climate Change (IPCC) and a sociopolitical framework through the SDGs and the Paris agreement. The IPCC relates the climate insights in [10] to the SDGs, and it links the specific needs for the energy transitions with the SDGs. A core message is that humanity must reduce the CO2 (and greenhouse gasses overall) to zero emissions by 2050 to stay at 1.5 °C temperature rise, as agreed in the Paris agreement [11]. The SDG framework sets the broader sustainability context for the energy transition.

1.2. What Makes the Energy Transition So Challenging

Earth is a natural system with complex properties and behavior. Even an organization like the IPCC, with many well-respected scientists, does not claim a complete understanding of our own planet.

The energy system is a System of System (SoS) to the power of n. Maier [12] defines SoSs as having managerial independence, operational independence, geographic separation, emergent behavior, and evolutionary development. Interestingly, constituent systems of the energy system have no full operational independence in the current grid architecture; however, they fulfill all other SoS characteristics. The energy system has many layers of systems, from (inter)continental grids, national grids with its power sources and controls, regional and local grids, individual buildings, and other platforms, down to personal power producing or consuming devices. The technologies and constituent systems of the energy system of the future are still under development or, worse, not yet invented. The 2050 target date is probably beyond most people's horizon.

The energy system is sociotechnical, with all the dimensions, political, economic, social, technical, environmental, and legal (PESTEL) [13] having relevance. Heleen de Coninck, one of the main authors of the IPCC report SR1.5 [14], states in [15], "*Technologies are always part of a cultural, social, economic, legal, and policy-based system. With a fundamental change in the energy system, they also change with it. Or maybe they even have to change first to make the technological change possible.*"

Lastly, the number of stakeholders is huge, as is their variety of roles, power, and knowledge. The stakeholder field is complex, due to the psychosocial and political nature of humans. Perceptions, emotions, personal, organization, and national interests make the problem domain more complex than the field of engineered systems, where systems engineering originates.

1.3. Introduction to the Town Best, Where the Research Takes Place

The approach of this research is to use action research to apply roadmapping and conceptual modeling at a local scale on specific parts of the energy transition. We will report on an initial roadmap and set of conceptual models that we made for the town named Best in the South of the Netherlands; see Figure 1. Best has about 30 thousand inhabitants. It is a town and municipality with a council and several aldermen as administration. For the energy strategy, Best is part of the metropolitan region Eindhoven [16]. In turn, the metropolitan region is part of the province Noord Brabant.

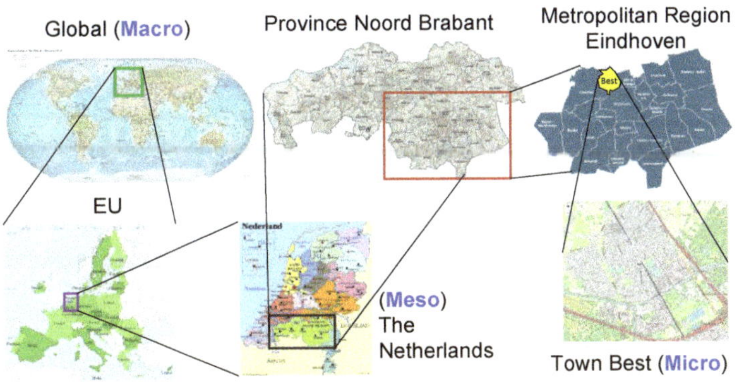

Figure 1. Best in Noord Brabant in the south of the Netherlands.

2. Methods from Systems Engineering

2.1. Theory of Roadmapping

Phaal [2] provides the basic framework of roadmaps with a horizontal timeline spanning short-term, medium-term, long-term, and vision. Vertical layers define why, what, and how. The more specific guidelines for sustainability roadmaps [17] explicitly add policy instruments and governance in the how layer. Figure 2 shows the starting point of the roadmapping based on [2,17].

Roadmapping is a well-researched field. Phaal [18] maintains a bibliography with over 1100 references. The website https://www.cambridgeroadmapping.net/ (accessed on 20 March 2021) provides examples, templates, research, and experiences of roadmapping over a broad set of domains, including global applications and organizations, such as the International Energy Agency (IEA) in the energy sector and United Nations organizations, such as the United Nations Framework Convention on Climate Change (UNFCCC) [19] and United Nations Industrial Development Organization (UNIDO) [20]. The UNIDO report [20] uses Foresight as another common name for methods to envision the future.

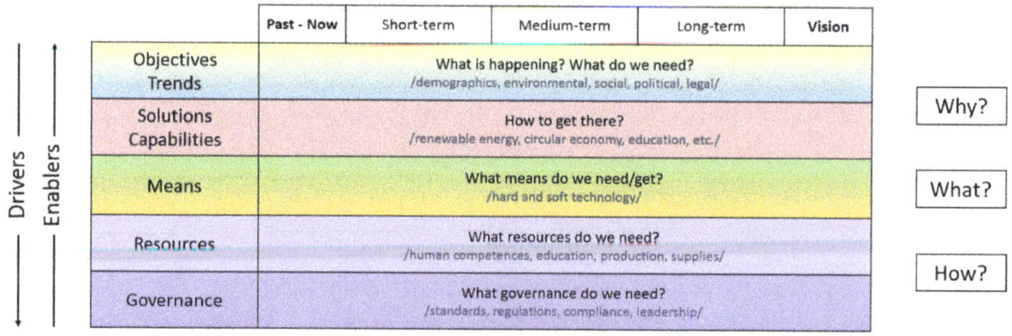

Figure 2. A 5-layer Roadmap Model.

2.2. *Theory of Conceptual Modeling*

Conceptual modeling has multiple origins, e.g., physics, for development of simulations, as a means for collaborative working, or as means for conceptual design in systems engineering. The common denominator between these various origins is the value of concepts to facilitate communication and thinking. In this subsection, we will discuss some of these origins.

There is a broad field of methods in *collaborative working* that use conceptual modeling more or less explicitly:

- Design Thinking [21], with a heavy emphasis on human interactions.
- Gigamapping, originating in systems-oriented design [22], employs many kinds of visualizations on a so-called gigamap to enhance communication. Gigamapping relates to design thinking in its interaction style.
- Neely et al. [6] propose collaborative conceptual modelling as a tool for transdisciplinarity. The workshops that this method [23] proposes are quite similar to gigamapping.

The field of simulations also provides much literature on conceptual modeling. Sargent [24] defines conceptual models as "the mathematical/logical/graphical representation (mimic) of the problem entity developed for a particular study." Many authors, e.g., [3,25,26] relate the use of conceptual models to simulation. Robinson [3,4] asserts, "It is almost certainly the most important aspect of a simulation project," and "conceptual modelling is more of an 'art' than a 'science' . . . ."

In the systems engineering field, there are many forms of conceptual modeling, as the various engineering disciplines all use their domain-specific conceptual models. Blanchard and Fabrycky [7] cover a wide variety of (conceptual) models. Tomita et al. [27] discuss the conceptual system design and propose to apply design thinking to lift the discussion from data and information level to knowledge and wisdom level. Montevechi and Friend [26] propose the use of *Soft Systems Methodology* (SSM) [5] in developing conceptual models. In the context of the energy transition, which is a challenge across technology, economy, society, and environment, the use of SSM makes a lot of sense. *Systems Thinking* is an essential competence when developing conceptual models. For example, Jackson [28] links systems thinking to address highly complex problems.

In this paper, we build on the author's earlier work [29]. Conceptual models bridge the first principle and empirical worlds. Empirical models provide a means to capture what we observe and measure without an understanding of what we observed. First-principle models are models that use the theoretical principles from science, such as the laws of physics, to explain the behavior of a property using the first principles. These models often take the form of mathematical equations and formulas. When we enter values in the formulas, then we can compute the resulting property for these values. A conceptual model explains observations and measurements using a selection of first principles. A conceptual model is a hybrid of empirical and first-principle models. It needs to be simple enough to understand and to reason, while it must be realistic enough to make sense.

3. Materials and Methods

A major challenge in researching these transdisciplinary problems is the number of relevant disciplines, each with its own frameworks and research methods. For instance, the research of heating options may require an understanding of thermodynamics, while the problem of acceptance by inhabitants requires social sciences.

Given the urgency and importance of the energy transition problem, this research adopts a pragmatic approach. First, we limit the research to a single, small, town. Next, we decompose the research into smaller research projects with a limited scope, such that the smaller research project can apply the specific research method suitable to the nature of that research topic.

The main research method is action research [30], all conducted in the context of this town. Each researcher visited the town and had access to the stakeholders and was given the opportunity to explore and test their findings in this real-world situation. The research took place in steps:

1. Preparation and project definition with the local stakeholders [31]
2. Creating the initial roadmap in cooperation with the local stakeholders [1,32]
3. Creating and operating a task force of representative stakeholders that discusses, maintains, and operationalizes the roadmap
4. Further studies of specific aspects of the roadmap (not yet published)
5. Continuous observations of the local situation

Steps 3 and 4 partially overlapped with the Covid-19 pandemic situation, which clearly affects the research, since the interaction with stakeholders was quite different.

The principal investigator is an inhabitant of this town and member of the sustainability cooperation. The other researchers came from outside the town and, in some instances, from outside the country. The double role of the principal researcher is clearly a dilemma in the research design. Benefits of such close ties are:

- The in-depth knowledge of the local situation.
- The access to relevant stakeholders.
- The a priori overview of the local situation.

Risks of such close ties are:

- Lack of objectivity.
- Bias from relations or from personal opinions and beliefs.
- Pressure from inhabitant stakeholders.

These risks exist for action research anyway [33], but the double role may exacerbate them. We expect that the benefits outweigh the risks. However, we acknowledge that the validity of the research is limited by this choice. The validity of this research is therefore limited in many directions. Generalizations are not possible, but the combined findings and methods applied offer insight for researchers in the domain.

4. Results

4.1. Step 2. Creating the Initial Roadmap

Laura Elvebakk, who studied Industrial Economy at University of South-Eastern Norway in Kongsberg (USN), helped to create an initial sustainability roadmap for Best during her master project in 2019 [1]. Figure 3 shows her approach, where she alternated interviewing stakeholders and synthesizing results in workshops. The initial interviews and workshop focused on the two top layers of the roadmap (the why), e.g., objectives and trends and possible solutions and capabilities. The second set of interviews and workshops elaborated the next layers (the what and how), e.g., what are the means, resources, and governance that we need to create the solutions and capabilities to achieve the objectives and cope with the trends? The time frames defined were short-term (2019–2021), medium-term (2022–2025), and long-term (2026–2030).

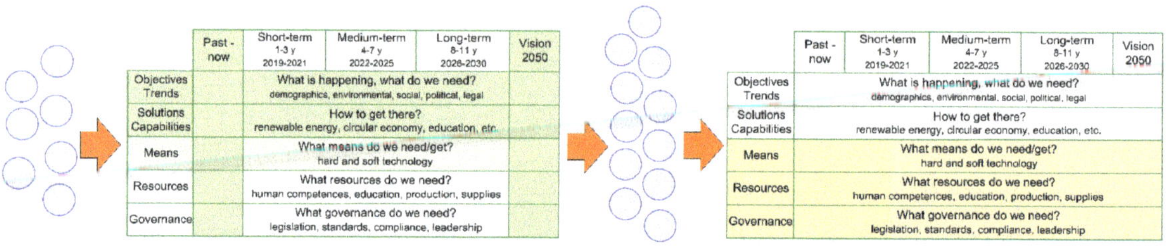

Figure 3. Interview and workshop progression [1].

The result was an initial roadmap as shown in Figure 4, and [32] discusses the roadmap contents. The main elements of the roadmap are

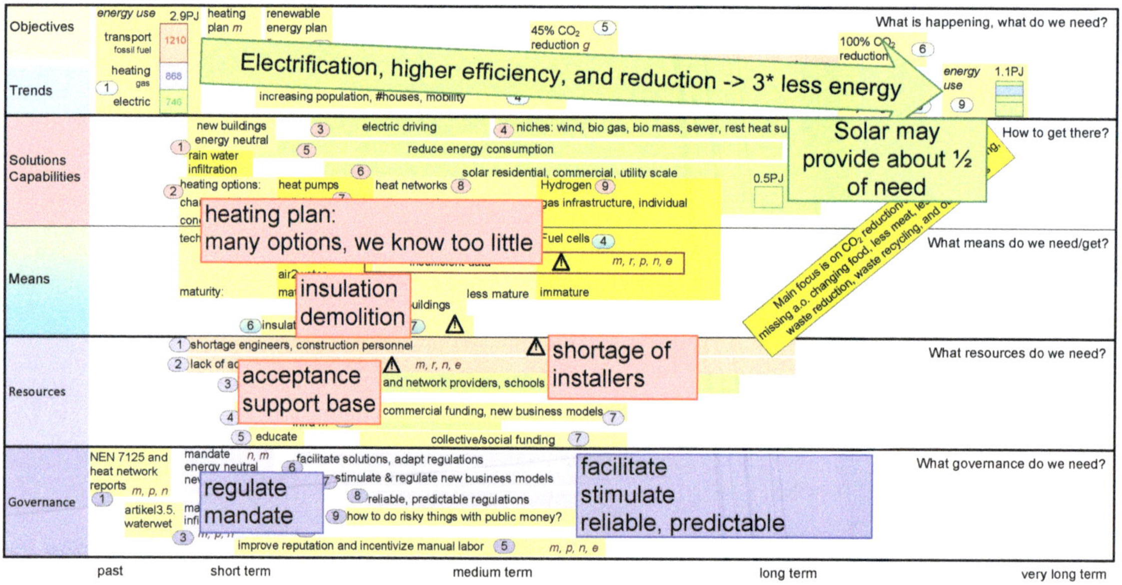

Figure 4. Summary of the initial Sustainability Roadmap for the town Best in the Netherlands [32]; the initial roadmap in A3 size is available at https://gaudisite.nl/BDRA3initialRoadmapBest.pdf (accessed on 20 March 2021).

- **Objectives and trends**: Reducing the current yearly energy consumption of about 3 PJ generated mostly with fossil fuels to about 1 PJ in the form of renewable sources and carriers.
- **Solutions and capabilities**: Using residential and industrial roofs, agricultural, and land close to motorways to install solar solutions will allow for the yearly generation of about 0.5 PJ. Other niche solutions may generate a fraction of solar. Main challenge is heating of buildings.
- **Means**: There are multiple technology options for heating, with rather varying degrees of maturity. An initial concept exploration is essential before the municipality can develop any meaningful policy. Obvious and urgent technologies that need policy and implementation are insulation and material choices (e.g., no concrete or traditional bricks) for buildings.

- **Resources**: A major bottleneck for execution of the transition is the installation and construction capacity of qualified installers. A major perquisite is support of the population. Stakeholders see motivation and incentives as better instruments than force or fear.
- **Governance**: A main challenge is creating the legislation and conditions to implement the solutions in such way that stakeholders are involved and motivated. Stakeholders emphasize the need for facilitation (incentivizing) and (long-term) predictability.

One of the challenging steps of roadmapping is the collection and ordering of supporting information. The roadmap itself should be limited, e.g., one sheet with the five layers. However, each word in a roadmap is the result of an analysis and discussion using various data sources. Here, the conceptual modeling appears. We can capture the underlying information in a number of conceptual models at the back-of-the-envelope level or the A3 level.

Figure 5 shows an example from [32] with an estimate of the amount of solar that municipality Best can generate. It analyzes commercial, residential, utility scale, and countryside solar options using available surface areas. Although this is a rough estimate, it helps to know what is possible and how various options compare. In hindsight, the countryside estimate is too low, due to underestimating the size of farm buildings.

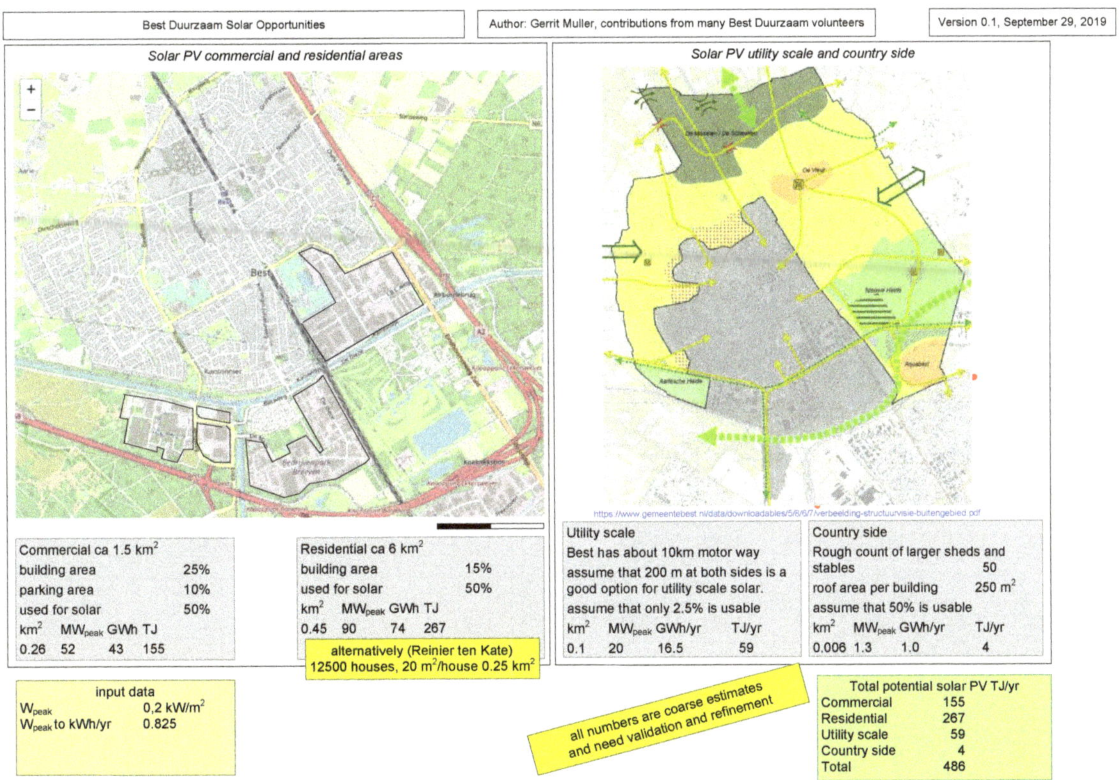

Figure 5. Example of underlying A3, elaborating how much solar Best can install [32].

4.2. Step 3. Creating and Operating a Task Force

The municipality is the formal owner of the transition plans and has the responsibility and mandate to set the policy and transform that into legislation, within the charters of the national and regional governments. The cooperation Best Duurzaam is a volunteer organization with members. In the preparation phase of the roadmap, Best Duurzaam identified the need for a consultative body of stakeholders. It took these two rather different organizations more than a year to set up a task force consisting of representatives of organizations that are stakeholders in the transition.

The objective of this task force is to involve and align the stakeholders of the transition. The idea is that a roadmap is a shared vision. However, stakeholders are not yet committed to that vision. The commitment follows in the elaboration of the roadmap into master plans per organization. The task force exchanges the master plans and discusses them. Organizations may adapt their master plan based on these discussions.

The task force has met three times since the creation of the initial roadmap. The task force clearly is in its early infancy. Various representatives have to build up mutual trust and to find their roles. The number of participating organizations is still small (municipality, sustainability cooperation, primary and secondary education, building cooperation, and agriculture sector). The idea is to grow the task force gradually.

4.3. Step 4. Further Studies

4.3.1. Seasonal Energy Storage

Since the roadmap identified the main energy source as a solar solution and the main energy consumption as heating, the energy supply and demand are out of phase. In the summer, we can harvest much energy, while in the winter, we consume most energy. A question is, how we will cope with these seasonal differences?

At the national level, the Netherlands will harvest a significant amount of energy via offshore wind. There is more energy potential for offshore wind in winter than in summer. At the national level, there are other energy consumers, especially in industry, that use significant amounts of energy, more or less constantly. This means that coping with seasonal energy fluctuations at the national level is easier than at the local level.

Another of the USN Norwegian master students, Erik Drilen, studied various concepts for seasonal energy storage. He made conceptual models for the most promising options. Figure 6 shows an example of a conceptual model for Power-to-Gas (PtG) [34].

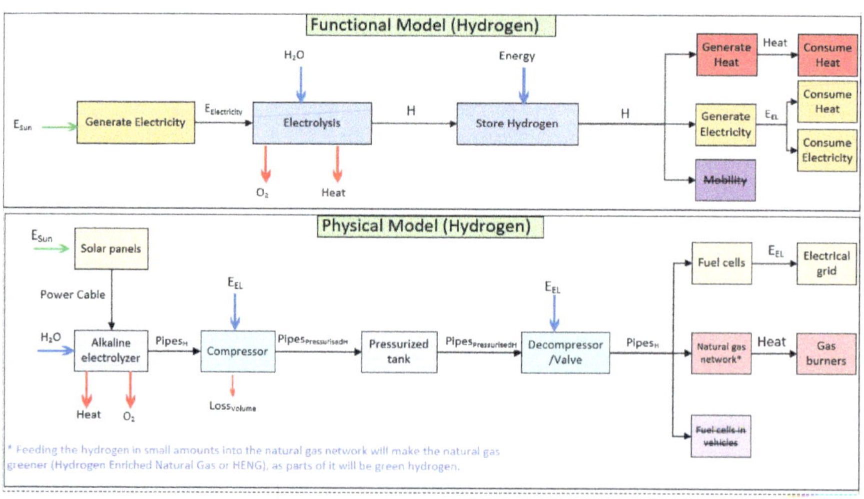

Figure 6. Power-to-Gas (PtG) functional and physical model. Reproduced with permission from [34].

Drilen then uses the functional and physical models to estimate the behavior and performance of the selected options. This estimate shows that a mostly electrified scenario using heat pumps is most energy efficient. The next step is that he uses that data to estimate the economic performance. These estimates indicate that storing the energy in the form of methane requires the least storage volume and has the lowest cost, despite the lowest end-to-end efficiency.

His paper ends with a reflection, which may be typical for conceptual modeling:

> *"The steepest part of [the] leaning-curve was at the late stage of the project. At this point, I started to get a better understanding of the technologies and the consequences related to them. With the increasing knowledge came an increasing understanding of what was missing or what I should im-prove. A learning from this is that with more knowledge, comes more questions. This created a chaotic final stage of the project. As I ended up with many new questions that I had to answer in a short period."*

4.3.2. Support for Sustainability in the Population

Vince Evers, a student from the Department of Sustainable Energy Technology & Innovation Sciences at the Eindhoven University of Technology, did his internship at Best Duurzaam. In that period, he studied the research question, "What factors, psychological or otherwise, influence the acceptance of sustainability measures and projects among the key stakeholders in the municipality of Best, and how do these influence implementation?" The trigger for this study is that district heating is one of the dominating options; however, it is poorly received by the citizens [35].

He developed a framework for acceptance, see Figure 7, based on the literature, among others [36–39]. **Financial Costs and Benefits** are probably obvious concerns for most stakeholders. **Values and Goals** relate to personal norms, values, motivations, and goals. A perceived lack of **Justice and Participation** may result in opposition to sustainability [36,39]. **Trust and Communication** is a major issue, as contemporary headlines confirm [40]. Lack of (perceived) **Efficacy and Feasibility** is eroding acceptance.

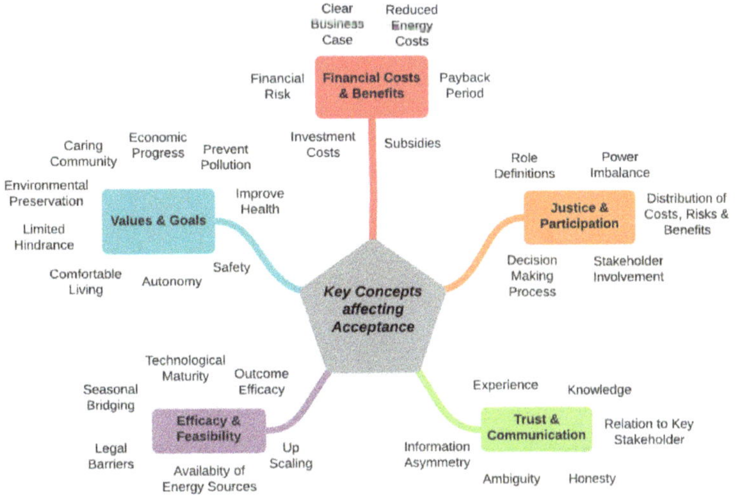

Figure 7. Conceptual framework for acceptance. Reproduced with permission from [35].

Figure 8 shows the results of a survey with 85 respondents. The centering of the bars is based on a net promotor score [41]. What is immediately clear is that "easy" measures, such as solar panels and insulation, are popular. These measures have a return on investments of several years with the current incentive schemes. However, crucial electrification, like heat pumps and electric vehicles, are not popular yet.

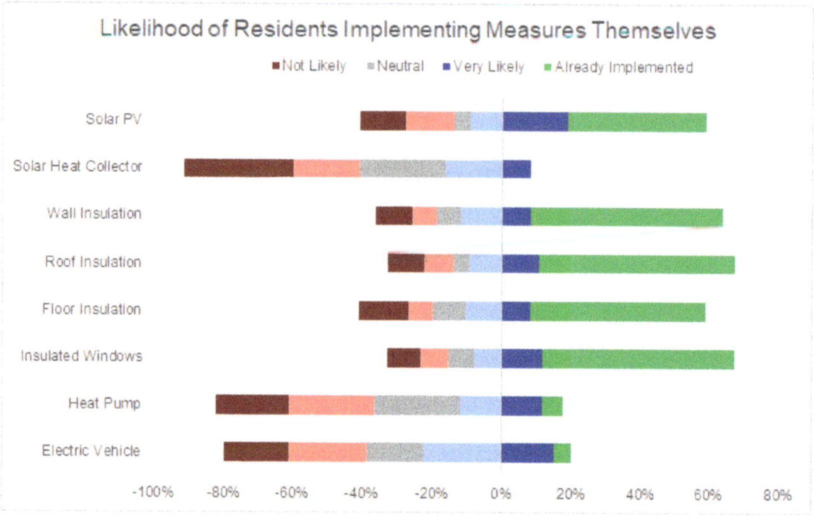

Figure 8. Survey results for sustainability measures under residents of Best for individuals. Reproduced with permission from [35].

Figure 9 is a continuation of the same survey; however, these questions address measures that the municipality may take. Only PV panels on public buildings get a rather positive reception. All other measures trigger significant opposition.

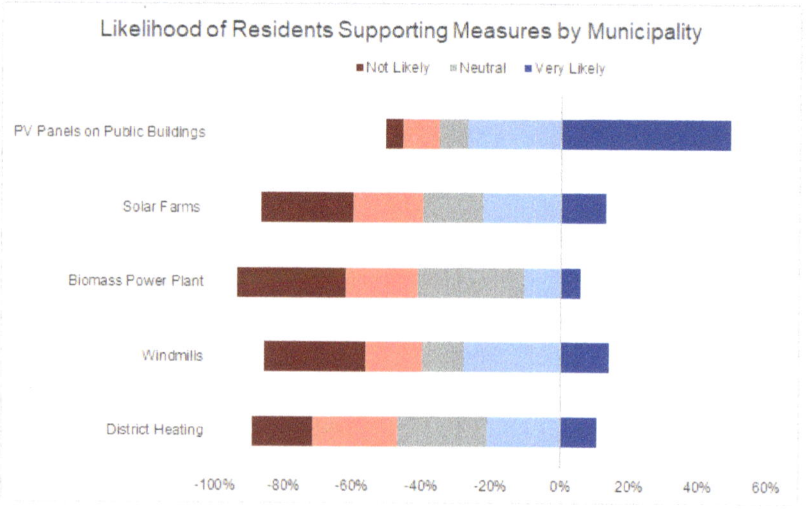

Figure 9. Survey results for sustainability measures under residents of Best for municipality measures. Reproduced with permission from [35].

4.4. Step 5. Continuous Observations of the Local Situation

One advantage of a co-located principal investigator is the possibility to collect continuous observations between research projects of events taking place in Best that relate to the roadmap.

a Heat storage in combination with solar collectors

The municipality developed a plan to harvest solar energy using solar collectors on the sport hall. The harvested heat would be stored in a large underground tank, and then several schools and the sport hall would use the stored heat for heating and hot tap water. Drivers behind the project were a high subsidy in solar collectors, a potential subsidy for frontrunner projects in heating, and the vision that local storage of heat would reduce the load on the electric network (when heating would change from using natural gas to using electricity). This plan was a mix of visionary aspects, such as using a fifth-generation heat network, and pragmatism, such as using planned maintenance of the sewer system to install a heat network in the streets.

The council rejected this plan, since its alignment with policy was unclear, while the financial risks were significant. The civil servants and the politicians who made this plan did not use the roadmap to explain how this project would help in achieving sustainability goals. The project preparation and discussion in the council were quite ad hoc.

b Municipality contribution for the 2030 renewable energy strategy

The process in the Netherlands to develop transition plans is highly distributed. There is a national agreement ("klimaatakkoord" [42]) that serves as a charter. Regions of tens of municipalities have to develop a Regional Energy Strategy. These regions ask the municipalities to make an "offer" of how much renewable energy they can produce in 2030. National government, regions, and municipalities then iterate until they have a fitting proposal. The alderman of municipality Best made an estimate very similar to the roadmap estimates to come to an offer. He consulted several stakeholders, including Best Duurzaam. This offer was in line with the roadmap.

c The request for commercial solar farms

Several commercial companies have proposed solar farms in Best. The national agreement states that half of the local renewable energy assets must be owned locally. There is a clear tension between the intent in the agreement of shared ownership to ensure support from the local population and commercial exploitation.

Concurrently with these proposals, the municipality was formulating the policy for environmental regulations. Best Duurzaam and several political parties pushed hard to ensure that the environmental regulations would not open the door for commercial exploitation without arranging how the shared ownership will work.

This is an example where "vertical" relations in the roadmap, in this case, solar farms as the solution, commercial companies and local population as resources, and environmental and exploitation regulations in governance, are coupled. A roadmap is a way to visualize such relations and to think of how to evolve to the desired state.

d Unrest in the countryside about the number of solar farms

The request from the regional energy board for a renewable energy offer triggered major unrest in the countryside. A news item on this offer suggested that a significant part of the countryside must be transformed into solar farms. Similar unrest played in other municipalities as well. An expert from a national university estimated solar on roofs can fulfill a significant part of such offer. However, the way the national system is set up more or less ignores small contributors. Hence, the risk is that, rather than using (unused) roofs, we would sacrifice agriculture or nature grounds. This is another example of lack of overview. The roadmap and its underlying conceptual models offer insight into what the options are and what the consequences of these options are.

5. Discussion

The discussion will first discuss how the constituent research projects affect the local sustainability efforts. Then, the discussion broadens to the larger scope, e.g., national and global. Finally, we discuss how the combined use of systems engineering methods roadmapping and conceptual modeling support the approach to the energy transition in the broader scope of the sustainability transition.

5.1. How the Research Projects Affect Local Sustainability Efforts

The 17 UN sustainability development goals articulate a significant global challenge. The energy transition, as part of them, is, in itself, a huge global challenge. We observe that, at national level in the Netherlands, there is an attempt to involve a wide variety of stakeholders. The so-called climate tables, where stakeholders discussed and negotiated how to achieve the energy transition, resulted in a national climate agreement. However, this agreement is a sociopolitical agreement, where the expertise of solving complex problems is missing. Many methods from systems engineering are useful in tackling this problem, especially roadmapping and conceptual modeling. In this research, in the limited scope of the municipality Best, we tried out several of the core methods to create an overview and to align a heterogeneous set of stakeholders.

We observed that most stakeholders were open for a constructive dialogue using a roadmap. However, we also observed that most stakeholders felt comfortable in their own niche (limited application scope, limited time horizon, concrete actions). Some of them were able and willing to connect their niche to the broader context. Other stakeholders preferred to do their own work, leaving the strategic work to others.

The municipality lacks systems engineering expertise. Most governmental organizations rely on external consultants for their expertise. In the heat storage case, the municipality had external expertise to make a design and a plan for the heat storage solution. The experts jumped at once to a feasibility study level without first exploring a range of options. Such exploration is required to understand the problem better, to analyze how solutions may fit, and to anticipate consequences.

The discussion and planning of solar farms shows the importance of providing overview and the need for communicating clearly. Estimating the required area for solar farms is relatively easy. However, the complexity of solar farms is in the socioeconomic model, which has impact on the support for such solar farms.

Although estimating the required area for solar is easy, we observed that current regulations trigger undesired rooftop sizing of solar. Residents get a good return on investment, as long as their production is similar to their consumption. However, overconsumption is not financially attractive. The consequence is that most people install "just enough" solar for their own consumption. In the wider perspective, we need to maximize solar on roofs, and to avoid that, we have to sacrifice agriculture or nature areas. Hence, we conclude that the current incentive scheme is counterproductive.

A recent news item [43] states that foreign investors own 79% of large Dutch solar farms. The subsidies to incentivize solar farms make solar farms rather attractive investments, rendering a yearly return up to 15%. This analysis makes clear that current legislation is not achieving the goals of the national agreement. Legislation at national and local level (e.g., environmental permitting) needs updating and alignment. This is a typical example of how roadmap layers connect.

The seasonal storage study illustrates how relatively simple conceptual models provide insight in strengths and weaknesses of various technology options. This type of modeling should take place before the municipality starts feasibility studies, such as the heat storage. We assert that the combination of roadmapping to set heat harvesting, storage, and distribution into a broader perspective and conceptual modeling to understand options and their impact will help to identify sensible options and to "sell" them to the council and other stakeholders. The acceptance study shows that securing support is a significant challenge, since the current opinion about such technologies is quite negative.

5.2. How Do These Methods Relate to the Broader Scope, e.g., National and Global?

Another future challenge is that we need similar methods and competence at the regional and national levels. Lack of overview and direction at these levels is frustrating the overview and alignment at local level. Moreover, solutions at the local level need scrutiny from regional and national (and European) levels. For instance, coping with seasonal variations is easier at national or European levels, since we then have more options to even out temporal and spatial variations.

5.3. How Does the Combination of Roadmapping and Conceptual Modeling Support the Approach to the Energy Transition?

This subsection addresses the research questions as far as the ongoing research allows this.

RQ1. How can we help a heterogeneous set of stakeholders to share an understanding of the current situation and of the options and their consequences to resolve them?

The conceptual models that the research used on A3s to build the initial roadmap helped in the stakeholder workshops. This is in line with [6,23]. These A3s helped in facilitating the stakeholders to have concrete and specific discussions in a problem field that is large and intangible.

Transferring these insights to a broader audience seems more difficult. Stakeholders who have not been part of the discussion missed the discussion and thereby the relevance of some facts, models, and their relationships. The step to politicians is especially challenging. Politicians have a rather different perspective. Could this be the reason that Stave and Hopper [44] position conceptual modeling so "low" in their taxonomy?

RQ2. How can such understanding help in reasoning about the solutions and, in that way, help in decision-making and preparing governance?

Several of the described events show that both conceptual models and the roadmap provide a reasoning framework for topics like local heat storage and positioning of solar farms. However, there is still a clear gap between understanding options for solutions and translating these into policy and legislation. There is a cultural and language mismatch between engineers and civil servants and politicians.

Fundamental to the decision-makers' role, and at the core of politicians, is that they cope with emotions and perceptions. As Westen [45] states, *"In politics, when reason and emotion collide, emotion invariably wins."* Thaler and Sunstein [46] describe how people make choices, indicating how important emotions are in decision-making. Lakoff elaborates on this in [47]. It is clear that just making technocratic conceptual models and roadmaps will not bridge the gap between experts and decision-makers.

RQ3. What is a manageable set of conceptual models to achieve understanding, reasoning, communicating, and decision-making?

A continuous challenge is to help stakeholders by offering them relevant information and sufficient context for a good discussion. The original roadmap workshop resulted in over 100 "issues." So many issues overwhelm humans. Experienced systems architects use about 10 views to capture an architecture [48]. Borches [49] proposes A3 Architecture Overviews (A3AO). A3 refers here to the standard A3 paper size of 297 × 420 mm. In this research, inspired by the A3AOs, we combined a limited set of conceptual models on a single A3 to make the information digestible for the stakeholders.

These A3s worked well in the workshops about the roadmaps. Some of the stakeholders still feel overwhelmed by these A3s. However, in these workshops, they felt involved, engaged, and informed. However, we see that the problem identified in RQ2, e.g., the gap between experts and decision-makers, has its equivalence for information that stakeholders can share. Just condensing information in roadmaps and conceptual models is insufficient to bridge the gap between these parties.

Our experiences also make clear that introducing these methods is challenging, in itself, due to, among others, a lack of systems engineering competence, while many stakeholders prefer to stay in their local scope. We see this challenge in the limited scope of industrial application of systems engineering. Increasing the scope to a transition scope makes this competence challenge significantly larger. Further research may help to find ways to introduce such competence.

6. Conclusions

The combination of conceptual modeling and roadmapping facilitates discussions between heterogeneous stakeholders on complex transition problems, such as the energy transition, at the local scale. However, there is a significant gap in the way of thinking and communicating between experts and decision-makers. Conceptual models and roadmapping may contribute to the societal challenge of the energy transition. However, humanity needs more means to bridge the transdisciplinary fields and the heterogeneous stakeholders.

Gerrit Muller worked from 1980 until 1999 in the industry at Philips Healthcare and ASML. Since 1999, he has worked in research at Philips Research, the Embedded Systems Institute, and TNO in Eindhoven. He received his doctorate in 2004. In January 2008, he became a full professor of systems engineering at University of South-Eastern Norway in Kongsberg (USN), Norway. He continues to work at TNO in a part-time position. Since 2020, he is INCOSE Fellow and Excellent Educator at USN.

Funding: This research received no external funding.

Institutional Review Board Statement: Not Applicable.

Informed Consent Statement: Not Applicable.

Data Availability Statement: Not Applicable.

Acknowledgments: The students performing their master projects and internships made this research possible. This paper builds on their publications or reports. The contributions of the stakeholders in Best are essential, too. The ownership of Best Duurzaam and the municipality Best for setting up the roadmap and the task force enabled this research. My colleagues Kristin Falk and Siv Engen supported me in writing this paper. Cecilia Haskins invited and encouraged me to write this paper.

Conflicts of Interest: The author declares no conflict of interest.

References

1. Elvebakk, L.; Muller, G. Creating a Roadmap to Capture a Vision for a Sustainable Community in Transition: A Case Study in a Dutch town Best. In Proceedings of the INCOSE, Virtual, 20–22 July 2020.
2. Phaal, R.; Farrukh, C.; Probert, D. Technology roadmapping—A planning framework for evolution and revolution. *Technol. Forecast. Soc. Chang.* **2004**, *71*, 5–26. [CrossRef]
3. Robinson, S. Conceptual modelling for simulation Part I: Definition and requirements. *J. Oper. Res. Soc.* **2008**, *59*, 278–290. [CrossRef]
4. Robinson, S. Conceptual modelling for simulation Part II: A framework for conceptual modelling. *J. Oper. Res. Soc.* **2008**, *59*, 291–304. [CrossRef]
5. Checkland, P. *Systems Thinking, Systems Practice: Includes a 30-Year Retrospective*; Wiley: Chichester, UK, 1999.
6. Neely, K.; Bortz, M.; Bice, S. Using collaborative conceptual modelling as a tool for transdiscipinarity. *Evid. Policy J. Res. Debate Pract.* **2019**, *17*. [CrossRef]
7. Blanchard, B.S.; Fabrycky, W.J. *Systems Engineering and Analysis*, 5th ed.; Pearson: London, UK, 2011.
8. United Nations. *The Sustainable Development Goals Report 2017*; UN: New York, NY, USA, 2017.
9. INCOSE. Systems Engineering. Available online: https://www.incose.org/about-systems-engineering/system-and-se-definition/systems-engineering-definition (accessed on 20 March 2021).
10. IPCC. *Summary for Policymakers*; World Meteorological Organization: Geneva, Switzerland, 2018.
11. UN. *Paris Agreement to the United Nations Framework Convention on Climate Change*; UN: New York, NY, USA, 2015.
12. Maier, M. Architecting principles for systems-of-systems. *Syst. Eng.* **1999**, *1*, 267–284. [CrossRef]
13. Gupta, A. Environment & PEST analysis: An approach to the external business environmen. *Int. J. Mod. Soc. Sci.* **2013**, *2*, 34–43.
14. De Coninck, H.; Revi, A.; Babiker, M.; Bertoldi, P.; Buckeridge, M.; Cartwright, A.; Dong, W.; Ford, J.; Fuss, S.; Hourcade, J.-C.; et al. Strengthening and Implementing the Global Response. In Global Warming of 1.5 °C; An IPCC Special Report on the Impacts of Global Warming of 1.5 °C above Pre-Industrial Levels and Related Global Greenhouse Gas Emission Pathways, in the Context of Strengthening the Global Response to the Threat of Climate Change, Sustai sustainable development, and efforts to eradicate poverty [Masson-Delmotte, V.; Zhai, P.; Pörtner, H.-O.; Roberts, D.; Skea, J.; Shukla, P.R.; Pirani, A.; Moufouma-Okia, W.; Péan, C.; Pidcock, R.; et al. (eds.)]. In Press. Available online: https://www.ipcc.ch/sr15/chapter/chapter-4/2018 (accessed on 20 March 2021).

15. Schut, G. Techniek is niet de oplossing (Technology is not the solution). *Technisch Weekblad*, 4 March 2021.
16. Metropoolregio Eindhoven. *Regionale Energiestrategie 1.0*; Eindhoven, The Netherlands, 2021. Available online: https://energieregiomre.nl/concept-res/default.aspx (accessed on 20 March 2021).
17. Miedzinski, M.; McDowall, W.; Fahnestock, J.; Muller, G.; Lopez, F.J.D. *Science, Technology and Innovation—Policy Roadmaps for the SDGs*. 2018. Available online: https://www.inno4sd.net/uploads/originals/1/inno4sd-pub-mgd-01-2019-fnl-sti-policy-roadmap-sdgs.pdf (accessed on 20 March 2021).
18. Phaal, R. *Roadmapping Bibliography*; Universityof Cambbridge: Cambbridge, UK, 2019.
19. Londo, H.; More, E.; Phaal, R.; Würtenberger, L.; Cameron, L. Background paper on Technology Roadmaps. In Proceedings of the Fifth Meeting of the Technology Executive Committee, Bonn, Germany, 26–27 March 2013.
20. UNIDO. Practice on Roadmapping. In Proceedings of the UNIDO, Vienna, Austria, 2–3 December 2008.
21. Plattner, H.; Meinel, C.; Leifer, L. *Design Thinking Research: Making Design Thinking Foundational*; Springer: Berlin/Heidelberg, Germany, 2015.
22. Wettre, A.; Sevaldson, B.; Dudani, P. Bridging Silos; A new workshop method for bridging silos. In Proceedings of the RSD8, Chicago, IL, USA, 13–15 October 2019.
23. Newell, B.; Proust, K. Introduction to Collaborative Conceptual Modelling; ANU Open Access Research. Available online: https://openresearch-repository.anu.edu.au/handle/1885/9386 (accessed on 20 March 2021).
24. Sargent, R.G. Verification and validation of simulation models. *J. Simul.* **2013**, *7*, 12–24. [CrossRef]
25. Balci, O.; Arthur, J.D.; Nance, R.E. Accomplishing Reuse with a Simulation Conceptual Model. In Proceedings of the Winter Simulation Conference, Miami, FL, USA, 7–10 December 2008.
26. Montevechi, J.A.B.; Friend, J.D. Using a Soft Systems Methodology Framework to Guide the Conceptual Modleing Process in Discrete Event Simulation. In Proceedings of the Winter Simulation Conference, Berlin, Germany, 9–12 December 2012.
27. Tomita, Y.; Shirasaka, S.; Watanabe, K.; Maeno, T. Applying Design Thinking in Systems Engineering Process as an Extended Version of DIKW Model. In Proceedings of the INCOSE, Adelaide, Australia, 15–20 July 2017.
28. Jackson, M. *Critical Systems Thinking and the Management of Complexity. Responsible Leadership for a Complex World*; Wiley: Newark, NJ, USA, 2019.
29. Muller, G. CAFCR: A Multi-View Method for Embedded Systems Architecting; Balancing Genericity and Specificity. Ph.D. Thesis, Technical University Eindhoven, Eindhoven, The Netherlands, 2004.
30. O'Brien, R. Um exame da abordagem metodológica da pesquisa (An Overview of the Methodological Approach of Action). In *Teoria e Prática da Pesquisa (Theory and Practice of Action Research)*; Richardson, R., Ed.; Universidade Federal da Paraíba: João Pessoa, Brazil, 2001.
31. Muller, G.; Elvebakk, L.; van der Velde, J.; Lean, F.M. Roadmapping for Sustainability; How to Navigate a Social, Political, and Many Systems-of-Systems Playing Field? A Local Initiative. In Proceedings of the SoSE, Anchorage, AK, USA, 19–22 May 2019.
32. Muller, G. A Roadmap for Sustainability for a Community in The Netherlands. In Proceedings of the SoSE, Virtual, 2–4 June 2020.
33. Falk, K.; Muller, G. Embedded Master's Students Conduct Highly Relevant Research Using Industry. *Technol. Innov. Manag. Rev.* **2019**, *9*, 54–73. [CrossRef]
34. Drilen, E.; Muller, G.; Syverud, E. Conceptual Modelling of Seasonal Energy Storage Technologies for Residential Heating in a Dutch town Best. In Proceedings of the INCOSE, Beijing, China, 12–13 April 2021.
35. Evers, V. *Intership Report*; Eindhoven, The Netherlands, 2020.
36. Huijts, N.M.; Molin, E.J.; Steg, L. Psychological factors in technology acceptance: A review-based comprehensive framework. *Renew. Sustain. Energy Rev.* **2012**, *16*, 525–531. [CrossRef]
37. Van der Lelij, B.; de Graaf, M.; Visscher, J. *Energievoorziening 2015–2050: Publieksonderzoek Naar Draagvlak voor Verduurzaming van Energie*; Ministerie van Economische Zaken: Den Haag, The Netherlands, 2016.
38. Bain, P.G.; Milfont, T.L.; Kashima, Y.; Bilewicz, M.; Doron, G.; Garoarsdóttir, R.B.; Saviolidis, N.M. Co-benefits of addressing climate change can motivate action around the world. *Nat. Clim. Chang.* **2016**, *6*, 154–157. [CrossRef]
39. Wüstenhagen, R.; Wolsink, M.; Bürer, M.J. Social acceptance of renewable energy innovation: An introduction to the concept. *Energy Policy* **2007**, *35*, 2683–2691. [CrossRef]
40. Dastagir, A.E. Trust no one? Americans lack faith in the government, the media and each other, survey finds. *USA Today*, 23 July 2019.
41. Reichheld, F. The one number you need to grow. *Harv. Bus. Rev.* **2003**, *81*, 46–55.
42. Anonymous. Klimaatakkoord. 2019. Available online: https://www.klimaatakkoord.nl/documenten/publicaties/2019/06/28/klimaatakkoord (accessed on 20 March 2021).
43. Laconi, P. Buitenlandse investeerders gaan er vandoor met miljoenensubsidies én winst van zonneparken. *Algemeen Dagblad*, 16 January 2021.
44. Stave, K.; Hopper, M. What Constitutes Systems Thinking? A Proposed Taxonomy. In Proceedings of the 25th International Conference of the System Dynamics Society, Boston, MA, USA, 29 July–3 August 2007.
45. Westen, D. *The Political Brain: The Role of Emotion in Deciding the Fate of the Nation*; PublicAffairs: New York, NY, USA, 2007.
46. Thaler, R.H.; Sunstein, C.R. *Nudge: Improving Decisions About Health, Wealth and Happiness*; Penguin Putnam: London, UK, 2009.
47. Lakoff, G. *The Political Mind*; Penguin: London, UK, 2009.

48. Muller, G.; Hole, E. Architectural Descriptions and Models. 2006. Available online: http://www.architectingforum.org/whitepapers/SAF_WhitePaper_2006_2.pdf (accessed on 6 March 2021).
49. P.D. Borches Juzgado, A3 Architecture Overviews: A Tool for Effective Communication in Product Evolution, Enschede: University of Twente. 2010. Available online: https://research.utwente.nl/en/publications/a3-architecture-overviews-a-tool-for-effective-communication-in-p (accessed on 20 March 2021).

sustainability MDPI

Article

An Approach to Sustainability Management across Systemic Levels: The Capacity-Building in Sustainability and Environmental Management Model (CapSEM-Model)

Annik Magerholm Fet [1,2,*] and Haley Knudson [2]

1 Department of Industrial Economics and Technology Management, Norwegian University of Science and Technology, 7491 Trondheim, Norway
2 Department of International Business, Norwegian University of Science and Technology, 6009 Ålesund, Norway; haley.knudson@ntnu.no
* Correspondence: annik.fet@ntnu.no

Abstract: A toolbox for assessing the environmental impacts of processes, products and services has been gradually developed over the last 30 years. The tools and methods place attention on a growing holistic concern to also consider stakeholders' views connected to impacts of the entire life cycle of products. Another change is the gradual increase in consideration of the economic and social dimensions of sustainability since the 1990s. This paper presents this development using two interlinked models that illustrate the changes from the scopes of time and system complexity. The two initial models are further merged into one, the Capacity-building in Sustainability and Environmental Management model (the CapSEM-model), which presents organizations a systemic way to transition to sustainability, seen from the scopes of system complexity and performance complexity. The CapSEM-model attempts to integrate the different dimensions of systems and of methodologies and their contribution to increased environmental and sustainability performance. The Sustainable Development Goals (SDGs) are further mapped onto the model as an example of how they can be useful in the transition to sustainability. The model is, therefore, a conceptualization and needs further development to specify accurate level boundaries. However, it has proven to be helpful for organizations that struggle to find a systematic approach toward implementing sustainability. This is described through a brief example from the manufacturing industry.

Keywords: environmental performance; management; sustainability; the SDGs; systems thinking; systems engineering; life cycle; capacity building in sustainability

Citation: Fet, A.M.; Knudson, H. An Approach to Sustainability Management across Systemic Levels: The Capacity-Building in Sustainability and Environmental Management Model (CapSEM-Model). *Sustainability* **2021**, *13*, 4910. https://doi.org/10.3390/su13094910

Academic Editor: Cecilia Haskins

Received: 31 March 2021
Accepted: 23 April 2021
Published: 27 April 2021

1. Introduction

The focus on methodologies to assess environmental aspects and connected impacts has increased tremendously over the last 30 years [1–7]. *Our Common Future*, also known as the Brundtland Report, was launched in 1987 by the World Commission on Environment and Development (WCED), and importantly linked the needs of the environment and development for future agendas [8]. Presented at the 1992 Rio Conference, the report in many ways initiated the quest for sustainable development (SD) on an international scale. It was a catalyst for nations, as well as for large international corporations and smaller organizations, to take responsibility in addressing their sustainability challenges through management of their environmental impacts. Twenty years later, at Rio+20, the foundational ideas of the report continued to influence global SD initiatives, and the development of the United Nations (UN) Sustainable Development Goals (SDGs). Adopted in 2015, the SDGs [9] are the present global call to action for nations and companies alike. Progress, however, is not on track to reach all prescribed targets, and has been further curbed by the coronavirus disease 2019 (COVID-19) pandemic [10,11].

In addition to the competitive advantage that comes from increased environmental management and sustainability consciousness [12–15], companies of all sizes have a duty to improve the sustainability of their organizations. They are an essential piece to solve the complex puzzle of global SD [10,11,16]. Nearly 35 years after the release of the Brundtland Report, this paper focuses especially on the environmental dimensions of SD. Furthermore, it presents the advances of life cycle based sustainability management tools over the period. It discusses how the tools relate to corporate practice, and how they have developed to expand thinking beyond firm level impacts to wider system level SD. It finally raises some critical questions to the extent the tools have advanced companies toward solving the challenges outlined in the SDGs.

To understand and manage the impacts of systems, the concepts of systems thinking and life cycle thinking are essential [17]. *Systems thinking* involves recognizing systems and subsystems, and the interactions within and between them, from a holistic perspective [18,19]. A *life cycle approach* to problem solving considers the material and resource inputs and resulting environmental, social and economic impacts across all phases of a product or service's life cycle [20]. It puts new demands on corporations as analytical requirements become increasingly complex, refined and demanding. In time, this will increase demand for specialized staff for monitoring and reporting. There is, in other words, a gap between the numerous and diverse analytical models for sustainability aspects, and organization capacity and practice. Furthermore, an overview of these methods and the knowledge needed to implement them is often lacking, especially in smaller companies with more limited resources [21]. As both internal and external requirements become more stringent to meet growing sustainability challenges, companies and organizations need a holistic toolbox to help them navigate the interacting systems of SD, from triple-bottom-line aspects, to geographic scope and long-term timelines.

2. Meeting Sustainability Challenges with Life Cycle-Based Environmental and Sustainability Management Tools

To clarify the toolbox of life cycle-based environmental management tools, sustainability challenges can be classified according to the systems in which they occur. For example, from pollution and environmental degradation caused by production processes, to resource depletion and impacts across different stages of products' life cycles, to a lack of awareness from the management side of companies and policy makers.

These challenges can be met by organizations with a combination of technological advancement and a change in procedures and strategies across different environmental performance levels that vary in temporal and environmental scope [22,23], for example:

1. Environmental engineering;
2. Pollution prevention;
3. Environmentally conscious design and manufacturing;
4. Industrial ecology; and
5. Sustainable development.

These environmental performance levels, or systems, are numbered and presented in Figure 1a and further explained in the following text. Several models for a systematic presentation of the development that has taken place since the early 1990s can be found in literature [24]. The models presented in Figure 1 are one way of illustrating the development of the field over time. They are also a way to demonstrate how the toolbox for environmental assessment and improvement can be used to assess the challenges of transitioning to sustainability and contributing to meeting the objectives set by the SDGs. Figure 1a, together with 1b, are the starting point of the Capacity-building in Sustainability and Environmental Management model (CapSEM-model), presented in Section 3. Each of the models has advanced the goal to guide companies and other organizations to systematically implement sustainability practices in their products and internal strategies while also building partnerships with the larger societal system.

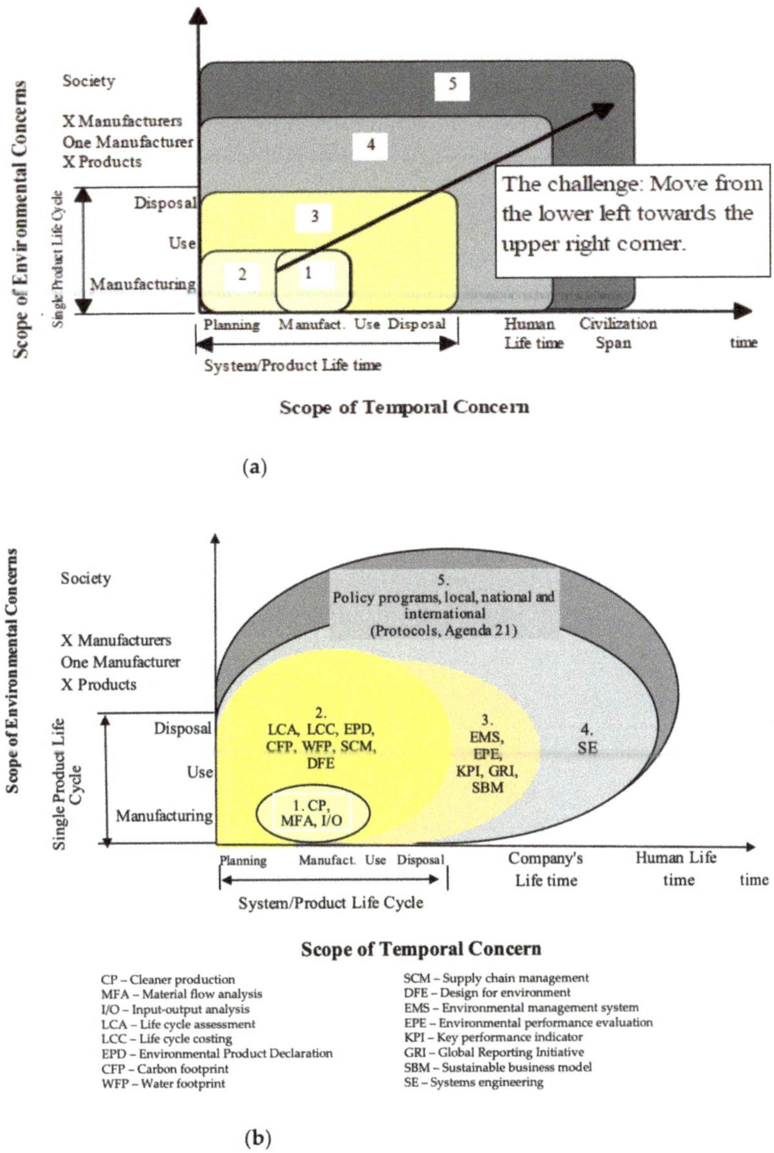

Figure 1. (**a**) Classification of environmental performance levels, [23] modified after [22]; (**b**) a classification of methods and tools for environmental performance improvements, modified after [23,25].

Area 1 in Figure 1a represents the perspectives related to environmental engineering strategies to reduce negative environmental impacts within production and manufacturing processes. This space takes a limited systemic scope in both time and environmental concern (only during the manufacturing process and life cycle stage).

Area 2 increases the temporal scope and involves pre-planning for the manufacturing phase to prevent pollution and negative impacts during the process. Pollution prevention strategies arose in 1992 through the initiatives launched by the Environmental Protection Agency (EPA) [26], with the objective to reduce the environmental impacts of products by identifying them in the design phase. This way, the impacts throughout the life cycle could be reduced through better planning of product design. For example, better planning might consider techniques for assembly and material selection to help avoid negative impacts in the use and dismantling phases later in the product's life cycle. So, even though this space has a limited system scope on planning and manufacturing only, it helps build an understanding of potential problems that may arise later in the life cycle. It can be seen as a prelude to the later consideration on the entire life cycle of a product.

Area 3 expands the scope from processes related to manufacturing to the product as a whole and considers design to reduce negative impacts across its complete life cycle. The increase in consciousness of environmental concerns is illustrated through the additional consideration of the use and disposal phases. The wider consciousness is also reflected in the expanding temporal scope related to the gradual knowledge development of how to address the entire life cycle of products [27].

Area 4 further broadens the system boundaries and understanding of impacts throughout the entire industrial system. This includes perspectives related to tracking material and energy flows according to principles of industrial ecology (IE), e.g., industrial symbioses and circular material flow models [28].

Finally, Area 5 represents the holistic consideration of environmental aspects over an extended timescale and beyond the firm and its network. This means considering aspects relevant for present and future generations and that address all stakeholders, and likely societal and political challenges over time.

Advancing Figure 1a, a model for a systematic approach to environmental performance improvements was developed [23,24]. This model is presented in Figure 1b and shows adaptations from the first model, most notably the addition of specific tools and methods for life cycle-based environmental assessment management mapped along environmental performance improvement levels.

Figure 1b suggests a series of environmental performance and management tools to be implemented for the purpose of moving to a higher level as indicated by Areas 1–5 presented in Figure 1a. The tools are further classified into a model for capacity building in sustainability and environmental management—the CapSEM-model. The application of the tools for the achievement of a transition towards sustainability is described in Figure 2. Readers should note that the models presented in Figure 1a and b focus mainly on environmental aspects of sustainability, and do not fully consider the needs of stakeholders and other social aspects. Systems engineering (SE) is, therefore, introduced as an overall process to better consider stakeholder opinions and involvement in a holistic transition process. SE can be viewed both as a discipline and as a process [23]. As a discipline, SE is about taking the holistic life cycle perspective and bringing in aspects from other disciplines when needed in a multidisciplinary context. SE as a process, is about bringing a system into being with an understanding of challenges to the system during its life cycle. A six-step SE-methodology is introduced by Fet [23], and suggests the following steps in the context of sustainability:

1. Identify stakeholders and their needs related to sustainability performance (of a system, hereunder also an organization or the society as a system);
2. Define the requirements for the achievements of stated needs;
3. Specify the current performances related to environmental, social and economic aspects;
4. Analyze and optimize the performances according to needs and requirements;
5. Suggest solutions according to stated needs and requirements;
6. Verify the suggested solutions against *1.* and *2.*

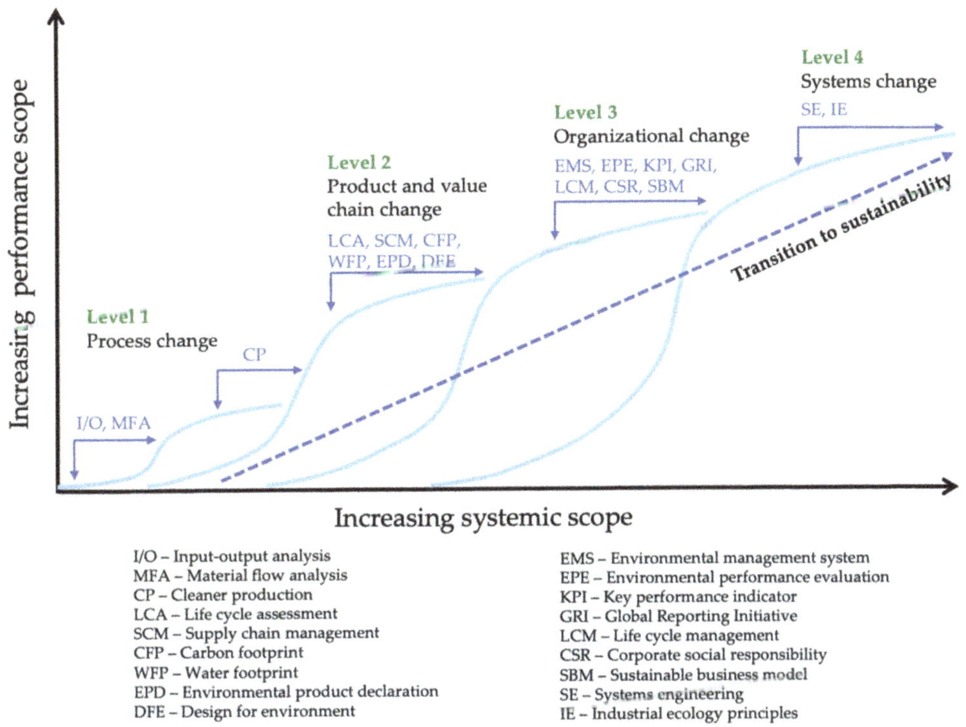

Figure 2. The Capacity-building in Sustainability and Environmental Management model (CapSEM-model)—a systemic approach towards sustainability, modified from [29].

This process can be used for each area in Figure 1. The complexity of stakeholder involvement, and thereby the sustainability aspects to be addressed along the road from the lowest to the higher levels, will increase. The steps in the SE-process can be performed in several cycles until the best solutions are achieved.

3. The Capacity-Building in Sustainability and Environmental Management Model (CapSEM-Model)

Based on the improvement of the models presented in Figure 1, the CapSEM-model was developed to illustrate the spectrum of environmental performance areas, here termed 'levels', ultimately reaching a holistic level of systemic sustainability. This requires that companies expand their environmental and sustainability management perspectives, extending the scope and number of impacts that are considered as they move toward more integrated sustainability. Figure 2 presents the CapSEM-model.

The waves in the CapSEM-model illustrate different levels of performance of the systems under study. A systematic use of the toolbox helps companies investigate the potential for appropriate actions to improve the environmental and sustainability performance related to production processes (Level 1), products and value chains (Level 2) and strategic organizational actions (Level 3). The highest level (Level 4) represents the larger societal system and a company's recognition of its place and responsibility within it. The term 'improvement' is used to mean the reduction of negative impacts and increase of, or replacement with, positive impacts—ultimately leading to strong, proactive and holistic sustainability as companies move toward the upper right of the model. As an organization traverses the levels, knowledge and tools from the previous levels are used as input to more extensive methods.

Each axis describes a change in scope. The horizontal axis shows the scope of systems and begins at the simple production process at Level 1. Furthermore, it extends to the set of processes within the value chain of a product at Level 2. Then, to the organizational level (Level 3), embracing concerns for production processes and products in addition to the integration of strategic management systems to implement sustainability consciousness in a more holistic manner. Within Level 3, aspects connected to economic and human factors should also be considered. The scope of the systems on Level 4 can be defined as the sector that the organization is a part of, or as wide as a societal system since all organizations are part of a larger system.

The vertical axis illustrates the scope of performance. Level 1 focuses on the environmental impacts of material flows, while in Level 2 the focus has broadened to the performance of the entire value chain where e.g., management of the supply-chain could contribute to an improvement of the value chain. Level 3 adds aspects to be considered from a strategic level, such as management systems that help organizations move to a higher level of performance over time. A broader range of sustainability aspects should also be considered at this level. Since Level 4 system scope depends on the context of the operation of the organization, a higher level of performance can be achieved under the holistic recognition of opportunities that come from improving system performance. From a systemic perspective, the different levels could be described as subsystems and system elements of a larger societal system.

As seen initially in Figure 1b, Area 1 contains the suggested tools of cleaner production (CP), material flow analysis (MFA) and input-output analyses (I/O) to monitor the environmental impacts during production and manufacturing processes. In the CapSEM-model, Level 1 encompasses process-related changes for environmental accounting and (more sustainable) performance (e.g., principles of eco-efficiency [30]). When setting objectives related to emissions, resource use and waste generation, companies must assess the current use and flows of materials in order to reduce consumption and waste in their production processes. The methods of I/O and MFA, therefore, fit in Level 1 as they measure baseline levels for defining improvement and resource efficiency [31]. CP is also located on this level, where source reduction is the objective rather than end-of-pipe solutions [32]—therefore moving its placement further along the scales of system scope and performance. The focus on resource efficiency is often driven by economic and/or policy incentives, as these methods provide for diagnostic comparison and benchmarking of companies. Focus only on environmental aspects means that the Level 1 system does not explicitly consider the wider impacts on society. Its system boundaries are drawn at the firm level around specific processes.

In Area 2 in Figure 1b, the tools for the purpose of environmentally conscious product development are life cycle assessment (LCA) [33], life cycle costing (LCC), supply chain management (SCM) [34], carbon footprint of products (CFP) and water footprint of products (WFP) [35], environmental product declaration (EPD) [36], and design for environment (DFE). By expanding from the boundaries of a single process, Level 2 in Figure 2 focuses on product- and value chain-related changes. This means a focus on a product or service and all activities and processes along its value chain. The methods in Level 2 include LCA, which quantifies material flows (from Level 1) across the full life cycle of a product. Results from an LCA are quantified and weighted in terms of environmental impact. The weighted criteria can then be used to implement changes for more sustainable SCM upstream in the value chain. In addition, the quantified impacts can be used to perform carbon- or water-foot printing of a product, or to reach certifications for acceptable levels of environmental impact, e.g., EPDs. The principles of DFE, e.g., design for recycling or dismantling, can transform the value chain, accounting and planning for reduced environmental impact through the full life cycle of the product and its materials. Social-life cycle assessment (S-LCA) could also be placed on Level 2, as a way to track social impacts through the life cycle of a product [37]. Such methods are younger in their methodological development and can be difficult to quantify. However, further developing both quantitative and qualitative indicators to measure social sustainability impact is essential to reach holistic sustainability as mandated in the SDGs.

Area 3 in Figure 1b presents tools to be used by companies to improve their strategic approach for being more environmentally conscious, e.g., by implementing environmental management systems (EMS) [38], environmental performance evaluation (EPE), key performance indicators (KPI), the Global Reporting Initiative (GRI) [39], and sustainable business model (SBM) frameworks [40]. To further increase the comprehensiveness and scope of aspects considered, Level 3 moves toward the implementation of methods for stronger sustainability within an organization's management systems and strategy. The transition from Levels 1 and 2 into Level 3 represents an important advancement of management and monitoring for sustainability, allowing the incorporation of more social aspects. The organization must now widen its view beyond the firm itself, or its associated value chains, and track and report on its impacts in relation to the past, to its competitors, and for its long-term survival.

To make and monitor strategic changes across a company's operations, tools and methods for organization-level changes help address more complex sustainability challenges. Meeting these challenges might include establishing management systems to monitor goals for reducing negative environmental impacts and engaging further with stakeholders and customers. It also means looking beyond the value chain for effects of the organization on its employees and global and local environments in the long-term. Level 3 tools, therefore, include EPE, life cycle management (LCM) and EMS for benchmarking, meeting goals and continuous improvement (e.g., through ISO14001). Corporate social responsibility (CSR) embraces the triple bottom line of sustainability and is one approach to stakeholder engagement [41,42]. Establishing KPIs is an essential step in setting these goals, and companies can use a range of indicator frameworks from national systems to large, standardized reporting and communication systems such as the GRI. Methods from Levels 1 and 2 can be used to collect the data required for measuring the KPIs—demonstrating the knowledge development path represented by the CapSEM-model. SBMs are also placed on this level as they can help firms conceptualize their current value flows (environmental, economic and social) and identify areas to innovate for sustainability [43].

To achieve sustainable development in the long-term perspective, Areas 4 and 5 in Figure 1b present the policy programs and international regulations that help to set goals for a larger societal system. The highest level in the CapSEM-model, Level 4 also focuses on systems-related changes. This includes the most comprehensive assessment of sustainability aspects, both environmental and social, and for the company to see itself as one actor in a complex network of actors. While Levels 1-3 focus mainly on environmental aspects, Level 4 (and the higher degrees of the Level 3) command the inclusion of stakeholders and their long-term needs. Here, systems engineering (SE) is suggested as a helpful methodology to address these challenges and to include the principles of industrial ecology, e.g., principles of industrial symbioses and circularity [44].

Just as discussed in relation to Figure 1, the six-step SE methodology, can be performed at each level of the CapSEM-model until the most sustainable performance has been achieved. For simplicity, SE is placed at Level 4 to illustrate that it yields to the lower levels, but also because the increased scope required for Level 4 represents the most advanced form of SE.

To summarize, the CapSEM-model shows a spectrum of tools and methodologies for transitioning towards sustainability. It does not mandate that a company place itself within one level. Rather, it shows the way the tools and perspectives are linked and build upon each other. Additionally, it provides an example toolbox of methods that can be applied for improved sustainability in an organization depending on its level of ambition or maturity.

4. Adding the Sustainable Development Goals (SDGs) to the CapSEM-Model

The SDGs were established to guide the global sustainable development agenda until 2030. They are an extension of the previous global development framework, the

Millennium Development Goals (MDGs), which laid out an agenda for global poverty reduction. Recognizing the limitations of the MDGs, the SDGs were developed in a participatory process involving stakeholders across the global south and global north and introduced a set of specific targets and indicators for national governments to measure and communicate progress [45]. The SDGs have two aims—the reduction of global poverty and the halting of climate change, and chiefly recognize the link between the two. Criticisms of the triple-bottom-line approach, for example [46,47], suggest replacing environmental, social and economic silos with a more integrated view that sees the dimensions in a nested system for SD and the SDGs, respectively. These factors combine to make the SDGs a systemic framework that dictates the recognition of the interconnections between the goals and their targets. The set of 17 goals must be seen as a whole to achieve SD on the system level.

Although the official SDG target and indicator framework is for national governments, the agenda depends on industry participation and commitment. Many companies today use the SDGs to guide and communicate their sustainability strategies. A number of organizations provide guidelines and frameworks for use in companies to set goals and indicators within their strategies and operations. The SDG Compass [48], a joint initiative between the World Business Council for Sustainable Development (WBCSD), UN Global Compact and the GRI, is one such guideline, and provides databases of business tools and indicators openly accessible to companies. Nonetheless, it can be challenging to navigate the 17 goals and their respective indicators.

Just as the CapSEM-model helps make sense of the plethora of methods to measure sustainability performance by grouping them in levels, it can also help companies understand how their activities contribute to each of the SDGs. This logic is explained through the exemplification of a company in the manufacturing sector. Figure 3 places the SDGs along the CapSEM-model and discusses them in relation to each of the levels. Although the goals are each placed on a single level of the model, this is only used to illustrate an entry point to their application. In parallel to Rockström and Sukhdev's 'Wedding cake model of the SDGs' [47], SDGs 6, 13, 14 and 15 are grouped in the environmental layer, SDGs 1, 2, 3, 4, 5, 7, 11, 16 and 17 within the social layer, and SDGs 8, 9, 10 and 12 on the economic layer. Even though the SDGs can be systematized this way, we stress that the systemic nature of the SDG frameworks also requires that they are considered on all levels. However, to incorporate the SDGs into company strategies, specific goals and targets must be prioritized as a starting point [49].

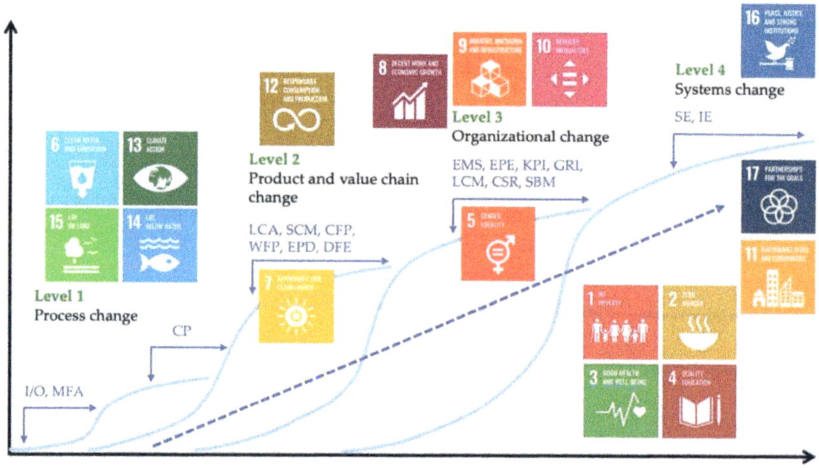

Figure 3. The CapSEM-model and the Sustainable Development Goals (SDGs), modified from [29].

Manufacturing involves several resource-consuming production processes (Level 1) where different materials, energy and chemicals are used, and resulting wastes generated. These wastes are typically disposed into air, land and water systems, and have contributed to the disruption of the Earth system. When considering improving sustainability in this sector, needs and requirements, therefore, include minimizing resource use, and avoiding pollution and the unnecessary expense and disposal of resources, especially into natural systems. I/O analyses can be used to quantify material flows within a production process or company system. Then, the quantified information can help inform decisions about the best solutions for designing new or adapting processes to reduce negative environmental impact. SDGs 6 (clean water), 13 (halting climate change), 14 (life under water) and 15 (life on land) have therefore been grouped on Level 1 in Figure 3 as their targets direct, for example, the increase in efficiency of water use (target 6.4) and the protection and restoration of water-related ecosystems (target 6.6). The selected goals and targets for improving sustainability in the manufacturing sector can be used to guide manufacturing companies in selecting indicators and making strategic decisions on how to reach them using the tools and methods at this level. The same process can be applied across the remaining levels and SDGs.

The move from Level 1 to Level 2 means that in addition to production processes, all other impacts related to the product and its value chain, e.g., the transportation of materials and components in the upstream life cycle of the product. In addition, downstream issues of distribution, maintenance and repair during the use phase and end of life treatment should be monitored for the entire life cycle of the product. Today, we see increased requirements for documentation of e.g., the carbon footprint of products. This means that the manufacturing company should take responsibility to achieve quantified information from the suppliers of materials, components and services across the life cycle. Based on the quantified information, optimized solutions for reduced GHG-emissions such as renewable energy sources should be achieved. SDGs 7 (clean energy) and 12 (responsible consumption and production) are therefore grouped on Level 2 to capture both upstream and downstream value chain sustainability improvements. SDG 12 places a focus on the entire value chain, and SDG 7 requires that products are designed and manufactured for cleaner energy systems. Because Level 1 can be seen as an input, or subsystem, to Level 2, the goals and targets at Level 1 must necessarily also be accounted for.

Pressure from public procurement and customer demands for products that support more sustainable living or help clean-up past damage, encourage manufacturing companies to report and communicate their progress toward improved sustainability. They must, therefore, develop their organizational strategies and practices (Level 3) in accordance with known guidelines and frameworks e.g., the SDGs. This requires trustful information from the companies across the other levels. For example, that all Level 1 processes are controlled and managed in a sustainable way, that systems for quantification of the carbon-footprints are in place at Level 2, and that the companies can present a management or certification system (e.g., ISO 14001) that supports the company in their annual assessment of improvements. The tools presented for Level 3, as well as Levels 1 and 2, should help the company to communicate the performance through a set of KPIs that give the stakeholders the information they need for an eventual approval of the sustainability performance or ranking of the company. SDGs 5 (gender equality) and 10 (reduced inequality) are placed on Level 3 and relate to the social aspects of e.g., equal employment and stakeholder inclusion to be mandated within the company's sustainability management systems and strategic organizational goals. SDGs 8 (decent work and economic growth) and 9 (industry, innovation and infrastructure) have also been grouped on the organizational level. This is because they pertain to the economic viability of a company and may further support its knowledge and innovation development relating to products that support a sustainable society.

Level 4 relates to the methods and tools that help drive systemic societal change and mandate the company view itself as one actor within a network of actors. SDGs 1 (no poverty), 2 (zero hunger), 3 (good health and well-being) and 4 (quality education) are placed at this level as they represent the basic criteria for thriving livelihoods. Without meeting these livelihood goals, sustainability will not be reached or maintained over time. They also require that companies consider all stakeholders in their actions. SDGs 11 (sustainable cities and communities), 16 (peace, justice and strong institutions) and 17 (partnerships for goals) are also placed on this level as they help companies recognize their place in the larger system, from communities and cities, to regional, national and global impacts. In a smart and sustainable city system, for example, there are increasing requirements to document the carbon footprint of subsystems, from furniture used in public spaces and private homes, to infrastructure that is designed for easier repair and that supports smart renewable energy systems. The need for take-back systems and sharing economy systems will also appear more frequently, and IE is one of the tools for developing symbioses within a circular economy. Similarly, SE is an important tool for seeing systems and their interactions from a holistic perspective. Level 4 embraces the underlying features of Levels 1, 2 and 3.

It is common to see the cherry-picking of select SDGs that neatly meet ongoing operations, ignore interactions between them or fail to reflect upon the system as a whole [34]. Clear company strategy is, therefore, needed for prioritization of areas for sustainability improvement and related SDGs and targets. The authors do not claim that the ordering of SDGs in Figure 3 is the absolute placement, but rather that it is one way to help a company identify the ways their operations initially relate to each goal. If companies better understand and engage with the goals, their ability to prioritize and make strong measurable contributions to their targets increases [49,50].

5. Discussion

The CapSEM-model demonstrates how the different dimensions of systems and of methodologies can be integrated to contribute to increased environmental and sustainability performance. Transitions can be achieved within organizations through the use of the tools presented first in Figure 1b and advanced since the early 1990s. The SDGs are further mapped onto the model as an example of how they can be useful in the transition to sustainability as entry-points to and objectives for action. The models in Figure 1a,b have their roots in the initiatives that were introduced in the 1990s. Work towards improved environmental consciousness in organizations has advanced since clean-up and pollution prevention were the main strategies. Over this period, a set of methodologies were developed and matured. For example, early versions of CP have contributed to the further development of standards for EPE and EMS. Similarly, the first versions of LCA were the foundation of other tools such as WFP, CFP, EPD and DFE, and the inclusion of the social dimension in S-LCA. Other tools have come later, or new versions of early pilots have been further developed under new names. The GRI framework is one such example. While indicators and reporting schemes were initially developed by different bodies, the GRI is now used as a common concept for reporting-systems and the use of performance indicators across different sectors. As methods and tools continue to be advanced, and new approaches or frameworks are initiated, the CapSEM-model will need to be updated to reflect changes in the toolbox and outlooks of organizations. The list of methods presented in the model is not exhaustive since new supportive tools are under continuous development.

Numerous scholars have suggested categorizations of environmental performance and sustainability methods (e.g., [51–53]). The CapSEM-model, however, classifies analytical methods and tools in a practical way that can serve as an entry- or positioning point for companies. Its development has paralleled the historical growth in concern for the environment and is a result of engagement with companies of various maturity levels and outlooks over the period.

As an organization moves between levels, tensions or limitations may be identified in relation to requirements or assumptions in methods at other levels. This may be due to the limited scope of certain methods that are unable to capture aspects across all SD dimensions. In many cases, tough decisions must be made between sustainability trade-offs and require that the organization has a clear strategy to guide their priorities.

In further research the CapSEM-model should be tested across different sectors and the different dimensions of sustainability. The systems studied at each level in the CapSEM-model should also be further described as they appear as different categories of systems, either as physical systems (e.g., production processes), theoretical systems (e.g., management systems), or geographical systems (e.g., for a societal study). Further development of the model is, therefore, encouraged, under a systematic approach to stakeholder involvement and actions for checking the achievements of initially formulated needs and requirements.

6. Conclusions

The purpose of this paper has been twofold. First, to illustrate how different initiatives of environmental consciousness and related monitoring and management tools have been developed over time and can be further systematized for the purpose of environmental and sustainability performance improvements at different system levels. Second, the paper demonstrates how these tools can be used in a systematic way for organizations in their transition to sustainability.

No matter what is the driver of sustainability improvement within an organization, SD is a wicked and complex problem (e.g., [54–56]), that requires transdisciplinary, collaborative and holistic thinking across triple-bottom-line principles, long-term systemic reasoning and wide stakeholder involvement. The CapSEM-model is a conceptualization of methods and approaches to help companies address this problem, and to identify opportunities within it. Although the CapSEM-model needs further development to specify accurate level boundaries, it has proven to be helpful for organizations that struggle to find a systematic approach toward implementing sustainability.

Author Contributions: Conceptualization, A.M.F. and H.K.; methodology, A.M.F.; writing—original draft preparation, A.M.F. and H.K.; writing—review and editing, A.M.F. and H.K. All authors have read and agreed to the published version of the manuscript.

Funding: This research received no external funding.

Institutional Review Board Statement: Not applicable.

Informed Consent Statement: Not applicable.

Data Availability Statement: Not applicable.

Conflicts of Interest: The authors declare no conflict of interest.

References

1. Hunt, C.B.; Auster, E.R. Proactive environmental management: Avoiding the toxic trap. *MIT Sloan Manag. Rev.* **1990**, *31*, 7–18.
2. Colby, M.E. Environmental management in development: The evolution of paradigms. *Ecol. Econ.* **1991**, *3*, 193–213. [CrossRef]
3. Dillon, P.S.; Fischer, K. *Environmental Management in Corporations: Methods and Motivations*; Center for Environmental Management, Tufts University: Medford, MA, USA, 1992.
4. Welford, R.; Gouldson, A. *Environmental Management & Business Strategy*; Pitman Publishing Limited: London, UK, 1993.
5. Sarkis, J. Manufacturing strategy and environmental consciousness. *Technovation* **1995**, *2*, 79–97. [CrossRef]
6. Berry, M.A.; Rondinelli, D.A. Proactive corporate environmental management: A new industrial revolution. *Acad. Manag. Perspect.* **1998**, *12*, 38–50. [CrossRef]
7. Soo Wee, Y.; Quazi, H.A. Development and validation of critical factors of environmental management. *Ind. Manag. Data Syst.* **2005**, *105*, 96–114. [CrossRef]
8. World Commission on Environment and Development (WCED). *Our Common Future*; Oxford University Press: Oxford, UK, 1987.
9. UN General Assembly. Transforming our world: The 2030 agenda for sustainable development. In Proceedings of the United Nations Summit, New York, NY, USA, 25–27 September 2015; Available online: https://sustainabledevelopment.un.org/post2015/transformingourworld/publication (accessed on 22 March 2021).
10. United Nations. *Sustainable Development Goals Report 2020*; United Nations: New York, NY, USA, 2020; Available online: https://unstats.un.org/sdgs/report/2020/The-Sustainable-Development-Goals-Report-2020.pdf (accessed on 22 March 2021).

11. Sachs, J.; Schmidt-Traub, G.; Kroll, C.; Lafortune, G.; Fuller, G.; Woelm, F. *Sustainable Development Report 2020: The Sustainable Development Goals and Covid-19*; Cambridge University Press: Cambridge, UK, 2020; Available online: https://sdgindex.org/reports/sustainable-development-report-2020/ (accessed on 22 March 2021).
12. Porter, M.E.; Van der Linde, C. Toward a new conception of the environment-competitiveness relationship. *J. Econ. Perspect.* **1995**, *9*, 97–118. [CrossRef]
13. Porter, M.E.; Van der Linde, C. *Green and Competitive: Ending the Stalemate*; Harvard Business Review: Brighton, MA, USA, 1995; pp. 119–134.
14. Klassen, R.D.; McLaughlin, C.P. The impact of environmental management on firm performance. *Manag. Sci.* **1996**, *42*, 1093–1227. [CrossRef]
15. Melnyk, S.A.; Sroufe, R.P.; Calantone, R. Assessing the impact of environmental management systems on corporate and environmental performance. *J. Oper. Manag.* **2003**, *21*, 329–351. [CrossRef]
16. Schaltegger, S.; Hansen, E.G.; Lüdeke-Freund, F. Business models for sustainability: Origins, present research, and future avenues. *Organ. Environ.* **2016**, *29*, 3–10. [CrossRef]
17. Fet, A.M. Environmental management and corporate social responsibility. *Clean Technol. Environ. Policy* **2006**, *8*, 217–218. [CrossRef]
18. Richmond, B. Systems thinking: Critical thinking skills for the 1990s and beyond. *Syst. Dyn. Rev.* **1993**, *9*, 113–133. [CrossRef]
19. Richmond, B. The "thinking" in systems thinking: How can we make it easier to master. *Syst. Think.* **1997**, *8*, 1–5. Available online: https://thesystemsthinker.com/the-thinking-in-systems-thinking-how-can-we-make-it-easier-to-master/ (accessed on 18 April 2021).
20. Rebitzer, G.; Ekvall, T.; Frischknecht, R.; Hunkeler, D.; Norris, G.; Rydberg, T.; Schmidt, W.P.; Suh, S.; Weidema, B.P.; Pennington, D.W. Life cycle assessment part 1: Framework, goal and scope definition, inventory analysis, and applications. *Environ. Int.* **2004**, *30*, 701–720. [CrossRef] [PubMed]
21. Perez-Sanchez, D.; Barton, J.R.; Bower, D. Implementing environmental management in SMEs. *Corp. Soc. Responsib. Environ. Manag.* **2003**, *10*, 67–77. [CrossRef]
22. Bras, B. Current educational status, interaction with industry, and the future of sustainable development. In *NTVA-Report 2: Industrial Ecology and Sustainable Product Design*; The Norwegian Academy of Technological Science: Trondheim, Norway, 1996.
23. Fet, A.M. Systems Engineering Methods and Environmental Life Cycle Performance within Ship Industry. Ph.D. Thesis, Norwegian University of Science and Technology, Trondheim, Norway, 1997.
24. Fet, A.M. Environmental management tools and their application: A review with references to case studies. In *Knowledge for Inclusive Development: International Series on Technology Policy and Innovation*; Conceição, P., Gibson, D.V., Heitor, M.V., Sirilli, G., Veloso, F., Eds.; Quorum Books, Greenwood Publishing Group: Westport, CT, USA, 2002; pp. 449–464.
25. Fet, A.M.; Aspen, D.M.; Ellingsen, H. Systems engineering as a holistic approach to life cycle designs. *Ocean Eng.* **2013**, *62*, 1–9. [CrossRef]
26. Environmental Protection Agency. *Facility Pollution Prevention Guide*; Environmental Protection Agency's Office of Solid Waste: Washington, DC, USA; The Risk Reduction Engineering Laboratory: Cincinnati, OH, USA, 1992.
27. Ehrenfeld, J.R. Industrial ecology: A strategic framework for product policy and other sustainable practices. In Proceedings of the Green Goods: The Second International Conference and Workshop on Product Oriented Policy, Stockholm, Sweden, September 1994.
28. Graedel, T.E.; Allenby, B.R. *Industrial Ecology*; Prentice Hall: Hoboken, NJ, USA, 1995.
29. Fet, A.M.; Knudson, H. Transdisciplinarity for sustainability management. In *Transdisciplinarity for Sustainability: Aligning Diverse Practices*; Keitsch, M., Vermeulen, W., Eds.; Routledge: London, UK, 2021; pp. 93–117.
30. Fet, A.M. Eco-efficiency reporting exemplified by case studies. *Clean Technol. Environ. Policy* **2003**, *5*, 232–240. [CrossRef]
31. Bringezu, S.; Moriguchi, Y. Material flow analysis. In *A Handbook of Industrial Ecology*; Ayres, R.U., Ayres, L.W., Eds.; Edward Elgar Publishing Limited: Cheltenham, UK, 2002; pp. 79–91.
32. Jackson, T. *Clean Production Strategies: Developing Preventative Environmental Management in the Industrial Economy*; Lewis Publishers: Boca Raton, FL, USA, 1993.
33. Nordic Council of Ministers. Product Life Cycle Assessment–Principles and Methodology. In *Proceedings of Nord 1992: 9th*; Nordic Council of Ministers: Copenhagen, Denmark, 1992.
34. Igarashi, M.; De Boer, L.; Fet, A.M. What is required for greener supplier selection? A literature review and conceptual model development. *J. Purch. Supply Manag.* **2013**, *9*, 247–263. [CrossRef]
35. Fet, A.M.; Panthi, L. Standards on carbon and water footprints and their implications for the maritime sector. In Proceedings of the International Conference on Maritime Technology, Harbin, China, 25–28 June 2012.
36. Fet, A.M.; Skaar, C.; Michelsen, O. Product category rules (PCR) and environmental product declarations (EPD) as tools to promote sustainable products. *Clean Technol. Environ. Policy* **2009**, *11*, 201–207. [CrossRef]
37. Huertas-Valdivia, I.; Ferrari, A.M.; Settembre-Blundo, D.; García-Muiña, F. Social life-cycle assessment: A review by bibliometric analysis. *Sustainability* **2020**, *12*, 6211. [CrossRef]
38. Fet, A.M.; Knudson, H. Environmental management from a systems perspective. In *Encyclopedia of Sustainable Technologies*, 1st ed.; Abraham, M.A., Ed.; Elsevier: Amsterdam, The Netherlands, 2017; pp. 165–173.
39. Fet, A.M.; Staniškis, J.K.; Arbačiauskas, V. Indicators and reporting as a driving tool for environmental activities in the region. *Environ. Res. Eng. Manag.* **2009**, *1*, 69–75.

40. Joyce, A.; Paquin, R.L. The triple layered business model canvas: A tool to design more sustainable business models. *J. Clean. Prod.* **2016**, *135*, 1474–1486. [CrossRef]
41. Skaar, C.; Fet, A.M. Accountability in the value chain: From Environmental Product Declaration (EPD) to CSR Product Declaration. *Corp. Soc. Responsib. Environ. Manag.* **2012**, *19*, 228–239. [CrossRef]
42. Carson, S.G.; Fet, A.M.; Skaar, C. A Nordic Perspective of Corporate Social Responsibility (CSR). *Etikk Praksis Nord J. Appl. Ethics* **2011**, *5*, 3–8. [CrossRef]
43. Evans, S.; Vladimirova, D.; Holgado, M.; Van Fossen, K.; Yang, M.; Silva, E.A.; Barlow, C.Y. Business model innovation for sustainability: Towards a unified perspective for creation of sustainable business models. *Bus. Strategy Environ.* **2017**, *26*, 597–608. [CrossRef]
44. Sopha, B.; Fet, A.M.; Keitsch, M.; Haskins, C. Using systems engineering to create a framework for evaluating industrial symbiosis options. *Syst. Eng.* **2010**, *13*, 149–160. [CrossRef]
45. UN General Assembly. Annex: Global indicator framework for the sustainable development goals and targets of the 2030 agenda for sustainable development. In *Resolution Adopted by the General Assembly on Work of the Statistical Commission pertaining to the 2030 Agenda for Sustainable Development (A/RES/71/313)*; UN General Assembly: New York, NY, USA, 2017; Available online: https://undocs.org/A/RES/71/313 (accessed on 22 March 2021).
46. Griggs, D.; Stafford-Smith, M.; Gaffney, O.; Rockström, J.; Öhman, M.C.; Priya, S.; Steffen, W.; Glaser, G.; Kanie, N.; Noble, I. Sustainable development goals for people and planet. *Nature* **2013**, *495*, 305–307. [CrossRef]
47. Rockström, J.; Sukhdev, P. Opening Keynote Speech. In Proceedings of the Presentation at the Stockholm EAT Food Forum, Stockholm, Sweden, 13–14 June 2016.
48. Global Reporting Initiative (GRI); UN Global Compact; World Business Council for Sustainable Development (WBCSD). SDG Compass: The Guide for Business Action on the SDGs. Available online: https://sdgcompass.org/ (accessed on 22 March 2021).
49. GRI; UN Global Compact. Integrating the SDGs into Corporate Reporting: A Practical Guide. Business Reporting on the SDGs, GRI and UN Global Compact. 2018. Available online: https://d306pr3pise04h.cloudfront.net/docs/publications%2FPractical_Guide_SDG_Reporting.pdf (accessed on 22 March 2021).
50. Mhlanga, R.; Gneiting, U.; Agarwal, N. Walking the talk: Assessing companies' progress from SDG rhetoric to action. In *Oxfam Discussion Paper*; Oxfam International: Oxford, UK, 2018. [CrossRef]
51. Robèrt, K.H.; Schmidt-Bleek, B.; Aloisi de Larderel, J.; Basile, G.; Jansen, J.L.; Kuehr, R.; Price Thomas, P.; Suzuki, M.; Hawken, P.; Wackernagel, M. Strategic sustainable development: Selection, design and synergies of applied tools. *J. Clean. Prod.* **2002**, *10*, 197–214. [CrossRef]
52. Singh, R.K.; Murty, H.R.; Gupta, S.K.; Dikshit, A.K. An overview of sustainability assessment methodologies. *Ecol. Indic.* **2009**, *9*, 189–212. [CrossRef]
53. Mura, M.; Longo, M.; Micheli, P.; Bolzani, D. The evolution of sustainability measurement research. *Int. J. Manag. Rev.* **2018**, *20*, 661–695. [CrossRef]
54. Lang, D.J.; Wiek, A.; Bergmann, M.; Stauffacher, M.; Martens, P.; Moll, P.; Swilling, M.; Thomas, C.J. Transdisciplinary research in sustainability science: Practice, principles, and challenges. *Sustain. Sci.* **2012**, *7*, 25–43. [CrossRef]
55. Brandt, P.; Ernst, A.; Gralla, F.; Luederitz, C.; Lang, D.J.; Newig, J.; Reinart, F.; Abson, D.J.; Von Wehrden, H. A review of transdisciplinary research in sustainability science. *Ecol. Econ.* **2013**, *92*, 1–15. [CrossRef]
56. Schaltegger, S.; Beckmann, M.; Hansen, E.G. Transdisciplinarity in corporate sustainability: Mapping the field. *Bus. Strategy Environ.* **2013**, *22*, 219–229. [CrossRef]

sustainability

MDPI

Article

Application of Systems Engineering and Sustainable Development Goals towards Sustainable Management of Fishing Gear Resources in Norway

Paritosh C. Deshpande [1,*] and Cecilia Haskins [2]

1 Department of Industrial Economics and Technology Management, Norwegian University of Science and Technology, 7491 Trondheim, Norway
2 Department of Mechanical and Industrial Engineering, Faculty of Engineering, Norwegian University of Science and Technology, 7034 Trondheim, Norway; cecilia.haskins@ntnu.no
* Correspondence: paritosh.deshpande@ntnu.no

Abstract: Commercial fishing is a critical economic sector for Norway, yet deficiency of scientific information, regulatory instruments, inadequate implementation, and lack of management infrastructure are among the significant causes of mismanagement of fishing gear (FG) resources. Mismanagement of FGs results in leakage of plastics through abandoned, lost, or discarded fishing gears (ALDFG), which is the most threatening litter fraction for marine wildlife. In EU-EEA states, the management of ALDFG is prioritized through a dedicated circular economy (CE) action plan. Historically, systems engineering (SE) methods are successfully applied for resource management studies. This study adopts and applies the SPADE method to evaluate sustainable management for the system of FG resources in Norway. SPADE comprises five problem-solving activities covering stakeholders, problem formulation, analysis, decision-making, and continuous evaluation. Each activity is accomplished by data collected through stakeholder interviews and literature analysis to establish an initial structure of problems and associated management strategies across FG's life cycle phases. The application of SPADE spanned across four years (2017–2020) and resulted in scientific outcomes aimed at the common goal of improving the system of FG resources in Norway within the framework of sustainable development goals and CE. SPADE's practice to integrate stakeholders at each step and provision for continual systems evaluation proved effective in building a holistic understanding of the complex system.

Keywords: systems engineering; SDGs; circular economy; recycling; waste management; ALDFG; fishing gear; plastic pollution; marine pollution; resource management

Citation: Deshpande, P.C.; Haskins, C. Application of Systems Engineering and Sustainable Development Goals towards Sustainable Management of Fishing Gear Resources in Norway. *Sustainability* **2021**, *13*, 4914. https://doi.org/10.3390/su13094914

Academic Editor: Just Tomàs Bayle-Sempere

Received: 30 March 2021
Accepted: 23 April 2021
Published: 27 April 2021

1. Introduction

Policies and programs for sustainable development (SD) have undergone significant shifts in focus since the introduction of the SD concept at the first global conference on the environment in Stockholm in 1972 [1]. With prospects of a rising global population, accelerating development, increased resource use, and the associated environmental impacts, the definition of sustainability has broadened from mere concern about pollution and biodiversity to the promotion of resource management [2].

The science of resource management involves generating a systemic understanding of the processes that lead to improvements in, or deterioration of, natural or anthropogenic resources. Management of resources is relatively more straightforward, especially when the resources and use of the resources by users can be monitored, and the information can be verified and understood in a non-complex way [3]. In resource management terminology, 'information' refers to the fundamental knowledge about stocks, flows, and processes within the resource system as well as about the human-environment interactions affecting the system [4]. Highly aggregated information may ignore or average out local data essential to identifying future problems and developing sustainable solutions [3].

Historically, local and regional governments are deemed responsible for managing resources through political instruments, and resource users are assumed incapable of reversing the deterioration of natural resources [5]. However, Johannes [6] provided a strong argument advocating the necessity to study the resource itself and the local ways, traditions, and knowledge associated with its use. As all humanly used resources are embedded in complex, social-ecological systems (SES) [7], one needs to incorporate ecological and sociotechnical knowledge of stakeholders or 'resource users' in describing the resource system.

In this paper, a system of "Fishing Gear (FG) Resources" is studied for developing sustainable strategies in the life cycle management of FG for the commercial fishing sector of Norway. A systems engineering (SE) based method, SPADE, is adapted and modified to analyze FG resources in Norway. The challenges and need for coining sustainable management of FG resources are elaborated as a background in Section 2. Section 3 provides a brief account of the SPADE method's activity steps as they were applied in the study. An application of the SPADE method to explore the strategies for sustainable life cycle management in commercial FG system in Norway is elaborated in Section 4. The Sustainable Development Goals (SDGs) and EU's Circular Economy (CE) strategies are used as a backbone in designing the improvement strategies for FG. Finally, the framework and application are discussed in Section 5. The adapted SPADE model aims to assist decision-makers and system stakeholders in comprehending otherwise complex and multi-dimensional FG management themes, sustainability and circular economy.

2. Background of the Fishing Sector in Norway

Norway is a Northern European country surrounded by water to the south (Skagerrak), the west (the North Sea and the Norwegian Sea), the north and northeast (the Barents Sea). With a resource-rich coastline of more than 25,000 km, Norway is the European leader in the commercial fishery and aquaculture sector [8]. Historically, commercial fishing has played a significant socioeconomic role, nationally and regionally, providing a foundation for settlement and employment along the Norwegian coast [9]. The commercial fishery sector includes all registered fishing companies in Norway that conduct fishing operations for economic benefits. The fishing sector is segmented into the coastal and ocean fishing fleet. A fishing fleet is an aggregate of commercial fishing vessels engaged in a particular type of fishing with selected FG. The coastal fishing fleet comprises smaller vessels operated by one to five crew and sizes range from 10–20 m [10]. the ocean fleet is known for its deep-water and sophisticated fishing practices, where fishing vessels are generally more than 28 m in size, and crew members vary from 20 persons or more [10,11].

In 2016–2017, 5946 vessels were registered in Norway, out of which approximately 93% are coastal vessels, and the rest belong to ocean fishing fleets [11]. The economically essential and primarily captured species include herring, cod, capelin, mackerel, saithe, blue whiting, and haddock. Additionally, fish species such as prawns, Greenland halibut and ling are caught in smaller quantities, but they have a high commercial value. Figure 1 shows the diversification of the fishing fleet concerning the number of vessels and type of FG they use. Although leisure fishers and foreign vessels also fish in Norwegian waters through quota agreements, only fishing activity through Norway's commercial fishing fleet was considered for assessment in this study.

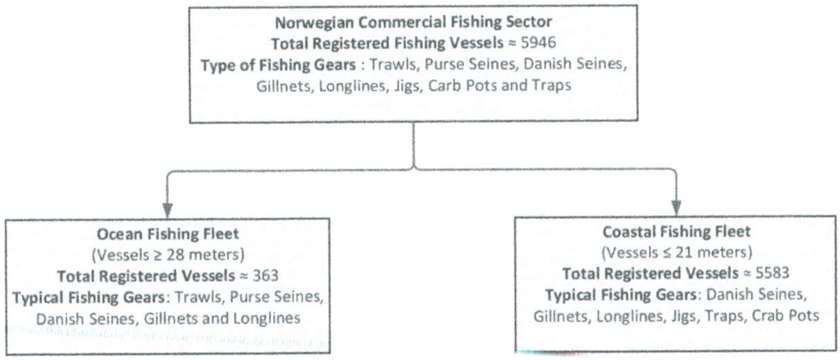

Figure 1. Structure of the commercial fishing fleet of Norway 2016–2017 [*modified after* 10].

The commercial FG resources are selected as a system in this study. FG are defined using an expansive definition proposed by FAO. According to FAO, FG are defined as "any physical device or part thereof or combination of items that may be placed on or in the water or on the seabed with the intended purpose of capturing or controlling for subsequent capture or harvesting, marine or freshwater organisms whether or not it is used in association with a vessel" [12]. Six major FG types, namely trawls, purse seines, Danish seines, gillnets, longlines, traps/pots and their associated ropes, are considered for this study. Polyethylene (PE), polypropylene (PP) and nylon (PA) are the primary building blocks of any FG [13]. Throughout the text, the term "plastics" includes PE, PP and PA. Although the FG unit contains other materials such as metals, lead, polyvinyl chloride (PVC) and wires, plastics constitute around 60% to 90% of any gear type [14]. Therefore, plastic polymers from FG are treated as resources in developing management strategies throughout this study.

Among the total plastic waste entering the oceans, *Abandoned, Lost or Discarded Fishing Gears* (ALDFG) are considered a particularly troublesome waste contribution that may continue to t trap, entangle and potentially kill marine life after all control of that gear is lost, which is defined as "ghost fishing" [15,16]. The amount, distribution and effects of ALDFG have risen substantially over past decades with the rapid expansion of fishing activity and fishing grounds and the transition to synthetic, more durable, and more buoyant materials used for some or all FG types [17,18]. In addition to the threat to marine ecology, the loss of fish stocks due to ghost fishing [19] and the additional cost of valuable resources on ALDFG also constitute significant economic setbacks [20].

The risk of ALDFG accumulation is relevant for countries characterized by a long and productive coastline. The geographic location and a strong dependence on fishing activity make Norway among the most vulnerable countries in the EU-EEA region to the detrimental effects of ALDFG pollution. Although ALDFG is the most dangerous fraction of marine litter, little or no information is available on the regional flows, sources and plastics from the fishing sector [14]. Consequently, there is a pressing need to build a holistic and systemic understanding of the fate, transport, sources, sinks and end-of-life (EOL) management alternatives of the fishing sector's regional plastic flow.

This paper aims to fill the knowledge gap by applying the SPADE method that facilitates the problem-driven, stepwise research to comprehend the sustainability dimensions for FG resource management in Norway. The adaption and application of SPADE are elaborated in Section 3.

3. Materials and Methods

Systems engineering is a discipline applied to structuring complex research problems where decision making is involved. The system of FG resource management presents a multifaceted, complex problem. Studying the entire system demanded detailed research spanning four years (2017–2020). The research is designed according to the SPADE framework of SE and elaborated here.

Systems Engineering (SE)

SE methods are characterized by their ability to structure and scope complex research problems [21]. The SE process involves a series of steps accomplished in a logical sequence to consider a holistic view of the total system, its life cycle and other interrelated life cycles (e.g., material, cash flow) [22]. SE was invaluable to help design and maintain the scope and boundaries of the research. Four critical characteristics of SE support a holistic understanding of a given problem.

i. A top-down approach where the system as a whole can be viewed.
ii. A life-cycle orientation where all phases of the system are addressed.
iii. A thorough identification of system requirements.
iv. An interdisciplinary collaborative approach to ensure that objectives are met in an effective manner.

Principles of SE were used in developing methods aimed at resolving the complex problems related to resource management and sustainability. FG resources' system is characterized by the unavailability of scientific information and the need to rely on stakeholders' knowledge as a major and, at times, only available source for developing an in-depth understanding of the system and interacting elements [23]. Consequently, an essential SE method, SPADE, proposed by [24], is used to structure the problem. SPADE is characterized by its ability to keep stakeholders at the forefront in defining the problems and identifying potential system improvements [24].

The first activity (S) involves defining system stakeholders and their needs. In the second, (P) the problems associated with the system under study are scoped and identified. The identified problems are analyzed using applied research methods in (A). After analyzing system performance, the decision-making (D) activity aims to identify solutions/alternatives to improve system performance. Finally, in (E), continuous evaluation and monitoring of suggested performance strategies are conducted for continual improvement of system performance. The SPADE model has been applied effectively in solving complex problems involving transdisciplinary research methods on decision making for sustainability in ship acquisition [25], creating an eco-industrial network [26] and solving design problems for offshore fish cages [27].

Figure 2 illustrates the adaption of the SPADE model to structure the research steps for the system of FG resources in Norway. The research spanned four years and included two research studies, hereafter referred to as study-1 [14] and study-2 [28], directly adhering to the SPADE model's activities presented as a stepwise progression. Both qualitative and quantitative research methods are applied to address the specific questions related to the case study of Norway's commercial fishing sector. Here, the SPADE framework and the rationale behind method selection for each step are elaborated. The research reported in this paper builds on two studies [14,28] that contribute to Problem definition (P), Analysis (A), Decision making (D) and Evaluation (E) parts of the SPADE model and shows how it can be used to structure long-term and multifaceted research.

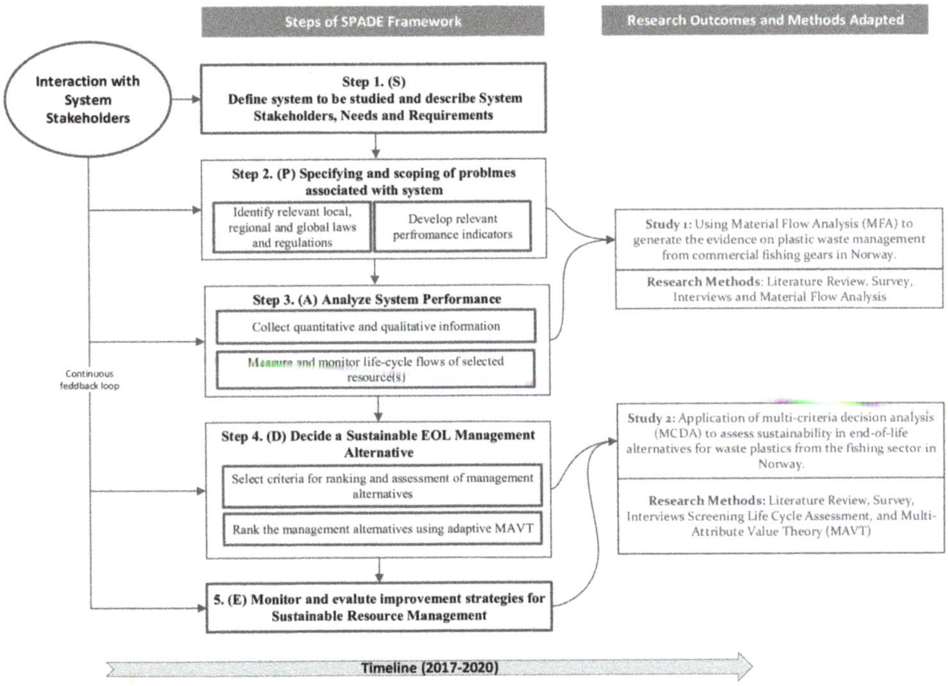

Figure 2. Research methods adapted in following SPADE framework for the management of FG resources.

The results of the literature review were used primarily to define the system and to map system stakeholders (step-1). The literature review and stakeholder interactions then were used to classify relevant life cycle processes of commercial FG and define associated problems across the life cycle phases (step-2). The industrial ecology (IE) tool, material flow analysis (MFA), is applied in study-1 to analyze system performance (step 3) by mapping and quantifying plastics' flows throughout the FG life cycle. In the absence of information to conduct MFA, questionnaires, semi-structured interviews and stakeholder surveys were used to collect the information necessary to validate the MFA model. Further, in study-2, a multi-attribute value theory (MAVT) method was adapted to determine sustainable management strategies for FG resources (step-4). The essential information for MAVT model inputs was obtained through site visits of waste management and recycling facilities and relevant expert stakeholders' questionnaires. The data collections routines and analysis of MAVT are elaborated in the study [28].

Finally, in the last step (step-5), literature review, expert opinion and insights gained through field visits were used to recommend strategies for continuous evaluation of the circular economy's targets for the FG resource management in Norway.

4. Results

SPADE is used here to organize the results regarding the system of commercial FGs in Norway.

4.1. System Stakeholders (S)

The first step taken is an identification of system stakeholders and their needs. There are wide ranges of groups that may be considered stakeholders in managing FG resources. Users and other stakeholders are individuals or groups of individuals who use the resource system in diverse ways for sustenance, recreation or commercial purpose [7]. The classification and mapping of stakeholders have been carried out in several ways based on the applicability and relevance to the problem. Here stakeholders are classified based on their relation to phases of the FG system life cycle. Purchase, use and end-of-life are the three main life cycle phases of FG. Stakeholders directly involved in one or more life cycle phase, their relationships and roles are presented in Table 1.

Table 1. List of stakeholders and their relevance to the life cycle stages of the FG system.

Stakeholders'	Pre-Use (Purchase)	Use-Phase	EOL Phase	Other	Category
Directorate of Fishery			X		Regulatory
Directorate of Environment			X		Regulatory, Environmental
Ports and harbours		X	X	X	Regulatory
Fishers and fishermen associations	X	X	X	X	Economic
Fishing Gear Producers/Suppliers	X				Economic
Relevant NGO's	X		X	X	Social, Environmental
Research & consultancy companies and Academia			X	X	Social, Environmental
Waste management companies			X		Economic, Environmental
Waste collection and recycling companies			X	X	Economic, Environmental

4.2. Problem Definition (P)

Problem formulation is among the most critical tasks for any research project. An in-depth understanding of the problem statement helps build a comprehensive analytical framework for the system under consideration. Figure 3 depicts the significant problems associated with the life cycle stages of FG resources. The problem identification was made by interacting with the indicated system stakeholder group(s). During the study period (2017–2020), a semi-structured questionnaire, interviews and a literature review were used to extract necessary information from relevant actors (FG producers, fishers, waste managers, beach clean-up volunteers, Directorate of fisheries, NGOs working on FG pollution, recyclers and landfill operators). Information was collected to build holistic problem definition for the study. The questions asked for each stakeholder groups across the life cycle of FG are presented in Supplementary Information Material (SI).

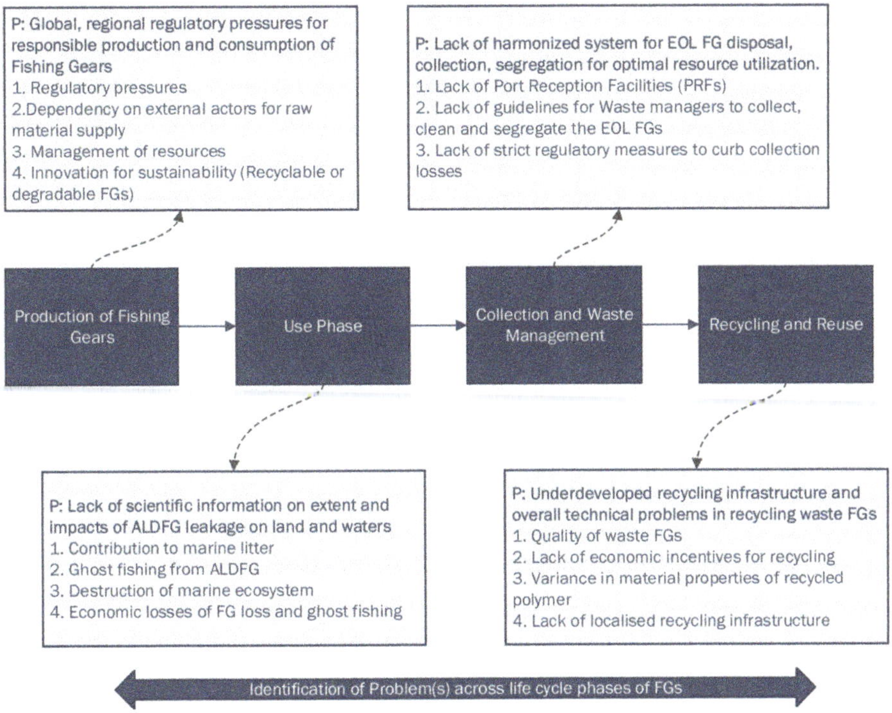

Figure 3. Problems associated across the system life cycle phases of fishing gear resources.

4.2.1. Problems across Production Phase

The design and production of FG vary considerably depending on the fishing community, type of target fish species, fishing grounds, size of the fishing vessel and local fishing regulations. Typically, FG are predominantly made of plastic polymers (PE, PP and PA) and other materials, such as metals, lead, polyvinyl chloride (PVC) and wires, only constitute around 10–30% of any FG type [14]. Intricate gear design, inability to track or trace FG once they get lost in the operation phase, and the dominance of durable plastic polymers resulting in ghost fishing are among the significant concerns associated with the current FG design. The use of three polymers and metal wires make it one of the most challenging products from which to recover material upon the end of useful product life. Additionally, the FG producers in Norway import most plastic polymers from Asian countries, causing dependencies on external actors [16]. Such dependency on external actors and the absence of in-house capacity to source raw materials had emerged as critical challenges for business firms during the current pandemic (COVID-19) times when import-export was significantly impacted. Consequently, developing local eco-industrial networks and improving self-reliance by recycling/reusing waste material could be explored as a long-term and sustainable alternative.

Furthermore, a rise in marine plastic pollution depletes fish stocks due to ghost fishing from ALDFG. These pressures on the marine ecosystem pose additional regulatory pressures on FG producers to innovate and develop biodegradable or recyclable designs for commercial FG. Hence, material resources and recovery underpin the fundamental problems identified by the regional FG producers in this study.

4.2.2. Problems across Use Phase

In the use-phase, fishers deploy FG in the ocean to catch a target species. Deployed FG, or their parts, may get lost during operation due to various reasons listed by Macfadyen [16]. FG have been lost, abandoned, or otherwise discarded in all seas and oceans since fishing began [13]. Moreover, in coastal countries like Norway and Iceland where fishing is among the primary economic contributors, impacts from ALDFG are detrimental [23]. Several studies document the damaging impacts of ALDFG on fish stocks and potentially harmful impacts on endangered species and benthic environments [16]. Apart from ghost fishing, ALDFG also increases ecological and economic threats, including navigational hazards and associated safety issues [29].

Commercial fishing is regulated in Norway through national, international and regional instruments. The Supplementary Information Material (SI) provides a brief overview of available policy instruments regulating the capture fishery practices. Norway's Marine Resources Act (6 July 2008) [30] is the primary fisheries legislation in Norway. This legislation prohibits the dumping of FG, moorings and other objects in the sea that may injure marine organisms, impede harvesting or damage gear. The act also mandates fishing vessels that lose FG to remove the object from the sea. If this is not possible, this loss must be reported to authorities. These lost-gear reports help the coast guard effectively plan the annual clean-up campaigns.

Although ALDFG is the most dangerous fraction of marine litter [13], little or no information is available on the regional flows, sources, and plastics' fate from the fishing sector. Jambeck [31] identifies this knowledge deficiency about plastic flows from fishing activities. Lack of scientific evidence resulted in strong dependence on precautionary principles or conservative methods to manage FG resources in coastal countries. Consequently, the problem associated with the use phase implies a need to build a holistic and systemic understanding of fate, transport, sources, sinks of ALDFG in land and waters.

4.2.3. Problems across End-of-Life Phase

At the end of gears useful life, fishers dispose of FG in the waste management facilities (WMF) as mandated in the Norwegian Pollution Control Act of 1981. The Pollution Control Act is modified to enforce the Port Reception Facility (PRF) directive by the EU. The Norwegian act states that littering is prohibited, which applies both on land and at sea [32]. Sound waste management is of crucial importance in preventing and reducing marine litter. Under the act, municipalities are responsible for collecting and treating household waste, while business and industry are responsible for properly handling and treating their waste.

The Norwegian Directorate of Environment and the Norwegian Directorate of Fisheries conduct programs to collect ALDFG from the ocean to minimize threats of ALDFG to the marine ecosystem. Similarly, ALDFG from shores are collected through annual beach clean-up operations in Norway. Waste FG collected through ocean and beach clean-up activities are deposited to the nearest WMF. Additionally, the waste generated through FG repair facilities end up in the WMF. At the WMF, this waste is segregated into fractions suitable for recycling, fractions for landfilling and fractions for incineration and then transported to respective facilities [14].

Semi-structured interviews during the site visits of waste management companies and waste recyclers in Norway for study-1 highlighted the problems in collections of EOL FG. The key challenge lies in the overall lack of PRF infrastructure across the Norwegian ports. The EU Directive 2000/59/EC dictates all EU-EEA member states to safeguard a PRF's availability and provide a waste management plan on all ports. PRFs are defined as *'any facility, which is fixed, floating or mobile and capable of receiving ship-generated waste or cargo residues'* [33]. According to the European Free Trade Association (EFTA) court's recent judgment, Norway has failed to fulfil the EU directive's obligations. Only one-third of the total registered ports in Norway contain a dedicated PRF or waste management plan. Lack of PRF can lead to an inappropriate collection of FG related waste, and may give rise to illegal dumping, burning or stockpiling of waste on ports hindering the waste collection regime from recovering valuable material [34].

4.2.4. Problems in Closing the Loop for FGs

Capacity building and technical support are critical to extracting value from waste based on CE principles. Currently, there exist numerous challenges in closing the loop for plastics from waste FG. The EOL collection, segregation, capacity, and availability of recyclers are among the critical concerns in realizing the economic benefits of material recovery. Hence, to establish opportunities for regional recycling or circular business models, it is imperative to know the quantity of waste plastic available for recycling from the fishing sector. Furthermore, while pursuing the goals of a circular economy, it is essential to assess the sustainability of the proposed EOL management strategies.

Considering all the collected information in the first two steps of the SPADE method, the following fundamental problems are defined in the second step of the SPADE framework that are subsequently answered in the last three steps SPADE method.

- What information is essential to aid system performance analysis and improvement? (answered in study-1)
- How much plastic is available at waste management facilities from the fishing sector in Norway? (answered in study-1)
- Which methods are available to assess sustainability in implementing circular economy strategies in managing FG resources? (answered in study-2 and partially in the current study)
- What measures are needed to ensure continual and sustainable improvement in system performance? (current study)

Additionally, the problem associated with plastic's value chain includes the absence of industrial recycling infrastructure. Lack of recycling facilities resulted in transboundary export of recyclable materials out of Norway, thereby missing an opportunity to recover material locally. If not handled responsibly, the exported waste may result in landfills or contribute to the pollution of marine or riverine ecosystems through leakages. Plastics polymer PE from waste FG can be processed through mechanical recycling to yield HDPE (high-density polyethylene) and LDPE (low-density polyethylene), which can be used further to replace virgin polymers in the products made by injection molding technology [35,36]. The PAs from waste FG retain their properties post-recycling, making them an economically attractive byproduct of recycling.

Therefore, the SPADE method's analysis must be articulated to understand opportunities and barriers in realizing sustainable closed-loop strategies for FG within Norway.

4.3. Analysis (A)

The analysis was conducted mainly through the IE-based MFA method and presented in study-1 [14]. The data and findings from study-1 are used here to define the analysis, and the lessons learnt from MFA are deliberated further. In analyzing the system's performance, it was essential to understand the physical flows and stock of mass of plastics (MoP) from the FG system in Norway.

Accordingly, in study-1 [14], a static MFA model was developed from 2016 to 2017 and accounting of MoP through purchase, use and post-use processes of FG life cycle were mapped. In MFA, static models provide insight into systems at a specific time, allowing holistic assessment of their current state [37]. Based on data from gear producers, fishers, collectors and recycling and waste management companies, an MFA model was established to quantify the annual stocks and flows of plastic polymers from six commonly used FG for commercial fishing by the Norwegian fishing fleet. The summarized results reported that the average MoP in the form of newly purchased FGs is estimated to be 2626 $\pm$ 143 tons per year.

Additionally, 1755 $\pm$ 681 tons of MoP are purchased as FG parts for replacement during significant repairs. Fishers reported the associated risk of damaging FG and losing part of or the entire gear upon deployment in the ocean. Such FG and/or parts leaked into the marine environment as deployment loss are estimated to be 400 tons per year. The beach and ocean clean-up efforts at the national scale contribute to removing/recovering cumulatively 100 tons/year of MoP from lost FG. Finally, if not lost during operation or repaired, fishers must dispose of FG at their end-of-life. The MFA estimates that annually around 4000 tons of MoP are collected at Norwegian WMFs for EOL treatment from the commercial fishing sector, and more than 50% of the collected waste fraction is sent out of Norway for further processing and recycling.

Building upon the problem definition stage, analysis uncovered the loopholes in the system of FG resources in Norway. Results suggest that although there are ample policy instruments available in regulating the fishing sector, leakage of plastics to the marine ecosystem, and an underdeveloped infrastructure to collect, handle and treat waste are causing system malfunction. Finally, the analysis can be used to develop additional insights through processing the data gathered from the system stakeholders and to design a strategy to ensure sustainable management of plastics from FG.

4.4. Decide (D)

Until the end of 2017, industrial-scale recycling of waste plastic was unavailable in Norway [28]. Consequently, the entire fraction of recyclable material was sent out of Norway for mechanical recycling of PP, PE and PA from EOL FG. However, industrial-scale recycling for obsolete plastics from the fishing and aquaculture sector began in Norway in the latter half of 2017. Although recycling began in Norway, a still significant fraction of waste is sent abroad for further processing and recycling. Therefore, decisions must be taken to assess the sustainable EOL management alternative for FG resources, ensuring optimum material or energy recovery. These decisions are studied and presented in Study-2 [28], and the findings are summarized here.

In study-2, to assess the sustainable EOL management alternative for FG resources, four scenarios were selected: landfilling, incinerating, recycling (inland) and recycling (export). Sustainable management is defined as "The ability of EOL management alternatives to manage 4000 tons of waste FG annually through maximizing environmental and economic and social benefits while minimizing the negative effects" [28].

Based on the qualitative and quantitative data from relevant system stakeholders, MAVT was adapted to rank the EOL alternatives based on their ability to manage 4000 tons of waste plastics from FG in Norway within the defined sustainability criteria. The application of multi-criteria decision analysis (MCDA) to address the decision-making problem is presented in the study [28]. Sustainable Development Goals (SDGs) and targets are considered helpful in assessing the three dimensions (environmental, economic and social) of sustainable development proposed by [38]. SDGs are preferred as they address some of the systemic barriers to sustainable development (SD) and contain better coverage of and balance between the three dimensions of SD and their institutional/governance aspects [39]. The assessment criteria are chosen to reduce the uncertainty, increase the understanding of the FG system, and measure the EOL alternatives' performance against a defined goal.

Figure 4 presents the value tree developed for the decision analysis problem in study-2. The top of the value tree diagram presents the decision problem, which was assessed to attain sustainable strategies for FG management as per the definition. SDGs are used to define the assessment criteria under environmental, economic and social objectives, and the four alternatives are evaluated. The first target under SDG 14 is to prevent and significantly reduce marine pollution of all kinds, particularly from land-based activities, including marine debris and nutrient pollution, by 2025. Under OSPAR convention's *Marine Litter Regional Action Plan 2014-2021*, Norway has adopted the goal of reducing litter inputs that have negative impacts on coastal waters, the sea surface, the water column or the seabed. Accordingly, the environmental criteria are chosen to minimize the risk of marine pollution and climate change.

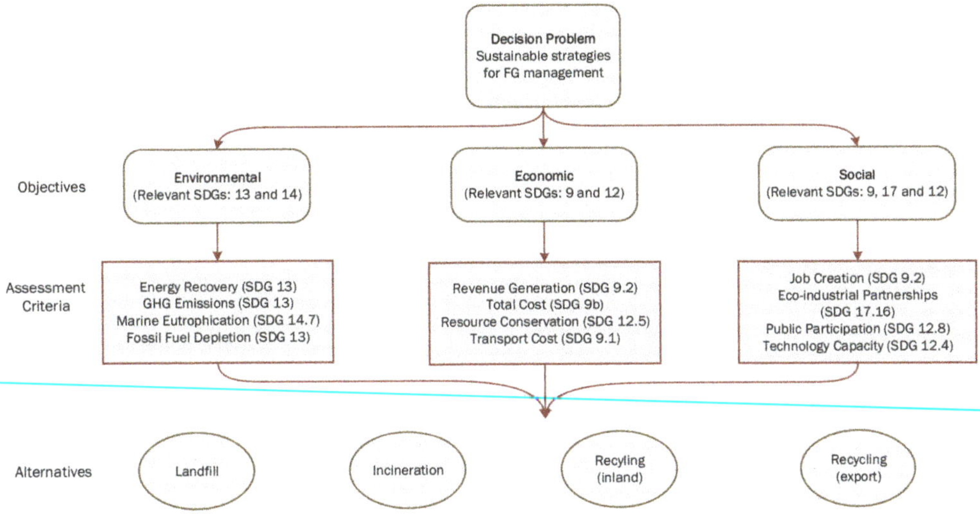

Figure 4. MCDA model for proposed alternative evaluation in selecting sustainable EOL management alternatives for FG and ropes.

The analysis was performed using DECERNS (Decision Evaluation in ComplEx Risk Network Systems) software [40]. The three core criteria, environmental, economic and social, and associated sub-criteria, are chosen through expert judgment. Additionally, the performance of each of the four alternatives against the selected sub-criteria was calculated using data from site visits and personal interviews of waste managers, recyclers. The rationale behind the selection of environmental, economic and social assessment criteria, and their relation to SDGs, data collection and analysis, are detailed in study-2 [28].

The adaptions in MAVT were made for addressing the research questions to facilitate the participatory process and gain flexibility in processing the available information. In a typical MAVT study, stakeholder opinion, qualitative and quantitative information is processed to find the "best" or "most-suitable" solution [41]. Hence MAVT was chosen to incorporate qualitative and quantitative information obtained from the expert stakeholders.

The results of MAVT decision-making analysis are detailed in study-2, which strongly suggest the importance of the location for recycling waste. The recycling operations shows the potentially maximum positive effects on the environment and society with additional economic benefits from resource conservation and energy recovery. In contrast, the overall sustainability scores were lowest for exporting recyclable FG waste material out of the country [28].

These decisions helped to determine the sustainable alternative for managing plastic resources from commercial FG in Norway. Adaption and application of MAVT provided flexibility in using available information in the assessment. Stakeholders are the key source of inputs throughout this research. In the MAVT study, 31 expert stakeholders participated, consisting of regional waste managers, recyclers, collectors, NGOs, academic experts and regulatory agencies. Involving these actors within the MAVT framework enhanced deliberated discussions on several aspects of EOL management that helped build a holistic understanding of EOL strategies during the decision-making process. Building on Analysis and Decision-making, the SPADE method's evaluation activity aims to encourage a continual evaluation of employing sustainable alternatives.

4.5. Continuous Evaluation (E)

Evaluation as an explicit activity of the SPADE method offers the opportunity to reflect on the progress of the research and the findings. The data collection steps applied for the previous studies yielded information for a comprehensive understanding of the system life cycle for commercial FG. The literature review on existing policy instruments (presented in SI) and the current status of system performance studied through the earlier steps were used to derive the elaborate evaluation strategies for improvements in system performance.

In improving system performance, sustainability and resource conservation following the circular economy principles are prioritized. CE is primarily at the forefront in Norway as the first analysis of the Circularity Gap Report Norway 2020 reveals that 97.6% of materials consumed each year in Norway never return to the Norwegian economy [42]. At only 2.4% circular rate, Norway has enormous potential and urgent need to explore various strategies to improve circularity within the region. CE is a priority of the European Green Deal through the *Circular Economy Action Plan* adopted in March 2020. Since the EU is among Norway's key trading partner, adopting CE principles is strategically important for Norway.

Additionally, CE strategies provide a new incentive by promoting local eco-industrial networks, creating new jobs and improving national self-reliance by recycling/reusing waste material. Therefore, the continuous evaluation of problems identified across system life cycle phases of commercial FG is an essential component to ensure follow-up of CE and SDG targets.

Based on information obtained from literature, site visits and interaction with system stakeholders for the data collection phase of earlier studies, a continuous evaluation strategy is proposed for enabling principles of CE and sustainability within the life cycle of commercial FG. Figure 5 presents a brief overview of evaluation strategies across the system life cycle of FG resources.

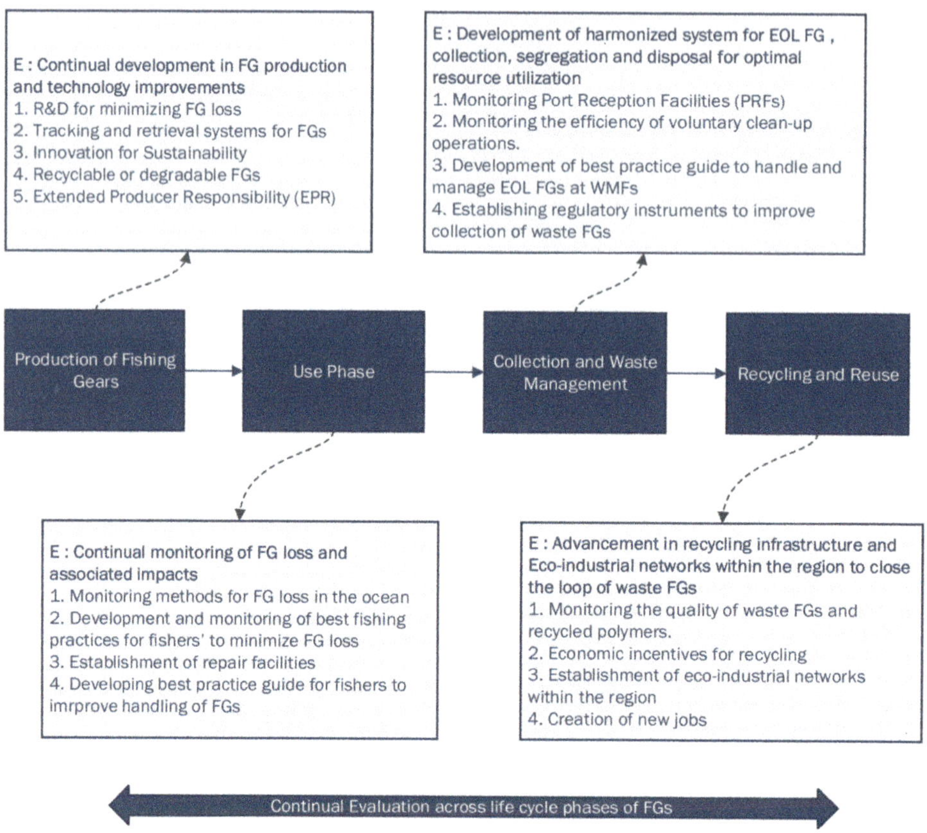

Figure 5. Continuous evaluation strategies suggested across the system life cycle phases of fishing gear resources.

4.5.1. Production Phase

Continuous evaluation is one way to improve the FG production material and technology. The research on alternative raw materials for FGs must be emphasized. The principles of eco-design, design for recycling and design for sustainability may contribute to create FG from material that allows efficient and profitable recovery upon end-of-life without hampering the product's effectiveness. In its current strategy, the EU invites innovation and business solutions across the member states to facilitate the transition towards a circular economy with a particular focus on marine plastic waste from FG [35]. Accordingly, the strategies such as gear marking, recycle friendly gear design and biodegradable FG should be explored further and monitored continuously to ensure the sustainability of improvement strategies.

Extended producer responsibility (EPR) is considered an efficient strategy to hold the manufacturer/producer responsible for the end-of-life treatment of their products, including FG. In effect, this removes the inconvenience and cost factors associated with waste management from the fishers. By linking the producer to the product's EOL stage, the scheme can also indirectly encourage more life cycle focused product design [43]. These schemes can also help trigger infrastructure development to support the EOL collection process. In Norway, the Ministry of Climate and Environment has announced its goal to introduce a producer responsibility scheme for fishing and discarding marine equipment from the aquaculture industry [44]. Together with EPR, gear marking or gear identification is identified as an essential strategy for responsible fishing and for controlling the ALDFG problem. The Fisheries Department of the FAO published systematic guidelines encouraging member states to incorporate gear marking policies. According to these guidelines, gear marking aids in understanding the location, scale and nature of FG in the water [12].

Careful introduction and incorporation of these strategies may improve the state of circularity in the system of FG resources. However, continuous monitoring is essential, as advised by SPADE, to ensure successful implementation and performance improvement.

4.5.2. Use Phase

The science of resource management involves generating a systemic understanding of the processes that lead to improvements in or deterioration of natural or anthropogenic resources. For the system of FG resources, management strategies are based mainly on highly segregated, generally outdated, non-uniform and unscientific estimates on FG life cycle phases resulting in an overall absence of system information. Therefore, to evaluate the system performance, continuous monitoring of flows and stocks of plastics throughout the life cycle phases of FG is essential. In order to establish scientifically sound mapping and monitoring systems for macro and microplastics in the marine environment, internationally agreed definitions, standardized quantitative methods and regionally suitable indicators are needed. Several European states are taking coordinated action to develop a common standard for analyzing and mapping microplastics and investigating the impacts of microplastics on the marine environment and seafood. These monitoring strategies need continual evaluation to understand better the fate and transport of plastics in the environment.

Additionally, ghost fishing's environmental, economic and social impacts must be studied and communicated to the system stakeholders. Fishers and other resource users are the critical stakeholders in the use phase of FG. Consequently, fishers' knowledge of fishing grounds and responsible gear use must be documented to develop best practice guides for young fishers to minimize FG loss or prolong the operational life span of FG. The performance of selected improvement strategies and collected information through monitoring must be evaluated continually to ensure the FG system's betterment.

4.5.3. End-of-Life Collection

The analysis (study-1) proved that in Norway, the logistics transporting EOL FG to respective recycling industries are immature or nonexistent [28]. On the contrary, findings from study-1 show the dominance of transboundary export of EOL FG from Norway, facilitated by the availability of organized actors collecting, segregating and transporting recyclable fractions of plastic FGs out of Norway for further treatment. Therefore, a continuous evaluation is needed towards harmonizing and regulating the network of actors responsible for EOL handling and management of waste FG. Additionally, monitoring the status of PRF is vital to improving the collection of waste gears. The recent judgment by the EFTA court highlighted Norway's apparent failure in fulfilling the obligations of establishing PRF across all the ports [34]. The gigantic coastline and numerous ports put an enormous burden on regulatory actors in managing the landing sites. Harmonized collection of waste FG is imperative for improving the material recovery in Norway, therefore, updating the PRF infrastructure is an urgent need for avoiding material loss.

Apart from the lack of collection facilities, material loss occurs within the WMF as well. The collected waste FG are often laden with dirt, rotten biomass and needs pre-treatment (segregation, cleaning, cutting) before proceeding further with material recovery. Intricate gear designs and presence of metal parts makes cutting and segregating recyclable plastic even more difficult. However, personal communication with a regional waste manager revealed that biomass laden waste is often subjected to incineration or landfills and completely bypasses any opportunity for material recovery. Minor changes in the waste handling practices at WMF and developing best-practice guides for handling the FG related waste could improve the recycling of FG in Norway.

4.5.4. Closing the Loop of Waste from the Fishing Sector

Previous research shows that commercial fishing practices generate an estimated 4000 tons of waste plastics from EOL FG collected at WMF in Norway. In addition to the commercial fishery, leisure fishing, fish farming, and inland fishing generate similar plastic composite material accessible for recycling. In addition, mechanical and chemical recycling technologies are available to recover material from waste FG. Mechanical recycling of waste FG generates HDPE and LDPE polymers that can be used as a raw material in the plastic products manufactured through injection molding technology. Pilot tests are underway to assess the feasibility of replacing virgin plastic polymers with recycled plastic pellets of HDPE and LDPE in the production of brackets and walkways for the aquaculture sector [45].

Positive results from the pilot test could boost the opportunity for realizing circular business models in Norway. Such product-to-product recycling may also reduce the dependence of plastic industries on foreign actors responsible for the supply of virgin polymers. Regional recovery of plastic polymers may provide flexibility in the supply chain, and additional positive social impetus by creating new jobs. In addition, Vildåsen [45] lists cost-cutting and reduced environmental impacts as other factors motivating regional plastic industries to aim for circular strategies.

The semi-structured interviews with regional recyclers pin-pointed the ambiguity in Norway's waste regulation in Chapter 9 [46], which states that *"All waste must be treated before landfilling, and landfilling is allowed if the processing and treatment of waste fraction are socio-economically non-viable."* According to the regional recyclers, the latter half of the statement results in landfilling as a preferred alternative over recycling because the lack of segregation and transport facilities for EOL FG makes it economically burdensome to recycle.

The availability of waste material as a resource for the technology to recycle plastic contents of FG indicated the opportunity to realize the circularity for the FG in Norway. However, establishing an eco-industrial network between the fishing and plastic industries demands assurance of quality and quantity of the eventual recycled polymers. Changes in policy drivers for waste management are necessary to promote material recovery over landfilling. Stakeholder awareness and involvement are vital in raising the demand for environmentally friendly products. Stabilizing all the factors may influence the market acceptance of products with recycled polymers and may result in elevated demands for such products.

Finally, a constant dialogue with system stakeholders, regulatory support through visionary policy instruments, research to advance technological feasibility and reduce operational challenges of FG recycling are among the areas that need continual evaluation to realize the sustainable management of FG resources in the Norwegian context.

5. Reflection and Conclusions

5.1. *Reflections on SPADE Method*

In this paper, we have explored an application of a SE framework, SPADE, to a case of the Norwegian commercial fishery sector, intending to develop strategies for the sustainable handling and life cycle management of FG resources. The science of resource management demands a transdisciplinary approach for studying complex socioecological systems. Consequently, a systems approach provides holistic coverage as it replaces the notion of resources as discrete entities in isolation from the rest of the ecosystem and social system. Here, the SPADE methodology was adapted to facilitate the problem-driven, interdisciplinary research to comprehend the sustainability dimensions in managing a system of FG resources in Norway.

The case study of FG resources is unique. There is an overall lack of scientific information to steer the meaningful policy or technological inputs for managing the potential plastic resources from FG. Qualitative research methods such as structured, semi-structured interviews, questionnaires and focus group workshops were used throughout the study period (2017–2020) to involve stakeholders. Involving stakeholders so early in the study helped build a dialogue with the actors, which eventually led to capturing several unnoticed or undocumented aspects of FG resources in Norway.

Although resource users develop a comprehensive knowledge of the resources they use/consume and associated environments, this knowledge is rarely collected systematically and therefore cannot be used scientifically. SPADE's application provided a scientific template to collect such knowledge in highly structured formats, resulting in the collection of large amounts of information on the FG resources and identification of associated problems. The data captured in the earliest phases also helped define the management problems for FG across their life cycle processes.

The continual interactions with several stakeholder groups further helped gain quantitative information on stocks and plastics flows from commercial FG in Norway. The mass of plastic flows across the system life cycle of FG provided a critical science and technology contribution for the Environmental Directorate and the Fishery Directorate of Norway, supporting the formulation of policies to monitor and abate the plastic pollution from the commercial fishing sector. Additionally, the reported annual quantities of plastic waste collected in the end-of-life stage are considered vital evidence for regional recyclers and waste managers looking for reasons for closing the material loop from FG resources in Norway. Simultaneously, the MAVT study results highlighted the improvement potential in realizing sustainable management's overall goal derived from the global framework of CE and SDGs.

Continual involvement of stakeholders' opinions in all the stages also allowed researchers (authors) to build a solid practical knowledge base for suggesting management strategies. Although proposed outcomes are limited to the case of Norway's commercial fishing sector, the knowledge can be adapted and exchanged with similar ecosystems elsewhere.

In retrospect, SPADE's structured simplicity and flexibility became increasingly valuable in selecting research methods to achieve a common goal. Nonetheless, it took four years (2017–2020) to address and apply the entire SPADE methodology, providing holistic coverage of all the life cycle processes of FG resources and enabling associated learning. Additionally, SPADE's logical and iterative workflow allowed the co-existence of qualitative and quantitative research methods throughout the four-years without hindering the achievement of the study's overall goal.

5.2. Integrating Sustainable Development Goals

In this study, the UN's Sustainable Development Goals and the EU's Circular Economy framework were used to define and outline the potential for improvement in the management of FG resources in Norway. Given the all-inclusive scope of the SDGs, inputs from science are deemed significant for policymakers to assess the economic, social and environmental implications of their strategies in an integrated way over the long term. Here, adaptations in the SPADE method were made to facilitate synergy between the SDGs and policymakers. The synergy is illustrated by identifying relevant SDGs, targets and further developing a set of assessment criteria to monitor and ensure sustainable management of FG resources. In study-2, the sustainable decision making at the fourth step was achieved through linking the assessment criteria to SDGs in the MAVT method. Application of relevant SDGs and targets aided in producing a focused, measurable and all-encompassing coverage of sustainability's triple-bottom-line aspects.

Additionally, the SDGs ensured better communication and understanding of the criteria, as stakeholders were familiar with the goals. Unlike the SDGs predecessor, the Millennium Development Goals (MDG), the SDGs explicitly focus on businesses to apply their creativity and innovation to solve sustainable development challenges. The elaborated targets and set of indicators to assess the growth of selected SDGs were proven helpful in devising strategies in line with global sustainability priorities.

5.3. Limitations and Application

The SPADE method proposed in this paper helped translate and operationalize the SDGs in managing FG resources in the fishing sector; however, this is not necessarily the only approach applicable for other sectors. Selections of relevant SDGs, targets and following assessment criteria are essentially subjective and might vary while reproducing the research elsewhere. Moreover, the current study's scope focuses on the recovery and management of plastics from fishing gears, excluding other marine litter sources, while developing a set of improvement strategies. Additional analysis and deliberations with experts in marine ecology must be organized to create sustainable management of other marine litter sources in Norway and the planet. Furthermore, the case study only illustrates possible connections between SDGs, targets, indicators and improvement strategies. System improvement is a continuous process that involves the active participation of a range of stakeholders. It also demands a strong political will to transfer scientific findings into policy instruments followed by constant monitoring of implemented strategies. The cyclic and iterative steps needed for continual system improvement are a vital part of SPADE, yet, maintaining and monitoring the iterative progress, in reality, may be challenging.

Lastly, the stepwise SE framework presented here advocates the need to incorporate stakeholders across all the life cycle stages of a resource to ensure the sustainable management of anthropogenic resources under consideration. The adapted SPADE method is transferable, especially in the cases of data-less or data-limited resource management. Nevertheless, it is essential to apply it in another context with local adaptations to validate its robustness. As more evidence is required to create informed strategies for managing FG across all the fishing nations, the proposed SPADE model is currently being applied by the author(s) to understand plastic pollution across European waters.

Supplementary Materials: The following are available online at https://www.mdpi.com/article/10.3390/su13094914/s1, A separate Supplementary Information File (SI) is available with this paper.

Author Contributions: Writing, Conceptualization and case study analysis, P.C.D. Method and review, C.H. All authors have read and agreed to the published version of the manuscript.

Funding: This research received no external funding.

Institutional Review Board Statement: Not Applicable.

Informed Consent Statement: Not Applicable.

Data Availability Statement: The data presented in this study are available on request from the corresponding author.

Acknowledgments: Authors kindly thank all the stakeholders who actively participated in the data collection and validation step throughout the study. Also, a special thanks to Gaspard Philis, a researcher at NTNU for assisting in data collection.

Conflicts of Interest: The authors declare no conflict of interest.

References

1. Ryan, C. Learning from a Decade (or so) of Eco-Design Experience, Part II: Advancing the Practice of Product Eco-Design. *J. Ind. Ecol.* **2004**, *8*, 3–5. [CrossRef]
2. Stø, E.; Throne-Holst, H.; Strandbakken, P.; Vittersø, G. Review: A multi-dimensional approach to the study of consumption in modern societies and the potential for radical sustainable changes. In *System Innovation for Sustainability*; Greenleaf Publishing Ltd.: London, UK, 2008. [CrossRef]
3. Dietz, T.; Ostrom, E.; Stern, P.C. The Struggle to Govern the Commons. *Science* **2003**, *302*, 1907–1912. [CrossRef]
4. Young, O.R. *The Institutional Dimensions of Environmental Change: Fit, Interplay, and Scale*; MIT Press: Cambridge, MA, USA, 2002.
5. Hardin, G. The tragedy of the commons. *Science* **1968**, *162*, 1243–1248. [PubMed]
6. Johannes, R. The case for data-less marine resource management: Examples from tropical nearshore finfisheries. *Trends Ecol. Evol.* **1998**, *13*, 243–246. [CrossRef]
7. Ostrom, E. A General Framework for Analyzing Sustainability of Social-Ecological Systems. *Science* **2009**, *325*, 419–422. [CrossRef] [PubMed]
8. Lawson, R. Mini-facts about Norway. In *Affairs*; Statistics Norway's Information Centre: Oslo, Norway, 2015.
9. Fisheries and Aquaculture Department. Fishery and Aquaculture Country Profiles. Norway [Internet]. 2013. Available online: http://www.fao.org/fishery/facp/NOR/en (accessed on 7 May 2018).
10. Deshpande, P.C.; Brattebø, H.; Fet, A.M. A method to extract fishers' knowledge (FK) to generate evidence for sustainable management of fishing gears. *MethodsX* **2019**, *6*, 1044–1053. [CrossRef] [PubMed]
11. Fiskeridirektoratet. Norwegian fishing vessels, fishermen and licenses. Available online: https://www.fiskeridir.no/English/Fisheries/Statistics/Fishermen-fishing-vessels-and-licenses (accessed on 3 April 2019).
12. Food and Agriculture Organization. *Report of the Expert Consultation on The Marking of Fishing Gear, Report No. 978-92-5-109275-0 Contract No. FIAO/R1157*; Food and Agriculture Organization of The United Nations: Rome, Italy, 2016.
13. Brown, J.; Macfadyen, G. Ghost fishing in European waters: Impacts and management responses. *Mar. Policy* **2007**, *31*, 488–504. [CrossRef]
14. Deshpande, P.C.; Philis, G.; Brattebø, H.; Fet, A.M. Using Material Flow Analysis (MFA) to generate the evidence on plastic waste management from commercial fishing gears in Norway. *Resour. Conserv. Recycl. X* **2020**, *5*, 100024. [CrossRef]
15. Laist, D.W. Impacts of Marine Debris: Entanglement of Marine Life in Marine Debris Including a Comprehensive List of Species with Entanglement and Ingestion Records. In *Marine Debris*; Springer: Berlin/Heidelberg, Germany, 1997; pp. 99–139.
16. Graeme Macfadyen THaRC. *Abandoned, Lost or Otherwise Discarded Fishing Gear*; Programme Food and Agriculture Organization of the United Nations: Rome, Italy, 2009; ISBN 978-92-5-106196-1.
17. Derraik, J.G. The pollution of the marine environment by plastic debris: A review. *Mar. Pollut. Bull.* **2002**, *44*, 842–852. [CrossRef]
18. Gilman, E. Status of international monitoring and management of abandoned, lost and discarded fishing gear and ghost fishing. *Mar. Policy* **2015**, *60*, 225–239. [CrossRef]
19. Smolowitz, R.J. Trap design and ghost fishing: An overview. *Mar. Fish Rev.* **1978**, *40*, 2–8.
20. Deshpande, P.C.; Aspen, D.M. A Framework to Conceptualize Sustainable Development Goals for Fishing Gear Resource Management. In *Handbook of Sustainability Science and Research*; Filho, L.W., Ed.; Springer International Publishing: Cham, Switzerland, 2018; pp. 727–744.
21. Blanchard, B.S.; Fabrycky, W.J.; Fabrycky, W.J. *Systems Engineering and Analysis*; Prentice Hall: Englewood Cliffs, NJ, USA, 1990.
22. Fet, A.M. Systems engineering methods and environmental life cycle performance within ship industry. *Dr. Ingeniøravhandling* **1997**, *21*.
23. Deshpande, P.C. Systems Engineering for Sustainability in the Life Cycle Management of Commercial Fishing Gears. Ph.D. Thesis, Norwegian University of Science and Technology, Faculty of Economics and Management, Trondheim, Norway, 2020.
24. Haskins, C. Systems Engineering Analyzed, Synthesized, and Applied to Sustainable Industrial Park Development. Ph.D. Thesis, Norwegian University of Science and Technology, Department of Industrial Economics and Technology Management, Trondheim, Norway, 2008.
25. Aspen, D.M.; Haskins, C.; Fet, A.M. Application of systems engineering to structuring acquisition decisions for marine emission reduction technologies. *Syst. Eng.* **2018**, *21*, 388–397. [CrossRef]
26. Haskins, C. Multidisciplinary investigation of eco-industrial parks. *Syst. Eng.* **2006**, *9*, 313–330. [CrossRef]

27. Shainee, M.; Haskins, C.; Ellingsen, H.; Leira, B.J. Designing offshore fish cages using systems engineering principles. *Syst. Eng.* **2012**, *15*, 396–406. [CrossRef]
28. Deshpande, P.C.; Skaar, C.; Brattebø, H.; Fet, A.M. Multi-criteria decision analysis (MCDA) method for assessing the sustainability of end-of-life alternatives for waste plastics: A case study of Norway. *Sci. Total Environ.* **2020**, *719*, 137353. [CrossRef]
29. Hong, S.; Lee, J.; Lim, S. Navigational threats by derelict fishing gear to navy ships in the Korean seas. *Mar. Pollut. Bull.* **2017**, *119*, 100–105. [CrossRef]
30. Ministry of Fisheries and Coastal Affairs, Norway. Norwegian Marine Resources Act. 2008. Available online: https://www.regjeringen.no/globalassets/upload/fkd/vedlegg/diverse/2010/marineresourcesact.pdf (accessed on 20 January 2021).
31. Jambeck, J.R.; Geyer, R.; Wilcox, C.; Siegler, T.R.; Perryman, M.; Andrady, A.; Narayan, R.; Law, K.L. Plastic waste inputs from land into the ocean. *Science* **2015**, *347*, 768–771. [CrossRef]
32. Ministry of Climate and Environment, Norway. The Pollution Control Act. 1981. Available online: https://lovdata.no/dokument/NL/lov/1981-03-13-6 (accessed on 12 February 2021).
33. European Commission. *Directive of the European Parliament and of the Council on Port Reception Facilities for the Delivery of Waste from Ships, Repealing Directive 2000/59/EC and Amending Directive 2009/16/EC and Directive 2010/65/EU*; European Commission, Directorate-General for Mobility and Transport: Strasbourg, France, 2018.
34. EFTA Surveillance Authority v The Kingdom of Norway (Failure by an EFTA State to Fulfill its Obligations—Directive 2000/59/EC on Port Reception Facilities for Ship-Generated Waste and Cargo Residues): Hearing Before the European Free Trade Association (EFTA), European Union Law: Case E-35/15 Sess. Available online: https://eftacourt.int/download/35-15-judgment/?wpdmdl=1559 (accessed on 12 March 2018).
35. Directorate-General for Environment, European Commission. *Reuse, Recycling and Marine Litter: Final Report—Study. Report No. KH-04-18-802-EN-N Contract No. 978-92-79-93917-4*; Directorate-General for Environment, European Commission, Environment D-Gf: Brussels, Belgium, 2018.
36. Gu, F.; Guo, J.; Zhang, W.; Summers, P.A.; Hall, P. From waste plastics to industrial raw materials: A life cycle assessment of mechanical plastic recycling practice based on a real-world case study. *Sci. Total Environ.* **2017**, *601*, 1192–2207. [CrossRef]
37. Allesch, A.; Brunner, P.H. Material Flow Analysis as a Tool to improve Waste Management Systems: The Case of Austria. *Environ. Sci. Technol.* **2017**, *51*, 540–551. [CrossRef]
38. Elkington, J. Partnerships from cannibals with forks: The triple bottom line of 21st-century business. *Environ. Qual. Manag.* **1998**, *8*, 37–51. [CrossRef]
39. Costanza, R.; Daly, L.; Fioramonti, L.; Giovannini, E.; Kubiszewski, I.; Mortensen, L.F.; Pickett, K.E.; Ragnarsdottir, K.V.; De Vogli, R.; Wilkinson, R. Modelling and measuring sustainable wellbeing in connection with the UN Sustainable Development Goals. *Ecol. Econ.* **2016**, *130*, 350–355. [CrossRef]
40. Yatsalo, B.; Gritsyuk, S.; Sullivan, T.; Trump, B.; Linkov, I. Multi-criteria risk management with the use of DecernsMCDA: Methods and case studies. *Environ. Syst. Decis.* **2016**, *36*, 266–276. [CrossRef]
41. Van Herwijnen, M. *Multi-Attribute Value Theory (MAVT)*; Vrije-Universiteit: Amsterdam, The Netherlands, 2010.
42. Initiative CGR. Circularity Gap Report, Circle Economy and Circular Norway, Oslo, Norway. Available online: https://www.circularnorway.no/gap-report-norway (accessed on 25 February 2021).
43. Sherrington, C.; Darrah, C.; Hann, S.; Cole, G.; Corbin, M. *Study to Support the Development of Measures to Combat a Range of Marine Litter Sources*; Report for European Commission; DG Environment: Bristol, UK, 2016.
44. Sundt, P.; Briedis, R.; Skogesal, O.; Standal, E.; Johnsen, H.R.; Schulze, P.E. Underlag for å utrede produsentansvarsordning for fiskeri og akvakulturnæringen. Available online: https://www.miljodirektoratet.no/publikasjoner/2018/mai-2018/underlag-for-a-utrede-produsent-ansvarsordning-for-fiskeri--og-akvakulturnaringen/ (accessed on 22 April 2021).
45. Vildåsen, S.S. Lessons learned from practice when developing a circular business model. In *Designing for the Circular Economy*; Charter, M., Ed.; Routledge Publishers: New York, NY, USA, 2018.
46. The Ministry of Climate and Environment, Norway. Available online: https://lovdata.no/dokument/SF/forskrift/2004-06-01-930 (accessed on 18 January 2021).

sustainability

Article

Merging Systems Thinking with Entrepreneurship: Shifting Students' Mindsets towards Crafting a More Sustainable Future

Matthew Lynch [1,2,*], Gunnar Andersson [1] and Frode Ramstad Johansen [1]

[1] Department of Engineering, Østfold University College, 1613 Fredrikstad, Norway; gunnar.andersson@hiof.no (G.A.); frode.r.johansen@hiof.no (F.R.J.)
[2] School of Economics and Business, Norwegian University of Life Sciences, 1430 Ås, Norway
* Correspondence: matthewl@hiof.no

Abstract: The major challenges confronting humanity are systemic in nature: climate change, pollution, poverty, and inequality. Entrepreneurship fails to tackle these challenges, and 'creative destruction' is mostly just leading to the destruction of the natural world that we inhabit. The present economic, financial, and productive systems can and should be transformed to lead and power a shift towards sustainability. If we are to reverse the course of destruction that current capitalist systems are creating, we need to introduce more of a systems perspective into entrepreneurial education. This article addresses how merging systems thinking and entrepreneurship can be used to nudge students towards sustainability. Through a single case study, we argue that a practice-based pedagogy that combines perspectives from entrepreneurship and systems thinking can be used as a catalyst to bring about local changes in business models by making the business case go beyond the individual organization and seeing entrepreneurship as being about creating more sustainable business systems.

Keywords: systems thinking; entrepreneurship; sustainability; experiential learning

Citation: Lynch, M.; Andersson, G.; Johansen, F.R. Merging Systems Thinking with Entrepreneurship: Shifting Students' Mindsets towards Crafting a More Sustainable Future. *Sustainability* **2021**, *13*, 4946. https://doi.org/10.3390/su13094946

Academic Editors: Cecilia Haskins and Marc A. Rosen

Received: 30 March 2021
Accepted: 24 April 2021
Published: 28 April 2021

1. Introduction

The world faces many challenges—climate change, plastic pollution, and inequality, to name a few [1–3]. These challenges are systemic in nature, and are made up of a complex interconnecting web of stakeholders with differing interests and incentives [4]. These are the types of issues to which systems thinking has traditionally been applied [5]. Resolving these challenges will therefore require a systemic solution. The present economic, financial, and productive systems can and should be transformed to lead and power a shift towards sustainability [1]. The question then becomes how to do this.

Traditionally, entrepreneurship has played a key role in transforming markets through the introduction of innovative products and services [6]. The forefather of entrepreneurship as a field, Joseph Schumpeter, famously described this change making process as the perennial gale of creative destruction, as entrepreneurs and firms continue to act entrepreneurially through innovation in search of sustainable profitability [7,8]. While much has been written about the term, at the core of the message is that entrepreneurs are the catalyst that bring about the destruction of old economic patterns through the creation of new enterprises [7]. Yet, if we review entrepreneurial progress in the last decades, there might be cause for concern [2]. The largest successful entrepreneurial firms such as Facebook, Apple, Tencent, Alibaba, Microsoft, etc., have amassed huge amounts of wealth, symbolizing their popularity and their success as economic organizations. Yet, none of them have as a primary or even secondary focus the resolution of the serious social issues facing the world. With a certain amount of greenwashing involved, there might be arguments made for the good they are having on the world, although we would posit that this is merely lip service [9]. It seems the perennial gale of creative destruction is mostly leading to the destruction of the natural world that we inhabit.

There is a growing recognition of the need to change the way the current economic system operates in order to resolve the serious issues facing the world [1]. In referring to the current economic systems and capitalism in general, we acknowledge there are variations in capitalism across regions and countries. We use the terms economic system and capitalism interchangeably to refer to the largely neoliberal markets of developed countries, such as those in western Europe and North America, although we do not consider the issues mentioned to be solely with these types of capitalism. Capitalism has many benefits, as evidenced by its ability to lift more than a billion people out of poverty at the start of the millennium [10]. Yet, even the earliest thinkers around capitalism saw the issues with unbridled capitalism, with Adam Smith and Karl Marx discussing accumulation of wealth (inequality) and externalities (pollution and destruction of nature) as being logical outcomes of the system. While capitalism has changed its form considerably since these thinkers raised their concerns, the issues they mention are as relevant today.

As is often quoted in systems thinking, a system operates exactly as it is designed to, and the issues in the world we are currently seeing are the clear result of the economic system being designed to create such issues [11]. Therefore, if we want our economic systems to behave differently, then we have to begin to design them differently.

One of the greatest points of leverage for changing a system is to focus on changing the mindsets of those involved in the system [12]. If we are to reverse the course of destruction that current capitalist systems are creating, and instead replace it with something more sustainable, then we argue that we need to introduce more of a systems perspective into entrepreneurial education [13]. This will in turn transform students' mindsets to having a more holistic view of the role of entrepreneurship in creating a sustainable future. We use a teaching case to demonstrate what the merging of these perspectives would look like, in the hope that inspiring a shift in mindsets will lead students towards being catalysts and creators of new more sustainable economic systems [14]. The case demonstrates how a practice-based pedagogy combines these perspectives as an initial step towards bringing about local changes in business models by making the business case go beyond the individual organization.

In order to argue for the combination of entrepreneurship with systems thinking, we begin with a brief overview of both systems theory perspectives and entrepreneurship. We then provide a merged theoretical perspective of the two fields, before demonstrating through a case study of how this can be taught as a practical hands on pedagogy. Lastly, we follow this with a reflection on the challenges and limitations of such an approach, along with suggestions for future research.

2. Systems Thinking and Entrepreneurship

As Meadows pointed out, "hunger, poverty, environmental degradation, economic instability, unemployment, chronic disease, drug addiction, and war, for example, persist in spite of the analytical ability and technical brilliance that have been directed toward eradicating them" [15] (p. 5). These are systems problems that, as she addressed, will not disappear until someone finds "the courage and wisdom" to restructure these systems.

Systems thinking is rooted in the understanding that society is so complex that linear ways of thinking and analyzing problems are insufficient and even counterproductive. Systems thinking has been developed based on the understanding that it is not enough to study only parts and processes in isolation; instead, there is a need to solve problems while considering the dynamic interaction of parts [16,17].

> *Similar general conceptions and viewpoints have evolved in various disciplines of modern science. While in the past, science tried to explain observable phenomena by reducing them to an interplay of elementary units investigable independently of each other, conceptions appear in contemporary science that are concerned with what is somewhat vaguely termed "wholeness"* [16] (pp. 36–37).

Von Bertalanffy [16] describes how systems thinking is a collection of methods, concepts, and techniques that are used to understand a phenomena where there are strong interactions between different parts that together make up complex systems like those in entrepreneurial studies and sustainable development.

Classic systems thinking literature focuses on balancing forces and counter forces in order to uphold an equilibrium, which adequately explains how the current social issues already mentioned have come to be. Newer literature, such as Cabrera [18,19], point towards a better understanding of the complexity in the models and thus developing more enhanced mental models (we use mental models refer to 'deeply held internal images of how the world works' [11] (p. 174)). The challenge lies in the fact that when we change a system, it is difficult to predict the effects that will result from those changes. A singular change can have flow on effects, leading to second, third, and nth order consequences [20,21]. This is not something readily discussed in entrepreneurship literature, meaning entrepreneurs' mental models are not well equipped to see the consequences of any new service or product that they may launch. Entrepreneurship as a field does not train nascent entrepreneurs to see the totality of their choices.

Entrepreneurship may be understood as a system or network of interconnected actors, intimately related to today's complex societal challenges like sustainability. Yet, this is a topic that is rarely discussed within teaching environments. A systemic literature review of entrepreneurship education [22] makes no mention of 'sustainability' and only a passing mention of social entrepreneurship, while a more recent article [23] focusing on the role of UK entrepreneurship centers in higher education uses the word 'sustainability' only in relation to 'financial sustainability'. A surface review of three core journals focusing purely on entrepreneurship (Entrepreneurship Theory & Practice, Journal of Business Venturing, and the Journal of Small Business Management) reveals that since 2010, there have collectively been 38 articles with the word 'sustainability' in the title. We are conscious that a more thorough review would have revealed more articles, and that there are smaller niche fields such as 'sustainopreneurship' springing up (only 165 results in google scholar since 2010), but this term is totally absent from articles published in the three key journals mentioned. We do not mean to imply there is no literature linking sustainability and entrepreneurship. The point is that sustainability has not been part of the core discussion occurring within mainstream entrepreneurship journals and has traditionally been absent from the curriculum being taught to students of entrepreneurship [24].

However, attitudes have been rapidly changing, with new programs sprouting up that focus on the intersection of entrepreneurship and sustainability [25]. The point is that if we want students to begin to see the environmental consequences of the economic choices they make, then we need to begin to train entrepreneurs to think differently and more systemically [26], or at least to think more widely in their perspective. A systemic perspective on entrepreneurship indicates a need to explore and discuss more of what relates to or affects the whole of the system the entrepreneur is targeting [27,28].

The systemic perspective is embedded in the concept of sustainability introduced in the UN World Commission on Environment and Development report 'Our common future' [29], warning of the necessity of making progress toward an economic development that can be sustained without depleting natural resources or harming the environment. The UN Sustainable Development Goals, the Paris agreement, and the two degree scenario (International Energy Agency, 2017) all require policy actions and an array of public and private support across all innovation stages, from strategically directed research & development (R&D) and market creation and technology-specific support towards holistic support and market pull policies and system integration [30] (p. 10). The situation requires an entrepreneurial mindset that grows beyond an economic viable product and company and that embraces a systems-wide approach to change.

The idea of sustainable entrepreneurship is based on a development paradigm that recognizes that entrepreneurship makes an important contribution to environmental, social, and economic development. Yet, it still views the value creation unit at the level of the individual organization [31] and largely ignores the wider systemic implications. As such, it sticks with a more traditionally linear mindset that ignores the kind of 'wholeness' [16] we referred to earlier. Systems thinking can be used to provide a new perspective and tools to broaden the field of influence entrepreneurs consider when innovating.

3. Previously Disconnected Literature and Mental Models

The connection between systems thinking and entrepreneurship is not entirely novel, although literature from the two fields is seldom sighted together. Initial conceptual work examining the connection between social entrepreneurship and systems thinking posited that a lack of understanding of social systems caused many social enterprises to fail [32]. We agree with this point, but note that the authors are still focused on the success of individual organizations and less on the systemic impacts of these organizations and ensuring that those systemic influences are positive.

Social enterprises are those whose main purpose is not just economic gain, but rather that they try to bring about social 'good' at the same time. The rise in popularity of social enterprises has resulted in a plethora of terms such as 'social purpose venture', 'community wealth venture', 'non-profit enterprise', 'venture philanthropy', 'caring capitalism', and 'social enterprise' [33]. Although, not all efforts to be socially responsible have been equally genuine, with many companies carrying out 'pseudo branding' in their attempt to appeal to consumers as socially responsible [32]. We suggest that minor changes in business models towards less damage or mild positive impacts does not go far enough towards supporting sustainability.

While Triverdi and Misera [32] are focused on social entrepreneurship, many of their arguments apply equally to the more mainstream field of entrepreneurship. They criticize academia for conceptual ethnocentrism, whereby traditional business management concepts are given undue weight and lead to mental models that concentrate on personal gain. Indeed, economics encourages people to be wealth-maximizing individuals because it is considered the rational thing to do (although this doctrine has been giving way to more relaxed positions that include space for human 'irrationality'). In pushing these forms of narrative, the result is that there is a tendency in academia and in classrooms to promote the idea of entrepreneurship being closely connected to personal wealth generation.

Yet, studies examining the motivations of entrepreneurs and student entrepreneurs [34,35] have shown that financial rewards are not the main motivators of those with entrepreneurial aspirations. Instead, factors like freedom, challenge, and the pursuit of meaning are themes that are conveyed as motivating nascent entrepreneurs.

Wealth maximizing mental models taught (consciously or unconsciously) to students can cause them to see the value of a forest as being based only on the lumber that it can produce, instead of seeing the inherent value in natural systems. A basic tenet in systems thinking is that optimizing for a singular metric results in a suboptimal performance of the entire system [11,17]. Accepting that businesses are part of economic and environmental systems leads to the conclusion that optimizing for individual performance in individual organizations will come at the expense of the entire system (even if it benefits the individual business owners/entrepreneurs). The current mental models being taught to entrepreneurs contribute to the further destruction of the environment as entrepreneurs strive to optimize the performance of their enterprises.

The result of these mental models has seen the increase in commercial activity, with phrases like hypercompetition being coined [36]. As Schumpeter predicted, as this commercialization speeds up and hypercompetition takes place, it gets harder for firms to maintain a competitive advantage [7]. We do not wish to sound like the American Patent Office in 1899, who said there was nothing left to invent. There is of course still room for further creative destruction in the deliverance of new industries and the creation of more artefacts for consumption. Yet, there is also the question of how much more the natural world can tolerate of this development before triggering a crash of ecosystems.

We should not need to be reminded that it is the economic system that is dependent on the natural world for survival and not the other way round [37]. If we are to continue with our current trajectory of singular focus, then it would stand to reason that we would push our planet to the point that we risk our own destruction. In addition, the hypercompetition referred to puts pressure on margins, meaning there are incentives to choose the cheapest inputs, which often correlate with the most negative environmental outcomes, creating a re-enforcing feedback loop in the destruction of the planet.

It is therefore not sufficient to focus on businesses that have an incremental approach to improving environmental outcomes, although we commend those businesses that endeavor to limit their negative outcomes. Instead, we need to begin to redesign business as we know it to have a benefit to the environment, other businesses, and to multiple stakeholder groups connected or impacted by the business. Hence, there is a clear need to begin to provide aspiring entrepreneurs with new knowledge, such as a focus on 'wholeness'. We argue that theories of systems thinking can be used to shift the mental models of aspiring entrepreneurs towards a more sustainable orientation.

4. Applying Systems Thinking to Entrepreneurship

In aiming to combine the fields of systems thinking, entrepreneurship and sustainability, we encounter an issue at the level at which we are theorizing. Sustainability is an outcome, while systems thinking is a theoretical model. Entrepreneurship is about taking action under uncertainty [38], although is supported by a broader field of theory. We suggest that through shifting mental models of entrepreneurs by including systems thinking theory, we can create sustainable outcomes. Therefore, this section aims to assimilate the two disparate fields of entrepreneurship and systems thinking theory; the choice of which theory elements to join is entirely arbitrary. The decision as to which elements to discuss was based purely on the authors subjective perception of which elements seemed to have a natural cross-over. This is not an exhaustive review of either field, but rather an initial attempt to pair the two fields with the goal being to shift the mental models in aspiring entrepreneurs away from their singular focus on economic benefits.

4.1. Disconnected vs. Interconnectedness—Rather Than Seeing Individual Elements, Instead Focus on the Interconnection

In systems thinking, the focus is on the 'whole' [16], while what is meant by the whole can vary vastly depending on the level of analysis. To use an organizational example, you might consider a whole team, or a whole division or a whole business, or perhaps a whole industry or economy. Applied to entrepreneurship, we might be able to see that the normal unit of analysis is that of the organization that is trying to bring a new product or service to market. Occasionally, this whole might be within the context of an industry and considering the competitive positioning of a new offering and how this might create a product–customer fit.

Applying a more holistic perspective to entrepreneurship might mean seeing the role of a new venture in relation to value networks, at the level of societal impacts, or the roles that the introduction of the new firm has on its environment. These perspectives have already become part of the wider narratives of social entrepreneurship, but are somewhat underrepresented in mainstream entrepreneurship literature. For example, a review of 'Disciplined Entrepreneurship' [39], one of the leading textbooks on entrepreneurship, does not comment on the wider effects of a venture's sustainability or impacts on the environment. Lean Startup only mentions sustainability in the sense of an ever-growing business [40], and the same is true for some other well-known entrepreneurship books, such as the Startup Owner's Manual [41].

4.2. Linear vs. Circular—Focus on Feedback Loops and Reinforcing Nature, Not Just Linear Cause and Effect

Systems thinking has a large focus on feedback loops, reinforcing structures, and thinking about second-, third-, and fourth-order consequences of actions [5]. In this sense, it takes a longer-form perspective and considers how actions will play out over time.

This particular perspective has somewhat of a foothold in entrepreneurship already, however generally from a monopolistic perspective. Examples include the idea of a flywheel as described by Jim Collins, whereby a competitive advantage is gained through creating a self-reinforcing system that builds momentum over time and makes it virtually impossible for your competitors to catch up with you [42]. Alternatively, it is sometimes used in the sense of generating viral momentum through marketing a new venture by ensuring the lifetime value of a customer exceeds the cost of acquiring new customers [39]. In digital industries, the circular nature might be associated with stickiness and attention management, whereby product offerings are designed to retain users and keep them coming back for more (even if this is to the detriment of the customer).

Applying the systems tool in a more constructive way, from the viewpoint of external stakeholders, we could begin to introduce the idea of paying attention to the second-, third-, and fourth- order consequences of ventures and their offerings, in turn encouraging entrepreneurs to think in a wider timeframe that expands beyond sales for the quarter. A well-known negative approach is designing for obsolescence, whereby Apple and Samsung have been prosecuted for designing their phones to be obsolete faster than what they might have otherwise been [43]. This is profitable for the company, but given the raw materials required for the phone, there are few external stakeholders who would think of this as a positive outcome.

Instead, we can encourage entrepreneurs to think about how they integrate their products and services into a more circular offer. Additionally, we might be able to think about encouraging entrepreneurs to design their business models to have positive outcomes on external holders. Again, this has gained some traction in social entrepreneurship [33], but is still far from being normal in mainstream entrepreneurship. In mainstream literature, there is reference to the lifetime value of a customer [39], yet no reference to the lifetime impact of products and services on the environment. We could begin to encourage entrepreneurs to think of the cost of production and disposal on the environment, and not have them just consider the financial costs of production.

4.3. Silos vs. Emergence—Results Emerge from Complex Interrelationships

Complex systems are those that are beyond being able to be analyzed, and therefore are best understood through seeing the patterns that emerge over time [44]. Emergence is the concept that the results of a system emerge from the individual parts of a system, and the interaction that they have with one another. The concept of emergence has a space within the entrepreneurship community already and is the philosophical basis for effectuation [45]—a key theory in the field of entrepreneurship. It says that ventures cannot always be planned for, but rather are the result of differing resources and actors being combined, and with new ventures emerging out of this complex interaction [46].

The normative implication of this is that we cannot always plan our way to a new venture, but rather through making concerted efforts, a new venture can emerge. Additionally, a crucial concept of emergence is that small points of leverage can have big impacts on the final emergent outcome. As such, we need to begin to encourage entrepreneurs to think about the leverage points in their own ventures that can result in a more positive economic system, and combined with the earlier points have them see that their venture is not a singular economic unit, but is part of a bigger whole, and the creation of a new venture can have long term implications for society.

4.4. Parts vs. Whole

A downside of the analytic approach is that we tend to focus on the individual elements. Systems thinking encourages a focus on the whole, and is sometimes explained with the memorable analogy that two halves of a horse are not the same as a horse [11].

In entrepreneurship, we have seen this be applied in the form of 'customer journey mapping' and 'customer lifecycle analysis', where it is not just the purchase transaction that forms the focus, but the entire value offering to the customer [47]. Yet, this lifecycle analysis does not normally take into account disposal of products or environmental externalities of the products sold by a venture. As such, we could encourage nascent entrepreneurs to think more holistically about their value offering. We could encourage them to see that their organization is not a single unit, but to view themselves as a part of a whole network of actors, whose individual actions collectively have large and important impacts.

4.5. *Analysis versus Synthesis—The Idea of Synthesis Is the Creation of Something New through the Combination of Existing Elements*

In its nature, entrepreneurship is synthetic. It focuses on building ventures through the combinations of resources, with the goal being to create something new out of component parts. Synthesis is the lens through which effectuation and design thinking explain how individuals can create ventures.

What is less discussed is the idea of what entrepreneurs 'should be' synthesizing. The underlying (hidden) mental model is that they should be creating business models that focus on value creation for themselves and shareholders. If we shift the focus of entrepreneurs towards synthesizing firms with a broader perspective and a realization that they are part of a whole, then we can begin to shift their mindset towards having an overall more positive impact.

4.6. *Isolation vs. Relationships—Not about the Individual Elements, but Paying Attention to Interrelationships*

Systems thinking conveys that it is not the individual elements that matter, but rather the relationship between them that should be of crucial importance [5]. If we apply this to the context of entrepreneurship, we can begin to re-imagine what an economic business model might look like. Instead of the traditional competitive perspective, where there is a race to create a competitive advantage and beat your competition, we focus on more co-opetition or complimentary business models [48]. In particular, we could have students reimagine what the role of a company is in facilitating positive outcomes for society, so that entrepreneurship could be more focused on creating value networks or clusters of value in small economic micro-systems.

This might sound naive or overly simplistic, however, there is a growing pressure on ventures to find ways to continue to survive in a world of hypercompetition [7,36]. Indeed, evidence suggest that across the board, most firms are finding it harder to maintain an economically sustainable model, and that creative destruction is speeding up the process in which firms are replaced by new entries [7]. Noting that traditional monopolists have been able to maintain their monopolies where is it is uneconomic to build new infrastructures can be seen as aspirational. Perhaps the sustainable economic infrastructures of the future are those value networks that, due to their systemic conceptualization and co-location, have a collective competitive advantage that cannot be easily replicated.

Indeed, this is one of the key features of our case that we will now shift our focus toward discussing in order to show how these six systems perspectives can be actively applied to an entrepreneurial class.

5. Sparking Change in the Next Generation of (Systems?) Entrepreneurs

One of the challenges mentioned so far is that the global economic system of today has reached a status quo that, while obviously problematic, seems unlikely to change of its own free will due to the complex mix of stakeholders and power dynamics. This might seem an uninspiring starting point, but Meadows [12] provides some guidance on where best to intervene in such a system. She highlights 12 points, with the most crucial being that of transcending paradigms. This is a lofty goal, but given the alternative of continued planetary destruction, we suggest it is worth striving for.

However, a more attainable place to begin is at the second most powerful leverage point inside a system, and that is with shifting mindsets of those operating in the system. Meadows points out that change may be brought about by consistently and repeatedly pointing out the failures and assumptions of existing systems to those with an open mind. She discusses the needs to shift social paradigms through questioning assumptions and resetting these collective assumptions into a new socially accepted paradigm.

Education would appear to be a location where mindsets and collective paradigms can be shifted amongst young people with open minds [49]. It seems therefore fitting to focus on education as a tool through which we can begin to implement the leverage suggested by Meadows. A lot has been written in the field of entrepreneurship about shifting mindsets [50–53] and creating entrepreneurial mindsets [50,54–56]. While the exact definition of an entrepreneurial mindset is under debate [51], there is nothing that precludes nudging students towards having a broader environmental perspective when thinking of venture creation. Indeed, the original definition of 'mindset' was described as individuals having their mind set on finding solutions to the challenge at hand [55].

If students are tasked with creating entrepreneurial ventures that take into account a broader system picture, then their mental models will likely have this as a new baseline when they consider entrepreneurship in the future. In order to demonstrate the way students can be encouraged to have this wider perspective, we now turn to a case study.

6. Case Context

To illustrate the use of entrepreneurship from a systems point of view, we choose to highlight one possible method for communicating how to shift students' mindsets to take a more holistic viewpoint. The case is not intended to suggest that this is the best approach to teaching a systems perspective in conjunction with entrepreneurship, but rather one way that appears to have impacted students' awareness of the complexity of entrepreneurship and was experienced as a valuable learning experience for those involved.

Case studies are recognized as a relevant real-world research design [57] and are especially relevant for producing contextual knowledge [58] for maneuvering within complex processes [59,60]. Case studies are also appropriate for contemporary and emergent phenomena in a real-world context, including real-time and retrospect data [61,62].

The case relates to a corporate challenge, whereby a local land developer contacted our University College to inform us that they were planning to build a new business park. The corner stone tenant of this development was to be a giant datacenter. A common issue with datacenters is that the servers inside them need to be cooled in order to operate efficiently, as a result there is excess heat energy that needs to be removed from the datacenter and is commonly just extracted into the local environment in the form of warm air via air conditioning [63]. According to the information provided by the developer, it was expected that 40 megawatts (MW) of excess energy would be expelled into the air each year. The developer was concerned about the impact on the local environment as well as feeling like this heat would be wasted if it was just dispersed into the atmosphere. To put the figure into perspective, an average home in Norway uses 16,000 kilowatt-hours (kWh) (https://energifaktanorge.no/en/norsk-energibruk/energibruken-i-ulike-sektorer/ (accessed on 27 April 2021)), meaning the excess waste from the datacenter would be equivalent to the power usage of 2500 homes.

From an economic perspective, the development of the datacenter was profitable based on current economic conditions regardless of what happened with this excess heat. However, the developer pointed out that this heat seemed like an underutilized resource and was curious whether there was an opportunity to find a potential upside. In an initial meeting, we decided to see if we could use first year bachelor students to find potential opportunities for this heat resource. From a technical perspective, there is a clear solution as to how to use the heat, so the challenge is not one of technology. The heat can easily be transferred out of the datacenter via water cooling systems, and used to heat nearby homes and businesses. There are multiple examples of such solutions in the region. However, the heat in this case went far beyond what any nearby businesses would need in regards to normal heating. The distance between the business park and any nearby housing meant that building a pipeline to transfer the warm water was instantly ruled uneconomic due to the costs and time involved in negotiating with landowners whose land the pipe would traverse. The challenge then became what to do with the heat within the confines of the business park.

7. Application in a Teaching Setting

In order to address the challenge from the corporate partner, we decided to have all 52 students from a compulsory first year course in a bachelors level program address the challenge towards the end of their first semester. The students study a program called innovation and project leadership, which is a three-year Bachelor's focused on experiential learning through running a number of projects, including starting businesses, building bridges, organizing wilderness experiences, and providing innovation consultancy services to businesses to name a few of the projects. The program allows a degree of freedom for students to steer their own learning through choosing which particular projects they work on. As the students were approaching the end of their first semester, they had been introduced to a number of theories or schools of thought, and had applied these in varying degrees to small week-long projects. These schools of thought included circular economy, biomimicry, entrepreneurship, and design thinking. The theory associated with each of these schools of thought was introductory in nature and had not been taught in-depth to students.

As part of the challenge, students were aquatinted with the basic theories of entrepreneurship around business models, customer's needs, and finding a problem–solution fit [39,47,64,65]. In addition, students were invited to three question and answer sessions lasting approximately 2 h, each with knowledgeable individuals from differing fields, such as a local expert on water heat exchangers, another on renewable energy and innovation, and the developer running the project. The introduction to these differing perspectives helped shape students' perspectives on how to think about the task at hand.

In addition, daily coaching sessions were held with each of the eight teams working on the challenge. Students were allowed to choose their teams, and all teams had a mix of genders. These coaching sessions allowed the teacher (one of the authors) to steer the direction students were heading in, although the students retained a great deal of freedom to solve the challenge in a way they thought suitable. The teacher aimed to have students avoid any technical type approach (i.e., focusing on the particulars of how heat would be transferred) and instead had students focus on 'who' needed the heat, which types of businesses required the type of excess heat the datacenter would generate, and what were the requirements of this heat (such as consistency, timing, peak loads, etc).

8. Student Approach and Results

The students had no prior technical knowledge of the area of heat-recirculation, requiring them to go out and discuss the idea with a large number of companies to better understand the commercial needs of businesses. In some instances, students visited companies to receive tours and discuss with the companies' engineers the technical requirements the production facilities needed. They also discussed with the financial officers and CEOs the needs of the business to gain greater insights into whether there was a fit between the waste heat from the datacenter and what the individual companies needed.

The students worked in groups of four to six people on the challenge for 10 days before having to present their ideas to the developer. The developer stated after the student presentations that 'they were impressed by the suggestions of the students and that the work by the students went well beyond what was expected from them'. The developer then decided to initiate a longer-term collaboration with students on the project, which we do not discuss in detail here. The students provided the developer with a two-page summary of their findings, including further leads for the developer to follow up with regards to individual companies and types of industry players who might need this type of heat in their production process.

To draw on an example, one group visited a paper production facility and discovered that the facility needed around 14 MW of steam production each year. As part of this production, the paper company was paying USD 4.1 M per year in gas, USD 410,000 in carbon credits, and USD 3.4 m in electricity. The company did not reveal which percentage of the electricity bill was related to steam production, but the carbon credits and gas were almost entirely related to steam production, with electricity being used instead of gas when spot prices made it more economical. The students were thus able to identify paper production as a key industry that could benefit greatly from co-locating beside the datacenter. While the particular company the students visited did not express a willingness to move, having recently refurnished their plant, the business case for such a location was clear. This allowed students to begin to narrow their search to other paper production companies who might be willing to shift location. The business case was that the reduction in the cost of gas and carbon credits would be shared equally between the data center and the paper production company, with the commercial upside for the developers being that they could demand a greater price for their land given the benefit of locating there, and as well as being able to use positive publicity around the project to increase their profile.

A separate team focused on a more circular economy perspective, aiming to find a wider range of commercial players for whom waste heat energy might be of use. They identified as examples shrimp farming, algae growth, greenhouses, and insect production as types of commercial players who if co-located together could benefit from synergies associated with each other. The perspective focused on creating a value cluster that also made use of other infrastructure such as roading networks and being relatively closely located to major food distribution centers.

Neither idea is revolutionarily innovative, or might not even be considered innovative at all given that none of the ideas were new industries. Instead, the focus was a more systemic view of how businesses could cross-benefit from each other's waste products and create shared value through a non-traditional business model. These have not traditionally been the key focus areas of entrepreneurship education. Traditionally, education has focused on using resources to produce something new, and often in the process creating additional waste and pollution [2]. In this instance, the focus for the student groups was about using the resources currently available (or planned to be available) and finding a way to ensure that they were used instead of simply being wasted. Yet, this requires a reconfiguration in the way business is normally organized.

9. Discussion: A Systemic Approach to Entrepreneurship

We do not mean to argue that this case highlights the best way of teaching entrepreneurship and system thinking together. We do not even argue that this way is better than teaching the two subjects separately. Instead, we mean to highlight that it is possible to have students shift their mindset relatively early on in their education to have them consider a wider perspective that goes beyond narrow economic gain often associated with capitalism and entrepreneurship [13,14]. None of the students presented a case that focused on solely maximizing the developers profit. Rather, they had their focus on generating value across multiple business organizations. This represents a shift in mindset from singular optimization of an individual organizations profit to a focus on optimization of a cluster/network of organizations in order to maximize the system as a whole [48]. This represents a clear difference from discussions around sustainable entrepreneurship to date [66], which focus on improving the environmental outcomes of individual organizations (and/or their supply chains). In including systems thinking theory, we saw students' perspectives widen in number of ways, as discussed below.

9.1. Disconnected vs. Interconnectedness

The students were able to see that there was a valuable resource (heat) that could be shared across multiple business organizations. Instead of engaging in typical business development focused on the profit maximization of a single entity, they were able to shift their focus or analysis to the interconnectedness of businesses that could be co-located together in order to create value that was relevant for a cluster of companies.

They were also able to see the interconnectedness of a single business activity with downstream consequences, and to design solutions that minimized the impact on the surrounding environment. Viewing the interconnectedness of the business with the surrounding environment, and the potential for negative consequences, allowed them to see the need for a shift in existing ways of carrying out business and that the single metric of profit was problematic.

They were also able to see their own interconnectedness to the project and that they had the power to influence the final outcomes of the development that was to occur in their local environment. They therefore went from seeing themselves as passive observers to seeing the agency they had to push the project to head in a direction they considered positive.

9.2. Linear vs. Circular

Students became aware of the nature of feedback loops; for example, the developer expressed concern that if cold air was to be used to cool the datacenter, then the warm air that was expelled might potentially reduce the snowfall in the area (an unpopular outcome in a ski-obsessed nation), which would further lead to warming in the area (as snow reflects thermal heat), and that this feedback loop could have negative consequences with regard to ski conditions.

Students also saw how the 'process' of producing profit does not stop once something has been sold, and that there are downstream consequences that could be used in a constructive manner to design a more circular-based economic model. They, for example, were able to see the possibility that existed to use the waste heat to warm water for shrimp farming; the waste from shrimp farming could then be used for fertilizer, with the additional heat also being used for greenhouses, with the organic rubbish from the greenhouses being able to be used for feeding insects that could eventually be fed back to the shrimp. They could then see the second- and third-order consequences of setting up a synergistic value network of businesses.

9.3. Silos vs. Emergence

Students were able to gain experience with emergence as a concept connected to entrepreneurship, whereby a path became clear as they took action, which opened further possibilities and understanding of the situation in which they were trying to create value. In addition, they were able to experience emergence in the design of a business system [13].

In the beginning, many of the students were uncertain of how to proceed and felt insecure about their ability to come up with a solution. They were guided through a process with strong roots in emergence and sharing pedagogical roots with design thinking that pushed them to speak with potential relevant stakeholders, seek information, and create their own questions and sources of answers. In doing so, they experienced the emergence of their own understanding and with it the emergence of solutions and paths forward. A critical philosophical point of emergence is that small changes in starting factors can lead to extremely different outcomes in end scenarios. In this sense, they were able to see that their own small interactions in projects could play a large role in determining the outcomes in their local environment.

9.4. Parts vs. Whole

In a traditional sense, a business case involving a datacenter would focus on profit and only profit. This case allowed students to see that there was much more involved in developing a business park involving a datacenter. They were much more able to see the whole picture and how the different parts hung together. The typical example of two halves of a horse not being the same as a horse applies here too; they were able to see that half of a development of just building a profitable datacenter missed the opportunity to build the other half of the horse, which could be interconnecting businesses leading to a reduction in power (measured across all businesses), a reduction in Co2, and saved costs with regards to carbon credits. They could see this whole and how creating this whole was far more valuable than focusing on singular parts of the value chain.

9.5. Analysis versus Synthesis—The Idea of Synthesis Is the Creation of Something New through the Combination of Existing Elements

The project had strong synthetic elements to it; there was previously no value cluster and there were no plans to create one either. The students were then compelled to use the resources they had, such as their own knowledge, the knowledge of others, and their networks, to begin to generate interest from different stakeholders. In doing so, they carried out a process strongly resembling effectuation [45].

9.6. Isolation vs. Relationships—Not about the Individual Elements, but Paying Attention to Interrelationships

This case was a clear example of how to have students shift their mindset towards focusing on the relationship between business organizations in an attempt to create mutual value. The flow of resources between the planned businesses was the key theme for the project, and a crucial part was having students understand the nuances of these relationships as well. For example, students had to research the peak flows of heat in the datacenter compared to the timing of needs in the other organizations. It was important to understand the timing and needs of each business process and how these could be interlinked.

Another example, an initial idea that was later discarded by students, was the use of the datacenter heat to dry corn. While corn farmers use a lot of heat, they only need this in a single month of a year, meaning using the heat for such a short period did not make sense. Yet, another idea discarded early was the use of the heat to generate steam, run engines, and recycle this energy back into electricity. However, it showed that the heat from the data center was not suitable for this kind of steam generation. In this sense, students focused on the relationships and demands between the individual elements in their quest for creating solutions.

9.7. Differing from Sustainable Entrepreneurship

One particular criticism that might be levelled at the suggestion of blending entrepreneurship and systems thinking is the question of how is this different from existing fields such as sustainable entrepreneurship or social entrepreneurship, where the focus is on having a 'green' approach to business. We would argue that there is a significant and important difference. Sustainable entrepreneurship is still focused on singular entities and trying to reduce their particular negative impact. It is still analytic/linear in nature and often has a more incremental starting point. We do not mean that as criticism, and think sustainable entrepreneurship has a valuable place within a transition to a more sustainable future.

However, combining systems thinking with entrepreneurship is different, as it is less about individual entities and instead focuses on the creation of value and the use of resources in a more systemic way, that is, the creation of value clusters, for example, the design of the interaction between these individual organizations in which the value creation is intended to be maximized across the group, and in which value can also include the reuse of resources or the reduction of waste. In this sense, it moves away from viewing organizations as silos and instead views their individual value creation within a larger system and context, and in this sense aims at removing a kind of systems blindness that can account for the current negative business practices.

9.8. Shifting Mindsets

The original idea of mindset relates back to the work of cognitive psychologists in the early 20th century at the Würzberg School of Cognitive Psychology [67], where 'mindset' was a term used to describe on an individual setting their mind on finding a solution to a particular challenge. We can consider the goal here to shift students towards having their mind set on using a systemic perspective when engaging in entrepreneurial behavior with the desired outcome being a more sustainable set of business outcomes. Examining whether there is evidence that mindset shifts have occurred would require an in-depth study of students' perspectives before and after. We accept that this particular study is explorative in nature and does not reach the standard of being causative evidence of a mindset shift. The intention behind the case is to demonstrate that there is a coherent way that these three areas can be taught together in order to connect systems thinking, sustainability, and entrepreneurship.

Based on the observed work of the students and their deliveries, it would appear that there is ample evidence that they were able to manage the existing tensions between these three theoretical perspectives and find resolutions that might have been suboptimal for any one organization, but were optimal when considered holistically across multiple organizations. In this sense, students were able to shift their mindsets toward having agency in creating outcomes that impacted their local environment. As mindsets can be shaped through education, we hope this formative education experience helped students to have a mindset that was more sustainable while acknowledging the important roles that setting up commercial systems can have in benefitting everyone.

10. Going Forward

We have only begun to explore the concept here of joining the fields of systems thinking and entrepreneurship. Given this is a new perspective, it likely lacks nuance and therefore requires further exploration. As a single case study, this is intended only to begin exploring the overlap between the two fields and how it might be taught. We would welcome much more rigorous approaches to examining the ability of teaching to shift students' mindsets in a way that they begin to consider the holistic picture. We suggest a fruitful path for further research would be to examine students' reflections on their experiences of learning in this manner, and to begin to examine whether a shift in mindset has occurred.

In addition, the goal of shifting mindsets around how to begin to tackle global systemic challenges is obviously a daunting and ambitious task, but all movements start with minor steps. We encourage further exploration of this topic in order to examine the topic with greater rigor.

11. Caution in Systems Thinking

We advocate for developing a sense of agency in students so as to empower them to make what they consider to be positive changes in their environments. At the same time, influencing systems needs to be approached with a genuine sense of caution. As even the earliest systems thinkers pointed out, issues of social systems are incredibly complex and go far beyond individuals' capacity to analyze their way toward a solution [21].

Going a step further, there has been a great deal of harm done by what has been labeled 'naïve interventionalism' [20], where the intention to do good was clear, but the result was the opposite of that intended. As Meadows points out, people often know intuitively where leverage points are; however, they have a tendency to pull in the wrong direction [12]. We therefore reiterate the need for caution in encouraging students to begin to experiment with systems change, and they should be made equally aware that not all problems have a 'technical solution' and will therefore require trade-offs that enter into the realm of ethics and competing moral values [68]. As the age of the citations in this subsection testify, these are not new issues, but still ones we need to be cognizant of as we continue to wrestle with how to move society forward in a desired direction.

12. Conclusions

The article starts from the assumption that the current economic systems are creating a number of problems that are systemic in nature, and therefore require systemic solutions. Traditionally, entrepreneurs have played a crucial role in the shaping of systems and markets, but the current mental models of entrepreneurship are flawed in that they teach students to focus on wealth maximization (often at the expense of stakeholders and the environment), which leads to suboptimal solutions at a systems level.

Based on prior literature, we argued that one of the best ways to change a system is to shift the underlying mindsets or mental models of those inside it. We therefore argued that education should place greater focus on viewing the 'wholeness' of environmental and economic systems when engaging in entrepreneurship. In order to demonstrate how this can be done, we highlighted the theoretical arguments with a case study to demonstrate how entrepreneurship could be taught with a greater focus on seeing the 'wholeness' of the situation and not focusing on value creation at the individual organizational level.

There is still a lot of work to be done in order to more elegantly connect the fields of entrepreneurship and systems thinking if more students are to produce sustainable economic systems, and more work needs to be done to shift the mindsets of students interested in these fields. This article aims to take an early step toward achieving both of these, and we welcome and hope for further contributions from the field in this area. Potentially fruitful areas for further research would focus more on examining whether shifts in mindsets do occur from such teaching and whether this influences entrepreneurs' ventures when they finish their education. In addition, there are likely contributions to be made with regard to clarifying the theoretical overlap and contradictions between the fields.

Author Contributions: Conceptualization, M.L.; methodology, M.L.; validation, G.A. and F.R.J.; formal analysis, G.A. and F.R.J.; writing—original draft preparation, M.L.; writing—reviewing and editing, G.A. and F.R.J. All authors discussed the results and contributed to the final manuscript. All authors have read and agreed to the published version of the manuscript.

Funding: This research has received funding from the Interreg Øresund-Kattegat-Skagerrak program project Greenbizz.

Institutional Review Board Statement: Not applicable.

Informed Consent Statement: Not applicable.

Data Availability Statement: Data sharing not applicable.

Conflicts of Interest: The authors declare no conflict of interest.

References

1. United Nations Environment Programme. *Making Peace with Nature: A Scientific Blueprint to Tackle the Climate, Biodiversity and Pollution Emergencies*; United Nations Environment Programme: Nairobi, Kenya, 2021.
2. Coad, A.; Nightingale, P.; Stilgoe, J.; Vezzani, A. The Dark Side of Innovation. *SSRN Electron. J.* **2020**, *28*, 102–112. [CrossRef]
3. Brem, A.; Puente-Díaz, R. Creativity, Innovation, Sustainability: A Conceptual Model for Future Research Efforts. *Sustainability* **2020**, *12*, 3139. [CrossRef]
4. Sillitto, H.; Martin, J.; Griego, R.; McKinney, D.; Arnold, E.; Godfrey, P.; Dori, D.; Krob, D.; Jackson, S. A Fresh Look at Systems Engineering-What is it, How Should it Work? *Insight* **2018**, *21*, 44–51. [CrossRef]
5. Stroh, D.P. *Systems Thinking for Social Change: A Practical Guide to Solving Complex Problems, Avoiding Unintended Consequences, and Achieving Lasting Results*; Chelsea Green Publishing: Hartford, CT, USA, 2015.
6. Shane, S.; Venkataraman, S. The Promise of Entrepreneurship as a Field of Research. *Acad. Manag. Rev.* **2000**, *25*, 217–226. [CrossRef]
7. Wiggins, R.R.; Ruefli, T.W. Schumpeter's ghost: Is hypercompetition making the best of times shorter? *Strat. Manag. J.* **2005**, *26*, 887–911. [CrossRef]
8. Schumpeter, J. *Capitalism, Socialism and Democracy*; Harpers & Brothers: New York, NY, USA, 1942.
9. Giridharadas, A. *Winners Take All: The Elite Charade of Changing the World*; Knopf Doubleday: New York, NY, USA, 2018.
10. Towards the end of poverty. In *The Economist*; Economist: London, UK, 2013.
11. Senge, P. *The Fifth Discipline. The Art & Practice of Learning Organization*; Doupleday Currence: New York, NY, USA, 1990.
12. Meadows, D. Leverage Points: Places to Intervene in a System. 1999. Available online: http://donellameadows.org/archives/leverage-points-places-to-intervene-in-a-system/ (accessed on 12 January 2021).
13. Kickul, J.; Gundry, L.; Mitra, P.; Berçot, L. Designing With Purpose: Advocating Innovation, Impact, Sustainability, and Scale in Social Entrepreneurship Education. *Entrep. Educ. Pedagog.* **2018**, *1*, 205–221. [CrossRef]
14. Gladysz, B.; Urgo, M.; Gaspari, L.; Pozzan, G.; Stock, T.; Haskins, C.; Jarzębowska, E.; Kohl, H. Sustainable Innovation in a Multi-University Master Course. *Procedia Manuf.* **2018**, *21*, 18–25. [CrossRef]
15. Meadows, D.H. *Thinking in Systems: A primer*; Chelsea Green Publishing: Hartford, CT, USA, 2008.

16. Von Bertalanffy, L. The meaning of general system theory. In *General System Theory: Foundations, Development, Applications*; Braziller: New York, NY, USA, 1973; pp. 30–53.
17. Teece, D.J. Dynamic capabilities as (workable) management systems theory. *J. Manag. Organ.* **2018**, *24*, 359–368. [CrossRef]
18. Cabrera, D.; Colosi, L.; Lobdell, C. Systems thinking. *Evaluation Program Plan.* **2008**, *31*, 299–310. [CrossRef]
19. Cabrera, L. What Is Systems Thinking? In *Learning, Design, and Technology*; Spector, M., Lockee, B., Childress, M., Eds.; Springer: Basel, Switzerland, 2019.
20. Taleb, N.N. *Antifragile: Things that Gain from Disorder*; Random House Incorporated: New York, NY, USA, 2012; Volume 3.
21. Rittel, H.W.J.; Webber, M.M. Dilemmas in a General Theory of Planning. *Policy Sci.* **1973**, *4*, 155–169. [CrossRef]
22. Pittaway, L.; Cope, J. Entrepreneurship education: A systematic review of the evidence. *Int. Small Bus. J.* **2007**, *25*, 479–510. [CrossRef]
23. Jones, P.; Maas, G.; Kraus, S.; Reason, L.L. An exploration of the role and contribution of entrepreneurship centres in UK higher education institutions. *J. Small Bus. Enterp. Dev.* **2021**, *28*, 205–228. [CrossRef]
24. Wyness, L.; Jones, P.; Klapper, R. Sustainability: What the entrepreneurship educators think. *Educ. Train.* **2015**, *57*, 834–852. [CrossRef]
25. Cincera, J.; Biberhofer, P.; Binka, B.; Boman, J.; Mindt, L.; Rieckmann, M. Designing a sustainability-driven entrepreneurship curriculum as a social learning process: A case study from an international knowledge alliance project. *J. Clean. Prod.* **2018**, *172*, 4357–4366. [CrossRef]
26. Dzombak, R.; Mehta, C.; Mehta, K.; Bilén, S.G. The Relevance of Systems Thinking in the Quest for Multifinal Social Enterprises. *Syst. Pr. Action Res.* **2013**, *27*, 593–606. [CrossRef]
27. Rieckmann, M. Future-oriented higher education: Which key competencies should be fostered through university teaching and learning? *Futures* **2012**, *44*, 127–135. [CrossRef]
28. Sammalisto, K.; Sundström, A.; Von Haartman, R.; Holm, T.; Yao, Z. Learning about Sustainability—What Influences Students' Self-Perceived Sustainability Actions after Undergraduate Education? *Sustainability* **2016**, *8*, 510. [CrossRef]
29. World Commission on Environment and Development. *Our Common Future*; Oxford University Press: Oxford, UK, 1987.
30. International Energy Agency. *Energy Technology Perspectives*; International Energy Agency: Paris, France, 2017.
31. Haldar, S. Towards a conceptual understanding of sustainability-driven entrepreneurship. *Corp. Soc. Responsib. Environ. Manag.* **2019**, *26*, 1157–1170. [CrossRef]
32. Trivedi, C.; Misra, S. Relevance of Systems Thinking and Scientific Holism to Social Entrepreneurship. *J. Entrep.* **2015**, *24*, 37–62. [CrossRef]
33. Cannon, C.M.; Fenoglio, G. Charity for Profit. *Natl. J.* **2000**, *32*, 1898–1904.
34. Lynch, M.; Kristoffer, S.; Federico, L.; Martin, S.; Gunnar, A. Examining Entrepreneurial Motivations in an Education Context. In Proceedings of the 21st International Conference on Engineering Design (ICED 17) Vol 9: Design Education, Vancouver, BC, Canada, 21–25 August 2017; pp. 079–088.
35. Neck, H.M.; Greene, P.G. Entrepreneurship Education: Known Worlds and New Frontiers. *J. Small Bus. Manag.* **2010**, *49*, 55–70. [CrossRef]
36. D'Aveni, R.A. *Hypercompetition: Managing the Dynamics of Stratehic Maneuvering*; Free Press: New York, NY, USA, 1994.
37. Jensen, D. *Endgame*; Seven Stories Press: New York, NY, USA, 2006; Volume 1.
38. McMullen, J.S.; Shepherd, D.A. Entrepreneurial Action and the Role of uncertainty in the Theory of the Entrepreneur. *Acad. Manag. Rev.* **2006**, *31*, 132–152. [CrossRef]
39. Aulet, B. *Disciplined Entrepreneurship: 24 Steps to a Successful Startup*; John Wiley & Sons: Hoboken, NJ, USA, 2013.
40. Ries, E. *The Lean Startup: How Today's Entrepreneurs Use Continuous Innovation to Create Radically Successful Businesses*; Crown Books: New York, NY, USA, 2011.
41. Blank, S.; Dorf, B. *The Startup Owner's Manual: The Step-by-Step Guide for Building a Great Company*; John Wiley & Sons: Hoboken, NJ, USA, 2020.
42. Collins, J. *Turning the Flywheel: A Monograph to Accompany Good to Great*; Random House: London, UK, 2019.
43. Gibbs, S. Apple and Samsung fined for deliberately slowing down phones. In *The Guardian*; The Guardian: London, UK, 2020.
44. Bedau, M.A. Weak emergence. *Philos. Perspect.* **1997**, *11*, 375–399. [CrossRef]
45. Sarasvathy, S.D. Causation and effectuation: Toward a theoretical shift from economic inevitability to entrepreneurial contingency. *Acad. Manag. Rev.* **2001**, *26*, 243–263. [CrossRef]
46. Sarasvathy, S. *Effectuation: Elements of Entrepreneurial Expectation*; Edward Elgar: Cheltenham, UK, 2008.
47. Osterwalder, A.; Pigneur, Y.; Bernarda, G.; Smith, A.; Papadakos, T. *Value Proposition Design: How to Create Products and Services Customers Want*; John Wiley & Sons: Hoboken, NJ, USA, 2015.
48. Aksoy, L.; Calabretta, G.; Driessen, P.H.; Hillebrand, B.; Humphreys, A.; Krafft, M.; Beckers, S.F.M. Consumer perceptions of service constellations: Implications for service innovation. *J. Serv. Manag.* **2013**, *24*, 314–329.
49. Lynch, M.; Kamovich, U.; Longva, K.K.; Steinert, M. Combining technology and entrepreneurial education through design thinking: Students' reflections on the learning process. *Technol. Forecast. Soc. Chang.* **2021**, *164*, 119689. [CrossRef]
50. Neck, H.M.; Corbett, A.C. The Scholarship of Teaching and Learning Entrepreneurship. *Entrep. Educ. Pedagog.* **2018**, *1*, 8–41. [CrossRef]
51. Naumann, C. Entrepreneurial Mindset: A Synthetic Literature Review. *Entrep. Bus. Econ. Rev.* **2017**, *5*, 149–172. [CrossRef]

52. Shepherd, D.A.; Patzelt, H. *Entrepreneurial Cognition: Exploring the Mindset of Entrepreneurs*; Palgrave Macmillan: Cham, Germany, 2018.
53. Haynie, M.; Shepherd, D.; Mosakowski, E.; Earley, P.C. A situated metacognitive model of the entrepreneurial mindset. *J. Bus. Ventur.* **2010**, *25*, 217–229. [CrossRef]
54. Lynch, M. *Entrepreneurial Mindset: Defining the Concept, How to Measure It, How to Teach It and Its Role in the Venture Creation Process*; Norwegian University of Science and Technology (NTNU): Trondheim, Norway, 2020.
55. Mathisen, J.-E.; Arnulf, J.K. Competing mindsets in entrepreneurship: The cost of doubt. *Int. J. Manag. Educ.* **2013**, *11*, 132–141. [CrossRef]
56. Corbett, A.C. Experiential Learning within the Process of Opportunity Identification and Exploitation. *Entrep. Theory Pr.* **2005**, *29*, 473–491. [CrossRef]
57. Robson, C.; McCartan, C. *Real World Research: A Resource for Users of Social Research Methods in Applied Settings*; Wiley: Chichester, UK, 2016.
58. Flyvbjerg, B. Five Misunderstandings About Case-Study Research. *Qual. Inq.* **2006**, *12*, 219–245. [CrossRef]
59. Johansen, F.R.; Kerndrup, S.; Andersson, G.; Rubach, S. A view of clustering as emergent and innovative processes. *Ind. Innov.* **2020**, *27*, 390–419. [CrossRef]
60. Van de Ven, A.H.; Polley, D.; Garud, R.; Venkataraman, S. *The Innovation Journey*; Oxford University Press: New York, NY, USA, 1999.
61. Yin, R.K. *Applications of Case Study Research*, 3rd ed.; Sage: Thousand Oaks, CA, USA, 2011.
62. Hoholm, T.; Araujo, L. Studying innovation processes in real-time: The promises and challenges of ethnography. *Ind. Mark. Manag.* **2011**, *40*, 933–939. [CrossRef]
63. Li, Z.; Kandlikar, S.G. Current Status and Future Trends in Data-Center Cooling Technologies. *Heat Transf. Eng.* **2014**, *36*, 523–538. [CrossRef]
64. Osterwalder, A.; Pigneur, Y. *Business Model Generation: A Handbook for Visionaries, Game Changers, and Challengers*; John Wiley & Sons: Hoboken, NJ, USA, 2010.
65. Brown, T.; Katz, B. Change by Design. *J. Prod. Innov. Manag.* **2011**, *28*, 381–383. [CrossRef]
66. Deets, S.; Rodgers, V.; Erzurumlu, S.; Nersessian, D. Systems Thinking as a Tool for Teaching Undergraduate Business Students Humanistic Management. *Humanist. Manag. J.* **2020**, *5*, 1–21. [CrossRef]
67. Gollwitzer, P.M.; Bayer, U. *Deliberative Versus Implemental Mindsets in the Control of Action*; In Dual-Process Theories in Social Psychology; Guilford Press: New York, NY, USA, 1999; pp. 403–422.
68. Hardin, G. The Tragedy of the Commons. *Science* **1968**, *162*, 1243–1248.

sustainability MDPI

Article

Systems Engineering for the Energy Transition: Potential Contributions and Limitations

Arjan Kirkels [1,*], Vince Evers [1] and Gerrit Muller [2,*]

1 School of Innovation Sciences, Eindhoven University of Technology, 5612 AE Eindhoven, The Netherlands; v.h.m.evers@student.tue.nl
2 Faculty of Technology, Natural Sciences and Maritime Sciences, Campus Kongsberg, University of South-Eastern Norway, 3613 Kongsberg, Norway
* Correspondence: a.f.kirkels@tue.nl (A.K.); gerrit.muller@usn.no (G.M.)

Abstract: Systems engineering finds its origin in analyzing and exploring complicated technical systems. In this positioning paper, we set out to discuss the value and limitations of a Systems Engineering approach in its contribution to societal challenges, notably the energy transition. We conceptualize the energy system as a sociotechnical system. We specifically explore stakeholders and their roles, agency, and acceptance. We illustrate the relevance by a case at the municipal level that shows the relevance of acceptance, pluralism, distributed agency, context, and process aspects. The municipality is still in a phase of exploration and conceptualization. Systems Engineering can be of great value in this phase to explore the problem and solution space. However, to make the most of this requires that Systems Engineering addresses policy making, distributed agency, and complexity. We discuss the challenges this poses for the traditional Systems Engineering approach; we indicate several potential strategies to address these challenges, and we show two fields that can help clarify how to address these challenges: transition studies and sustainability assessment.

Keywords: systems engineering; energy transition; renewable energy; sociotechnical system; stakeholders; agency; acceptance; complex problems; transition studies; sustainability assessment

Citation: Kirkels, A.; Evers, V.; Muller, G. Systems Engineering for the Energy Transition: Potential Contributions and Limitations. *Sustainability* **2021**, *13*, 5423. https://doi.org/10.3390/su13105423

Academic Editor: Cecilia Haskins

Received: 30 March 2021
Accepted: 10 May 2021
Published: 12 May 2021

1. Introduction

Energy drives all activities and is strongly related to prosperity, production, consumption, and well-being [1–4]. As such, a good energy supply is crucial for achieving many sustainable development goals (SDGs) [2], and an urgent call for action is needed to ensure peace and prosperity for people and the planet, now and in the future [5,6]. Especially SDG 7 targets the energy supply: "Ensure access to affordable, reliable, sustainable and modern energy for all" [5,6].

Achieving the goals as stated in SDG 7 requires a transition, a radical shift in the energy system [1], from a fossil fuel-based system to one based on renewable energy sources. Especially in developed countries, reducing greenhouse gas emissions is a crucial driver behind recent developments in the energy sector [7,8]. The complexity, scale, and urgency of the required transition make it one of the grand societal challenges for the 21st century. Over the last few decades, this problem has received increased attention in academic [9–11] and policy circles [12–14].

The energy supply has all characteristics of a system: primary energy sources are mined in nature and are converted by an apparatus to secondary energy sources, which in turn are converted to energy services. All elements are interconnected by infrastructure and transport [15]. Traditionally, System Engineering has had substantial contributions to the sector, with a specific focus on understanding and assessing the techno-economic feasibility of these engineered systems [16]. In addition, for the future, there is a significant potential for contributions from the field of Systems Engineering, given the enormous changes required for the energy transition.

However, traditionally, Systems Engineering has been focused on exploring and analyzing technical systems and complicated problems. Although that is a valuable contribution by itself, we will argue in this positioning paper that the energy transition, and societal challenges in general, require a different approach. In essence, by their nature, energy systems are sociotechnical systems. These systems include infrastructure and conversion technology and markets, cultural aspects, regulatory paradigms, and consumer behavior [1,15,17]. As such, they require unpacking socioeconomic factors, including stakeholders' roles, normativity, and agency. These are issues that are hardly addressed by traditional Systems Engineering that would rather see normativity and agency either as clients' goals and customers' needs and requirements or as contextual factors for the engineering system. In the traditional approach, normative and agency aspects are typically addressed in a rather instrumental and technocratic way, focusing on how to model or how to manage [18–20].

Discussions and developments in the energy transition show that many struggles to embrace and accelerate the energy transition relate to the sociotechnical nature of the transition. One challenge is translating the generally accepted policy goals into specific roadmaps, action plans, and policy approaches. These serve to align stakeholders' interests and visions, empower stakeholders, and build up momentum to build up a robust approach. However, change invokes resistance. With the recently increased diffusion and (larger scale) implementation of solar PV and onshore wind, the acceptance of renewables has become a prominent issue [21–30] that can severely affect the support and direction of the energy transition.

In summary, we conceptualized the energy system as a sociotechnical system in transition. Our goal for this positioning paper was to indicate the relevance of Systems Engineering approaches, to illustrate the problems that arise due to the misfit with sociotechnical systems, to discuss strategies to overcome that misfit, and to indicate related fields from which lessons can be learned. As such, we contributed to the goal of this special issue to focus on Systems Engineering's role in contributing to grand societal challenges and providing policy advice [16,31].

We specifically focused on the diversity of stakeholders, their roles, and their agency, including societal acceptance issues. To do so, we drew upon an illustrative case study on the heating transition for the small city of Best in the Netherlands. This approach is analytically helpful as it reduces complexity: Goals, boundaries, and involved stakeholders can be clearly identified, dynamics can be easily understood. Over recent years, Dutch municipalities have been assigned a more substantial role in the energy transition, e.g., by making local plans for the heat transition in the built environment and specifying local potential [32]. However, municipalities typically lack experience and expertise in this field and struggle with questions about a regional energy transition [33,34]. The city of Best is especially relevant as it committed itself to far-reaching goals [35–37]; there is an active energy cooperative, Duurzaam Best, indicating a willingness to act among citizens; but despite this, a project to implement a heating network in the Naastenbest neighborhood recently failed [38,39]. As such, it provides an excellent illustrative case to study both stakeholders and broad aspects of acceptance. Finally, a systems architecting study had already been conducted for Best [40,41], which offered some prior knowledge.

In the next paragraph, we introduce sociotechnical systems and acceptance concepts more thoroughly based on relevant literature. Next, in the methodology section, we discuss the approach for the illustrative case, followed by the empirical results. Finally, we present a conclusion and discussion on the potential for contribution and the challenges in Systems Engineering and how to overcome these challenges.

2. Sociotechnical Systems and Acceptance

We argue that complex societal problems, such as the energy transition, require a sociotechnical system approach. A sociotechnical system has been introduced and developed in the field of science and technology studies (STS). A sociotechnical system has been defined as *"a configuration of technologies, services, and infrastructures, regulations, and actors (for example, producers, suppliers, policymakers, users) that fulfills a societal function"* [42]. It assumes that a broader set of stakeholders is involved in developing and implementing technology and that the social and technical aspects co-evolve. It builds on the notion of a coupled development of technology with social practices, rules, and normative settings [15,17,43]; the distributed agency over a variety of stakeholders and moments, i.e., the capacity to act and influence [44]; and the complexity of the problems, characterized by normative pluralism, inherent uncertainty, and emerging behavior [45–47]. As such, a broad set of stakeholders is relevant as are their roles, preferences, practices, and processes of interaction and decision-making. Relevant issues in this field include justice, public engagement, social construction, expectations, transforming innovations, and governing complex transitions [48]. A traditional limited stakeholder analysis focusing on power and interest will not cover all these aspects.

Recently the energy transition started to take shape. It moved from an idea or concept to an actual realization by implementing, among other things, larger scale solar PV and onshore wind. Consequently, the impact and hindrance of these technologies became clear and started to affect more extensive groups of stakeholders. The lack of social acceptance of renewables has become a prominent issue [21–30] that can severely affect the support and direction of the energy transition.

Acceptance, or the lack thereof, is a typical example of sociotechnical aspects of the energy transition. Acceptance is a broad concept that can refer to people's ability to "tolerate", "regard something as inevitable", or "approve of something or to encourage it". It can manifest itself as attitudes and actions on a spectrum of involvement and agency, i.e., the capacity to make choices [21,49]. Other crucial questions are who is accepting and what is being accepted. When it comes to social acceptance, Wüstenhagen et al. [29] differentiate between sociopolitical, market, and community acceptance: Sociopolitical acceptance refers to the acceptance of general goals, technologies, and policies by the public, key stakeholders, and policymakers. Market acceptance relates to the adoption and diffusion of innovations in the market by companies, consumers, and investors. Finally, community acceptance refers to residents' and local stakeholders' acceptance in siting decisions and renewable energy projects. This differentiation can help to increase understanding. For example, the not in my backyard (NIMBY) phenomenon that is seriously hampering the local energy transition [22,23,25,30] is a combination of large sociopolitical acceptance with low community acceptance.

The three types of acceptance as identified by Wüstenhagen et al. are influenced by five key factors related to the stakeholders involved and local factors, as determined from the literature (see Figure 1): financial costs and benefits; values and goals; justice and participation; trust and communication; and efficacy and reasibility. *Financial costs and benefits* refer to the initial investment, the payback time, and the expected return on investment. Low investment costs, short payback times, and a reduction of monthly energy costs are considered critical preconditions for community acceptance. Simultaneously, a solid business case is more decisive for market and sociopolitical acceptance [50–52].

Values and goals refer to an individual's preferences based on personal norms, values, motivations, and goals. When the intended effects of sustainable measures are perceived to align with these values and goals, the actions are more likely to be considered favorably [53]. Common goals influencing community acceptance are the improvement of public health, the stimulation of (economic and scientific) advancement, and the creation of moral and caring communities [50,53], as well as safety, autonomy to choose, little hindrance, a comfortable living environment, and preservation of the natural environment [51].

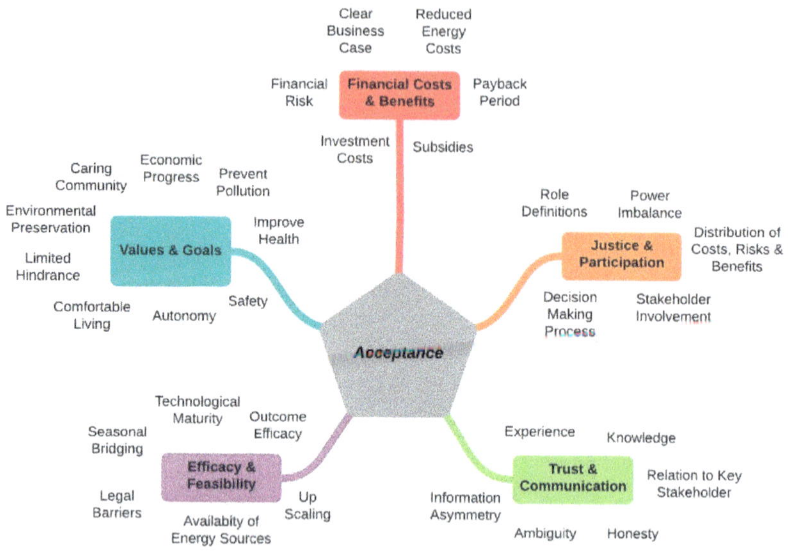

Figure 1. Conceptual map of key concepts affecting different types of acceptance.

Justice and participation both relate to the process of decision-making. Distributional justice relates to a perceived (im)balance in the distribution of costs, benefits, and risks among critical stakeholders. Procedural injustice is often experienced as a result of a perceived imbalance of power [23]. A perceived lack of justice and participation can result in damaged stakeholder relations and opposition [29,50,54]. As such, it is especially relevant for community acceptance and sociopolitical acceptance.

Trust and communication both relate to the stakeholders involved and their interaction. When knowledge of sustainable measures and technologies is limited, community acceptance depends on trust in experts and decision-makers [23,24,29,30,50]. As such, sociopolitical acceptance (or lack thereof) may trickle down to community acceptance. Trust is shown to affect acceptance most strongly through information relating to costs, benefits, and risks. Ambiguous or dishonest communication negatively affects trust in responsible stakeholders and increases perceived injustice. The importance of trust is also well described in other literature, e.g., in business [55,56], in governance [57–59], and for new large and risky technologies [60,61]. We argue that it is crucial for all three forms of acceptance. Due to recent developments, trust issues have become even more critical (e.g., due to individualism, alternative facts, distrust of central governments, and the role of social media) [62,63].

Finally, the *efficacy and feasibility* of proposed measures are essential. Individuals are less accepting if an action is considered unrealistic or when the effect of implementation on a final goal (outcome efficacy) is considered small [30]. Questions of feasibility are often related to technological maturity, the availability of energy sources, the matching of supply and demand of energy, the scalability, and the ownership of proposed solutions [52,64].

The more recent literature on acceptance is more critical and elaborates further on the justice and participation aspect. It primarily focuses on injustices and inequalities, and as such, aligns more with an STS agenda. It concentrates on the role of time and history in the energy transition, adopts a more relational framework (studying the production and consumption side and local and national level at the same time), takes into consideration power relationships (e.g., the relationship between expert political and lay systems), and the role of sociopolitical contexts. It moves away from positivist and individual frameworks towards the analysis of socially constructed discourses [65].

3. Methodology

We set out to explore stakeholders involved, their roles, and the issue of (social) acceptance for the case of Best, more specifically for the failed Naastenbest project. We studied directly involved key stakeholders to understand better their roles, the sociopolitical context, the process, and the differences and tensions between the political, expert, and lay systems. In addition, we studied the social acceptance of the local public and how that was affected by different factors.

The most relevant (directly involved, influential) stakeholders in the Naastenbest project were identified by the initial study of academic and grey literature and preliminary interviews. The municipal council has the formal responsibility for the energy transition in the municipality. The municipal officials support this council. The housing cooperation owns and builds rental houses as a delegated government responsibility and is especially relevant for initiatives in the built environment, as was the case for the Naastenbest project. Installation companies play a central role in engineering and installing energy and heating infrastructure and equipment. Best Duurzaam is an association of citizens who strive for sustainability. It engages with the societal debate and local initiatives.

In the next step, semi-structured interviews were held with each of these stakeholders. This approach matches the research's explorative nature and the level of involvement of the stakeholders [49]. Probing questions are employed to elicit additional information on potentially sensitive topics, assist respondents in recalling the information, and explore issues that were not previously identified as relevant [66]. Based on selective sampling, 32 stakeholders were approached, of which 10 agreed to the interview, covering all five relevant stakeholder groups, see Table 1. Interviews were conducted over the phone or by Skype (due to COVID-19 restrictions), following an open interview guide. Interviews typically lasted 45–75 min. Interviews were not recorded but notes were taken. Anonymized results were analyzed to identify key concepts and themes.

Table 1. Selective sampling of relevant stakeholder groups.

Stakeholder	# Interviewed
Municipality of Best	1
Municipal Council	2
Energy cooperation "Duurzaam Best"	4
Housing cooperation	1
Installation company	1

Finally, social acceptance in the community was assessed by an online survey that focused on broader acceptance of energy measures and policies. The survey consisted of closed questions and open questions (remarks) and was designed to assess (1) demographic characteristics of respondents; (2) the general attitude towards sustainability and the energy transitions; (3) perceived drivers and barriers for taking action; (4) the role of the municipality; and (5) the support for concrete measures, either individually or at the municipal level. Respondents were approached on multiple online platforms to increase the response rate and reduce potential sampling bias. These platforms included 'Plein Best' and 'Groeiend Best', as well as Facebook communities such as 'Best voor Best' and 'Best anders'. Moreover, fourteen area representatives were contacted to spread the survey among residents in their neighborhoods. In total, this resulted in 85 respondents.

4. Results of the Illustrative Case

In February 2020, officials of the Municipality of Best, responsible for the energy transition in Best, submitted a proposal for a district heating network including seasonal storage. This project targeted several national subsidies for the exploration of heating options. For the planning, the municipality cooperated with an installation company. The prospected users included a sports hall, a few schools, private individuals, and, most notably, tenants of the local social housing corporations. Members of Best Duurzaam, the local energy cooperative, reviewed and supported this proposal, albeit conditionally. Ultimately, the municipal council rejected the proposal later that year. Both the interviews and surveys were held after the rejection.

4.1. The District Heating Naastenbest Project

Differences in stakeholders' *values and goals* resulted in varying attitudes towards sustainability and district heating in Best. Both Best Duurzaam and the municipality were motivated by (inter)national climate goals (5 respondents), either intrinsically (Best Duurzaam) or by pragmatic political motivations (municipality). Members of the municipal council (2) stressed the importance of aligning their goals with their political party's program. The housing corporation and the installation company (2) were mainly motivated by the prospect of future financial gains. Most stakeholders (7) raised concerns about the *technological feasibility* with respect to the location of the system (3), the scalability of the distribution network (3), and the seasonal balance between generated and stored energy (1). Additional concerns were raised about uncertainty and the distribution between stakeholders of *financial costs and benefits*, e.g., uncertainties regarding (exploitation) costs and subsidies (5), unclarity on ownership of the network (3), uncertainty regarding (the potential for) upscaling and future financial returns (2), and doubts on the long-term risks of operation (4). Most stakeholders (8) acknowledged the importance of stakeholder *participation and justice* but were not satisfied with the participation process in the Naastenbest project (6). Both Best Duurzaam (4) and the municipal council (2) believe that they were inadequately involved in the decision-making process, as key information was shared too late. This strongly contributed to the discontinuation of the Naastenbest project. Most of the interviewees (6) partly attributed doubts and critiques to unclear *communication* by the municipality. A lack of *trust* among stakeholders was not explicitly mentioned, but frequently implied (6) through the questions that were raised concerning the municipal alderman's perceived lack of effort in passing the proposal (5), the trustworthiness of commercial stakeholders (3), and the involvement of the energy cooperative Best Duurzaam (3). An additional issue was the lack of *vision and policy* (8). Most stakeholders (7) indicated the absence of an overarching long-term vision on the energy transition at the municipal level. Long-term prospects are missing (2) and follow-up actions are undefined (2), making it hard to judge the strategic relevance of this project. In addition, a more precise definition of *stakeholder roles* was considered crucial for sound policy making (7). Opinions on the merit of the district heating proposal varied widely among stakeholders. Best Duurzaam members (4) indicated they were favorable towards the design despite their publicly expressed concerns. In contrast, the municipal council members (2) were less optimistic and expressed concerns that the (long-term) risks were not transparently and sufficiently covered.

4.2. Acceptance among Residents of Best

Overall, the survey revealed a variety of responses per question, indicating that the group is heterogenous: Different people have different drivers, preferences, and experience various barriers. There is broad support for sustainability and energy-saving measures within the sample. Many inhabitants have taken feasible and straightforward actions, such as insulation (over half) and domestic solar PV (about half), or are thinking of implementing them. Personal drivers are both pragmatic and value-based, e.g., saving money, improving comfort, reducing the impact on climate change, and protecting the environment are indicated to be relevant, as well as belief in progress and morality of choice. Community aspects (community connection, social obligation) are not perceived as relevant drivers. For many participants, previous bad experiences, lack of favorable financing, risks, and climate critique were not barriers for further implementation. Some concerns were raised with respect to costs and their contribution. More large-scale, intrusive, novel, expensive, or contested technologies received less support, e.g., heat pumps, electric vehicles, solar collectors for implementation by residents, solar farms, biomass power plants, wind turbines, and district heating by the municipality. Only the application of PV panels on public buildings received broad support. Most participants were neutral on the role and activities of the municipality. However, in the open questions, there was strong criticism. In particular, unclarity of information, lack of participation, and distrust towards the municipality were mentioned. Another frequently mentioned topic was the potential risk of large-scale projects on the environment.

5. Conclusions and Discussion

5.1. Lessons from the Case

Our primary interest was in exploring the diversity of stakeholders, their roles, and the acceptance of renewable energy as an illustrative case for the social struggles that affect the energy transition. The literature review shows that acceptance is a broad concept that manifests itself as attitudes and actions on a spectrum of agency and involvement. Acceptance also depends on who is accepting and what is being accepted. The literature differentiates between sociopolitical, market, and community acceptance. Different factors influence these types of acceptance: financial costs and benefits; values and goals; justice and participation; trust and communication; and efficacy and feasibility. In addition, they are affected by issues of power and sociopolitical context and are visible in and affected by socially constructed discourses.

Overall, the interviews showed the distributed agency in the Naastenbest project: Several stakeholders were involved in the process by steering, influencing, decision making, and acting over an extended period and in a sometimes ill-defined project. The lack of clear data and analysis were generally shared concerns among stakeholders, e.g., on technical and financial feasibility and the distribution of cost and benefits. Process aspects strengthened these concerns: the lack of clear communication, issues of good governance and participatory approaches, ambiguity in the process, stakeholder roles, and alignment with broader societal and transition goals. Most stakeholders are supportive towards addressing climate change (sociopolitical acceptance), but at the same time, are supportive of many other goals. It is not clear how stakeholders deal with trade-offs; or how to come to an approach that does justice to all stakeholders involved, clarifies their roles, and results in a shared agenda. An overarching long-term vision of achieving the energy transition on the municipal level (either focused on actions or process and roles) is lacking. This lack of a long-term vision hinders short-term decision making and an integrated assessment of the feasibility and potential contribution of the project. Representatives of the municipality see the pilot project as experimentation, which can be an excellent strategy to overcome complexity if it results in increased visioning, networking, and learning within the community and broader stakeholder groups [11,67]. In this specific case, this contribution was disputed by several stakeholders. Trust was a sensitive topic, especially considering the disagreement on goals and roles and the failed project.

The survey on community acceptance showed the heterogeneity of the group: different people have different drivers, preferences, and experience other barriers. Overall, there is broad support for sustainability and energy-saving measures. Many inhabitants are taking or are willing to take measures, especially feasible and straightforward measures such as insulation and solar PV installation. More large-scale, intrusive, novel, expensive, or contested technologies received less support. Expressed concerns included unclarity of information, lack of participation, and distrust towards the municipality.

The survey was distributed over multiple channels, resulting in residents' participation from all neighborhoods. Still, some districts, house owners, older people, and members of the energy cooperative were overrepresented. However, we feel that these issues have not seriously hampered our conclusions, as we aimed for exploratory research on an illustrative case. We think that an even more representative sample could have revealed more diversity and more considerable differences in goals, context, and preferences.

5.2. Systems Engineering of Sociotechnical Systems

The case revealed that the municipality is still in a phase of exploration and conceptualization of the problem and potential solutions. We argue that Systems Engineering can be of great value [16,31,68]; it offers a straightforward and well traceable approach, in which the problem space is made explicit, and the solution space is explored. It evaluates technologies for their potential contribution to foreseen systems, and it does so in a consistent way. It can help to make sense of possible solutions and can clarify the roadmap of how to get there. It helps to focus the decision making and discussion on real issues by identifying (lack of) potential contribution to the overall goals. It can help prevent a focus on only short-term actions that receive a lot of support.

In particular, the last argument seemed to be relevant to the case of Best. Politics is an opportunistic process, which looks for mandate by the community and support and acceptance by stakeholders involved, strategically acting by creating and using societal momentum. However, that approach only results in short-term and feasible actions, the win-win solutions. However, addressing climate change requires disruptive change, a fundamental change in the energy system and its underlying driving forces. The uneasy message that there is no silver bullet, no simple solution that will solve all problems and be beneficial to all is hardly heard in politics. The energy transition requires profound, fundamental changes: reinvestment, rebuilding, redefining winners and losers of the system, upscaling of renewable energy technologies, and at some point might require intense system interventions, e.g., the out-phasing of specific technologies, products, or energy carriers [43,69–71]. Systems thinking is critical to understanding the inertia of the current sociotechnical system [72], to understanding the need for change in underlying driving forces, feedback mechanisms, and paradigms [69,73], and for exploring coherent alternative systems. Systems Engineering might be of great value in this process [16,74–77] to ensure that final goals are anchored in solid policy-making.

The case also reveals several characteristics that can be challenging for a traditional Systems Engineering approach. There is not one decision-maker, and stakeholders' needs and requirements are diverse and fuzzy. Instead, several stakeholders are involved: companies, governments, consumers, citizens, NGOs. They are characterized by different roles and behavior, different interests, all of which are context dependent. That diversity is visible between and within stakeholder groups (e.g., citizens' preferences). It results in pluralism, multiple and sometimes conflicting views on problems, potential solutions, and the preferred process. Second, all these stakeholders have different levels of agency, the power to steer, decide, and influence. This agency is not only distributed over stakeholders but also over time. It makes sense making, goal setting, decision making, and steering a matter of processes, rather than discrete decision-making moments. Third, the involvement of stakeholders and agency results in socialware in addition to the traditional technology hardware, i.e., norms, rules, regulations, and practices that are an inherent part of the system. Social scientists talk of sociotechnical systems to indicate the intimate interdependencies and influences of the social and technical. The socioeconomic part is no longer considered the context of a technical system, but rather becomes an inherent part of that system. This has major consequences for how to conceptualize and analyze these systems. Finally, the behavior of this system is complex, resulting in inherent uncertainties. These are the results of the pluralism and agency by stakeholders and the many interactions between all sociotechnical elements and between systems (systems-of-systems), resulting in emerging behavior.

We argue that to understand or predict system behavior and to be of more value for decision making and governance, the traditional Systems Engineering approach needs to be adapted. Here, we indicate some potential strategies. Note that several of these strategies are non-exclusive and can be combined. First, introduce pluralism by taking into account multiple viewpoints or applying sensitivity analysis. Second, operationalize sociotechnical systems, including diversity of stakeholders and distributed agency. Third, conduct participatory Systems Engineering, as is done in conceptual modeling. Stakeholders know best their preferences and viewpoints, and it offers possibilities for shared learning. Fourth, put less emphasis on the preferred action or technology and more on learning about the deep problem structure, system behavior, and solution spaces. This way, it offers second-order learning and empowerment of stakeholders involved. Fifth, put more emphasis on sufficient and resilient solutions, rather than on optimal solutions. Finally, look at the potential role of System Engineering studies as boundary objects: an object that ties together scientific knowledge and societal goals and values. As a boundary object, the outcome of a Systems Engineering study enters the social arena of governance, public debate, and decision making.

We realize that these are not topics that come naturally to the average Systems Engineer. Most of these strategies are not simple, e.g., looking at different system boundaries or doing an additional task. They require bridging between different worldviews and research paradigms, between positivist approaches present in engineering sciences and constrained relativist approaches of value for transition thinking, policy-making, and governance [78]. However, initial efforts are visible in the field to start engaging with these challenges [79], e.g., by addressing sociotechnical systems (e.g., [77,80–84]), stakeholders' goals and agency (e.g., [18,85]), complex systems and systems-of-systems (e.g., [86–89]), conceptual modelling (e.g., [90,91]), and systems thinking (e.g., [75,76,82,92]).

However, Systems Engineering does not have to start from scratch as it can draw upon and learn from closely related fields [68,93,94]. Especially, the field of transition studies seems relevant. It builds on sociotechnical systems to understand and potentially influence transitions and diffusion of sustainable innovations. Its starting points are co-evolution of sociotechnical systems; multi-actor processes; embracing uncertainty and complexity; values and pluralism; normative directionality; and a focus on stability and change [9,10]. As such, it has a significant conceptual and theoretical contribution. However, in the field there is less focus on quantitative approaches and assessment [74], which offers room for contributions by Systems Engineering.

A second relevant field is sustainability assessment. It shares with Systems Engineering a strong focus on methodology, quantification, systems, and technology, which might be helpful as both fields speak the same language. However, the field seems to be ahead in addressing complex problem solving and being of policy relevance, e.g., by broadening of scope ([95–102]) to clarify the ambivalent and disputed concept of sustainability and assess it in all its dimensions; by strategic future orientation and dealing with uncertainties ([101–108]); by participative approaches and dealing with stakeholders, normative aspects, and plurality ([107,109–115]); by analyzing how assessments can support policy-making and governance ([78,106,114,116–119]); and by increasing reflexivity and making the normative basis of assessments explicit ([78,101,102,117,120]).

6. Final Conclusions

We set out to discuss the value and limitations of a Systems Engineering approach for contributing to societal challenges, in particular the energy transition. We did so by approaching the energy system as a sociotechnical system. We unpack the stakeholders' roles, agency, and acceptance in the case of the Naastenbest project in Best. It revealed the pluralism between and within stakeholder groups, the distributed agency, the relevance of trust and decision-making processes, and the context-dependency. Finally, we discussed the challenges this poses for the traditional Systems Engineering approach; we indicated several potential strategies to address these challenges; and identified two fields that can help clarify how to address these challenges: transition studies and sustainability assessment.

Author Contributions: A.K., Conceptualization; Formal analysis; Investigation; Writing–original draft; Writing–review and editing; Supervision. V.E., Conceptualization; Methodology; Formal analysis; Validation; Investigation; Resources; Data Curation; Writing–empirical chapters; Writing–review and editing; Visualization. G.M., Conceptualization; Formal analysis; Writing–review and editing; Supervision. All authors have read and agreed to the published version of the manuscript.

Funding: This research received no external funding.

Institutional Review Board Statement: Not applicable.

Informed Consent Statement: Not applicable.

Data Availability Statement: Not applicable.

Acknowledgments: We would like to thank two anonymous reviewers for their constructive comments.

Conflicts of Interest: The authors declare no conflict of interest.

References

1. Singh, H.V.; Bocca, R.; Gomez, P.; Dahlke, S.; Bazilian, M. The energy transitions index: An analytic framework for understanding the evolving global energy system. *Energy Strat. Rev.* **2019**, *26*, 100382. [CrossRef]
2. Fonseca, L.M.; Domingues, J.P.; Dima, A.M. Mapping the Sustainable Development Goals Relationships. *Sustainability* **2020**, *12*, 3359. [CrossRef]
3. Anderson, D.; Coelho, S.T.; Doucet, G.; Freudenschuss-Reichl, I.; Jefferson, M.; Jochem, E.; Williams, R.H. *World Energy Assessment Overview: 2004 Update*; United Nations Development Programme: New York, NY, USA, 2004; ISBN 978-92-1-126167-7.
4. Goldemberg, J. *World Energy Assessment. Energy and the Challenge of Sustainability*; United Nations Development Programme: New York, NY, USA, 2000; ISBN 92-1-126126-0.
5. United Nations. *The Sustainable Development Goals Report 2017*; United Nations: New York, NY, USA, 2017; ISBN 978-92-1-101368-9.
6. United Nations; Department of Economic and Social Affairs. The 17 Sustainable Development Goals. Available online: https://sdgs.un.org/goals (accessed on 5 May 2021).
7. IPCC. *Summary for Policymakers*; IPCC: Geneva, Switzerland, 2018; ISBN 978-92-9169-151-7.
8. United Nations/Framework Convention on Climate Change. *Adoption of the Paris Agreement, 21st Conference of the Parties*; Paris, France, 2015. Available online: https://unfccc.int/resource/docs/2015/cop21/eng/l09r01.pdf (accessed on 5 May 2021).
9. Markard, J.; Raven, R.; Truffer, B. Sustainability transitions: An emerging field of research and its prospects. *Res. Policy* **2012**, *41*, 955–967. [CrossRef]
10. Köhler, J.; Geels, F.W.; Kern, F.; Markard, J.; Onsongo, E.; Wieczorek, A.; Alkemade, F.; Avelino, F.; Bergek, A.; Boons, F.; et al. An agenda for sustainability transitions research: State of the art and future directions. *Environ. Innov. Soc. Transit.* **2019**, *31*, 1–32. [CrossRef]
11. Sengers, F.; Wieczorek, A.J.; Raven, R. Experimenting for sustainability transitions: A systematic literature review. *Technol. Forecast. Soc. Chang.* **2019**, *145*, 153–164. [CrossRef]
12. International Renewable Energy Agency. *REthinking Energy*; International Renewable Energy Agency: Abu Dhabi, United Arab Emirates, 2017.
13. EC-European Commission. *2020 Report on the State of the Energy Union*; EC-European Commission: Brussels, Belgium, 2020.
14. EC-European Commission. *Renewable Energy Progress Report*; EC-European Commission: Brussels, Belgium, 2020.
15. Bradford, T. *Energy System: Technology, Economics, Markets, and Policy*; MIT Press: Cambridge, MA, USA, 2018.
16. INCOSE. *A World in Motion. Systems Engineering Vision 2025*; INCOSE: San Diego, CA, USA, 2014.
17. Geels, F.W. From sectoral systems of innovation to socio-technical systems. *Res. Policy* **2004**, *33*, 897–920. [CrossRef]
18. Salado, A. A systems-theoretic articulation of stakeholder needs and system requirements. *Syst. Eng.* **2021**, *24*, 83–99. [CrossRef]
19. Kasser, J. 7.7.1 Getting the right requirements right. *INCOSE Int. Symp.* **2012**, *22*, 1005–1020. [CrossRef]
20. SEBoK Authors. Stakeholder Needs and Requirements. Available online: https://www.sebokwiki.org/w/index.php?title=Stakeholder_Needs_and_Requirements&oldid=59944 (accessed on 7 March 2021).
21. Batel, S.; Devine-Wright, P.; Tangeland, T. Social acceptance of low carbon energy and associated infrastructures: A critical discussion. *Energy Policy* **2013**, *58*, 1–5. [CrossRef]
22. Raven, R.; Mourik, R.R.; Feenstra, C.; Heiskanen, E. Modulating societal acceptance in new energy projects: Towards a toolkit methodology for project managers. *Energy* **2009**, *34*, 564–574. [CrossRef]
23. Perlaviciute, G.; Steg, L. Contextual and psychological factors shaping evaluations and acceptability of energy alternatives: Integrated review and research agenda. *Renew. Sustain. Energy Rev.* **2014**, *35*, 361–381. [CrossRef]
24. Steg, L.; Perlaviciute, G.; Van Der Werff, E. Understanding the human dimensions of a sustainable energy transition. *Front. Psychol.* **2015**, *6*, 805. [CrossRef]
25. Segreto, M.; Principe, L.; Desormeaux, A.; Torre, M.; Tomassetti, L.; Tratzi, P.; Paolini, V.; Petracchini, F. Trends in Social Acceptance of Renewable Energy Across Europe—A Literature Review. *Int. J. Environ. Res. Public Health* **2020**, *17*, 9161. [CrossRef] [PubMed]
26. Seidl, R.; Von Wirth, T.; Krütli, P. Social acceptance of distributed energy systems in Swiss, German, and Austrian energy transitions. *Energy Res. Soc. Sci.* **2019**, *54*, 117–128. [CrossRef]
27. Sonnberger, M.; Ruddat, M. Local and socio-political acceptance of wind farms in Germany. *Technol. Soc.* **2017**, *51*, 56–65. [CrossRef]
28. Krick, E. Ensuring social acceptance of the energy transition. The German government's 'consensus management' strategy. *J. Environ. Policy Plan.* **2017**, *20*, 64–80. [CrossRef]

29. Wüstenhagen, R.; Wolsink, M.; Bürer, M.J. Social acceptance of renewable energy innovation: An introduction to the concept. *Energy Policy* **2007**, *35*, 2683–2691. [CrossRef]
30. Heiskanen, E.; Hodson, M.; Mourik, R.M.; Raven, R.P.J.M.; Feenstra, C.F.J.; Torrent, A.A.; Brohmann, B.; Daniels, A.; Di Fiore, M.; Farkas, B.; et al. Factors Influencing the Societal Acceptance of New Energy Technologies: Meta-Analysis of Recent European Projects. 2008; ECN-E–07-058.
31. Sillitto, H.; Griego, R.; Arnold, E.; Dori, D.; Martin, J.; McKinney, D.; Godfrey, P.; Krob, D.; Jackson, S. A fresh look at Systems Engineering—What is it, how should it work? *INCOSE Int. Symp.* **2018**, *28*, 955–970. [CrossRef]
32. Rijksoverheid Klimaatakkoord. *Klimaatakkoord. Den Haag*. 2019. Available online: https://www.rijksoverheid.nl/documenten/rapporten/2019/06/28/klimaatakkoord (accessed on 5 May 2021).
33. Hoppe, T.; Miedema, M. A Governance Approach to Regional Energy Transition: Meaning, Conceptualization and Practice. *Sustainability* **2020**, *12*, 915. [CrossRef]
34. Vringer, K.; De Vries, R.; Visser, H. Measuring governing capacity for the energy transition of Dutch municipalities. *Energy Policy* **2021**, *149*, 112002. [CrossRef]
35. Van Bergen, T.; Van Neerven, M.; Van Oosterhout, M. *Beleidsplan Energie-en Materiaaltransitie*. 2011. Available online: https://repository.officiele-overheidspublicaties.nl/externebijlagen/exb-2019-24239/1/bijlage/exb-2019-24239.pdf (accessed on 5 May 2021).
36. Algra, C. Uitvoeringsprogramma Energietransitie 2017–2018. 2017. Available online: https://decentrale.regelgeving.overheid.nl/cvdr/xhtmloutput/Historie/Best/624284/CVDR624284_1.html (accessed on 5 May 2021).
37. Gemeente Best Samen aan de Slag met Duurzaam Wonen in Best. 2017. Available online: https://www.gemeentebest.nl/data/downloadables/7/4/6/3/nota-samen-aan-de-slag-met-duurzaam-wonen-in-best.pdf (accessed on 5 May 2021).
38. Gemeenteraad Best. Raadsvoorstel: Duurzame Warmtevoorziening bij Reconstructie Naastenbest Aanleiding. 2020.
39. Gemeenteraad Best. Motie Gemeenteraad Best: Duurzame warmtevoorziening Naastenbest (20015) 09-03-2020. Available online: https://best.notubiz.nl/modules/6/moties_en_toezeggingen/view (accessed on 5 May 2021).
40. Elvebakk, L.; Muller, G. Creating a Roadmap to Capture a Vision for a Sustainable Community in a Global Perspective: A Case Study in a Dutch Town Best. In Proceedings of the 29th INCOSE International Symposium, Orlando, FL, USA, 20–25 July 2019.
41. Muller, G.; Elvebakk, L.; Van Der Velde, J.; Lean, F.M. Roadmapping for sustainability; How to navigate a social, political, and many systems-of-systems playing eld? A local initiative. In Proceedings of the 14th Annual Conference System of Systems Engineering, SoSE 2019, Anchorage, AK, USA, 19–22 May 2019; pp. 317–322.
42. Schot, J.; Kanger, L.; Verbong, G.G. The roles of users in shaping transitions to new energy systems. *Nat. Energy* **2016**, *1*, 16054. [CrossRef]
43. Köhler, J.; Geels, F.; Kern, F.; Onsongo, E.; Wieczorek, A.; Alkemaade, F.; Avelino, F.; Bergek, A.; Boons, F.; Bulkeley, H.; et al. *A Research Agenda for the Sustainability Transitions Research Network*. 2017. Available online: https://pure.tue.nl/ws/files/101288346/STRN_Research_Agenda_2017.pdf (accessed on 5 May 2021).
44. Moncada, J.A.; Lee, E.H.P.; Guerrero, G.N.; Okur, O.; Chakraborty, S.; Lukszo, Z. Complex Systems Engineering: Designing in sociotechnical systems for the energy transition. *EAI Endorsed Trans. Energy Web* **2017**, *3*, 152762. [CrossRef]
45. Rittel, H.W.J.; Webber, M.M. Dilemmas in a general theory of planning. *Policy Sci.* **1973**, *4*, 155–169. [CrossRef]
46. Probst, G.; Bassi, A.M. *Tackling Complexity. A Systemic Approach for Decision Makers*; Greenleaf Publishing: Sheffield, UK, 2014.
47. Kurtz, C.; Snowden, D. The new dynamics of strategy: Sense-making in a complex and complicated world. *IEEE Eng. Manag. Rev.* **2003**, *31*, 110. [CrossRef]
48. Sovacool, B.K.; Hess, D.J.; Amir, S.; Geels, F.W.; Hirsh, R.; Medina, L.R.; Miller, C.; Palavicino, C.A.; Phadke, R.; Ryghaug, M.; et al. Sociotechnical agendas: Reviewing future directions for energy and climate research. *Energy Res. Soc. Sci.* **2020**, *70*, 101617. [CrossRef]
49. Hisschemöller, M.; Midden, C.J.H. Improving the usability of research on the public perception of science and technology for policy-making. *Public Underst. Sci.* **1999**, *8*, 17–33. [CrossRef]
50. Huijts, N.M.A.; Molin, E.J.E.; Steg, L. Psychological factors influencing sustainable energy technology acceptance: A re-view-based comprehensive framework. *Renew. Sustain. Energy Rev.* **2012**, *16*, 525–531. [CrossRef]
51. Van Der Lelij, B.; De Graaf, M.; Visscher, J. *Energievoorziening 2015–2050: Publieksonderzoek naar Draagvlak voor Verduurzaming van Energie*; Ministry of Economic Affairs: Den Haag, The Netherlands, 2016; pp. 1–81.
52. PAW Rapportage Reflectieve Monitor 2019. Voortgang & Leerervaringen. 2020. Available online: https://www.rijksoverheid.nl/binaries/rijksoverheid/documenten/rapporten/2020/01/22/voortgang-en-leerervaringen-27-proeftuinen-aardgasvrije-wijken/rapportage-reflectieve-monitor-2019.pdf (accessed on 5 May 2021).
53. Bain, P.G.; Milfont, T.L.; Kashima, Y.; Bilewicz, M.; Doron, G.; Garðarsdóttir, R.B.; Gouveia, V.V.; Guan, Y.; Johansson, L.-O.; Pasquali, C.; et al. Co-benefits of addressing climate change can motivate action around the world. *Nat. Clim. Chang.* **2016**, *6*, 154–157. [CrossRef]
54. Lennon, B.; Dunphy, N.P.; Sanvicente, E. Community acceptability and the energy transition: A citizens' perspective. *Energy Sustain. Soc.* **2019**, *9*, 1–18. [CrossRef]
55. Blois, K.J. Trust in Business to Business Relationships: An Evaluation of its Status. *J. Manag. Stud.* **1999**, *36*, 197–215. [CrossRef]
56. Gounaris, S.P. Trust and commitment influences on customer retention: Insights from business-to-business services. *J. Bus. Res.* **2005**, *58*, 126–140. [CrossRef]

57. Uslaner, E.M. Trust, Democracy and Governance: Can Government Policies Influence Generalized Trust? In *Generating Social Capital*; Springer: Berlin, Germany, 2003; pp. 171–190.
58. DiMaggio, P.; Braithwaite, V.; Levi, M. Trust and Governance. *Contemp. Sociol. A J. Rev.* **1999**, *28*, 731. [CrossRef]
59. Nye, J.S., Jr. In government we don't trust. *Foreign Policy* **1997**, 99–111. [CrossRef]
60. Midden, C.J.H.; Huijts, N.M.A. The Role of Trust in the Affective Evaluation of Novel Risks: The Case of CO_2 Storage. *Risk Anal.* **2009**, *29*, 743–751. [CrossRef] [PubMed]
61. Visschers, V.H.M.; Siegrist, M. How a Nuclear Power Plant Accident Influences Acceptance of Nuclear Power: Results of a Longitudinal Study Before and After the Fukushima Disaster. *Risk Anal.* **2012**, *33*, 333–347. [CrossRef] [PubMed]
62. Harrison, N.; Luckett, K. Experts, knowledge and criticality in the age of 'alternative facts': Re-examining the contribution of higher education. *Teach. High. Educ.* **2019**, *24*, 259–271. [CrossRef]
63. Newman, S. Post-truth and the crisis of the political. *Soft Power* **2019**, *6*, 90–108.
64. Westera, N. District Heating Ownership. Master's Thesis, Delft University of Technology, Delft, The Netherlands, 2018.
65. Batel, S. Research on the social acceptance of renewable energy technologies: Past, present and future. *Energy Res. Soc. Sci.* **2020**, *68*, 101544. [CrossRef]
66. Barriball, K.L.; While, A. Collecting data using a semi-structured interview: A discussion paper. *J. Adv. Nurs.* **1994**, *19*, 328–335. [CrossRef]
67. Smith, A.; Raven, R. What is protective space? Reconsidering niches in transitions to sustainability. *Res. Policy* **2012**, *41*, 1025–1036. [CrossRef]
68. Sataloff, R.T.; Johns, M.M.; Kost, K.M. *Complex Systems. Breakthrough Innovations Through People. The H-SEIF Project Booklet*; University of South-Eastern Norway: Notodden, Norway, 2020; ISBN 9781626239777.
69. Abson, D.J.; Fischer, J.; Leventon, J.; Newig, J.; Schomerus, T.; Vilsmaier, U.; Von Wehrden, H.; Abernethy, P.; Ives, C.D.; Jager, N.W.; et al. Leverage points for sustainability transformation. *AMBIO* **2017**, *46*, 30–39. [CrossRef]
70. Meadowcroft, J. What about the politics? Sustainable development, transition management, and long term energy transitions. *Policy Sci.* **2009**, *42*, 323–340. [CrossRef]
71. Fazey, I.; Schäpke, N.; Caniglia, G.; Patterson, J.; Hultman, J.; van Mierlo, B.; Säwe, F.; Wiek, A.; Wittmayer, J.; Aldunce, P.; et al. Ten essentials for action-oriented and second order energy transitions, transformations and climate change research. *Energy Res. Soc. Sci.* **2018**, *40*, 54–70. [CrossRef]
72. Turnheim, B.; Geels, F.W. The destabilisation of existing regimes: Confronting a multi-dimensional framework with a case study of the British coal industry (1913–1967). *Res. Policy* **2013**, *42*, 1749–1767. [CrossRef]
73. Meadows, D. *Leverage Points: Places to Intervene in a System*; The Sustainability Institute: Hartland, VT, USA, 1999.
74. Holtz, G.; Alkemade, F.; de Haan, F.; Köhler, J.; Trutnevyte, E.; Luthe, T.; Halbe, J.; Papachristos, G.; Chappin, E.; Kwakkel, J.; et al. Prospects of modelling societal transitions: Position paper of an emerging community. *Environ. Innov. Soc. Transit.* **2015**, *17*, 41–58. [CrossRef]
75. Boardman, J.; Sauser, B. *Systems Thinking: Coping with 21st Century Problems*; CRC Press: Boca Raton, FL, USA, 2008.
76. Checkland, P. *Systems Thinking, Systems Practice*; John Wiley and Sons: New York, NY, USA, 1999.
77. Rechtin, E.; Maier, M.W. *The Art of Systems Architecting*; Apple Academic Press: Cambridge, MA, USA, 2010.
78. Tukker, A. Philosophy of science, policy sciences and the basis of decision support with LCA Based on the toxicity controversy in Sweden and the Netherlands. *Int. J. Life Cycle Assess.* **2000**, *5*, 177–186. [CrossRef]
79. Akeel, U.U.; Bell, S.J. Discourses of systems engineering. *Eng. Stud.* **2013**, *5*, 160–173. [CrossRef]
80. Kroes, P.; Franssen, M.; Van De Poel, I.; Ottens, M. Treating socio-technical systems as engineering systems: Some conceptual problems. *Syst. Res. Behav. Sci.* **2006**, *23*, 803–814. [CrossRef]
81. Ottens, M.; Franssen, M.; Kroes, P.; Van De Poel, I. 8.1.1 Systems engineering of socio-technical systems. *INCOSE Int. Symp.* **2005**, *15*, 1122–1130. [CrossRef]
82. Davis, M.C.; Challenger, R.; Jayewardene, D.N.; Clegg, C.W. Advancing socio-technical systems thinking: A call for bravery. *Appl. Ergon.* **2014**, *45*, 171–180. [CrossRef] [PubMed]
83. Davis, K.; Mazzuchi, T.; Sarkani, S. Architecting technology transitions: A sustainability-oriented sociotechnical approach. *Syst. Eng.* **2012**, *16*, 193–212. [CrossRef]
84. Haskins, C. Using patterns to transition systems engineering from a technological to social context. *Syst. Eng.* **2008**, *11*, 147–155. [CrossRef]
85. Wixom, B.H.; Todd, P.A. A Theoretical Integration of User Satisfaction and Technology Acceptance. *Inf. Syst. Res.* **2005**, *16*, 85–102. [CrossRef]
86. Poller, A. Exploring and managing the complexity of large infrastructure projects with network theory and model-based systems engineering—The example of radioactive waste disposal. *Syst. Eng.* **2020**, *23*, 443–459. [CrossRef]
87. Maier, M.W. Architecting Principles for Systems-of-Systems. *Syst. Eng.* **1999**, *1*, 267–284. [CrossRef]
88. Gorod, A.; Sauser, B.; Boardman, J. System-of-Systems Engineering Management: A Review of Modern History and a Path Forward. *IEEE Syst. J.* **2008**, *2*, 484–499. [CrossRef]
89. Boardman, J. System of Systems—the meaning of of. In Proceedings of the 2006 IEEE/SMC International Conference on System of Systems Engineering, Los Angeles, CA, USA, 24–26 April 2006; pp. 118–123.

90. Neely, K.; Bortz, M.; Bice, S. Using collaborative conceptual modelling as a tool for transdisciplinarity. *Evid. Policy A J. Res. Debate Pr.* **2021**, *17*, 161–172. [CrossRef]
91. Newell, B.; Proust, K. *Introduction to Collaborative Conceptual Modelling*; Work. Pap. ANU Open Access Res.; 2012; p. 20. Available online: https://digitalcollections.anu.edu.au/handle/1885/9386 (accessed on 5 May 2021).
92. Muller, G. Who Does the Societal Systems of Systems Thinking?
93. Geels, F.W.; Berkhout, F.; Van Vuuren, D.P. Bridging analytical approaches for low-carbon transitions. *Nat. Clim. Chang.* **2016**, *6*, 576–583. [CrossRef]
94. McDowall, W.; Geels, F.W. Ten challenges for computer models in transitions research: Commentary on Holtz et al. *Environ. Innov. Soc. Transit.* **2017**, *22*, 41–49. [CrossRef]
95. Jeswani, H.K.; Azapagic, A.; Schepelmann, P.; Ritthoff, M. Options for broadening and deepening the LCA approaches. *J. Clean. Prod.* **2010**, *18*, 120–127. [CrossRef]
96. Onat, N.C.; Kucukvar, M.; Halog, A.; Cloutier, S. Systems Thinking for Life Cycle Sustainability Assessment: A Review of Recent Developments, Applications, and Future Perspectives. *Sustainability* **2017**, *9*, 706. [CrossRef]
97. Dijk, M.; De Kraker, J.; Van Zeijl-Rozema, A.; Van Lente, H.; Beumer, C.; Beemsterboer, S.; Valkering, P. Sustainability assessment as problem structuring: Three typical ways. *Sustain. Sci.* **2017**, *12*, 305–317. [CrossRef] [PubMed]
98. Sala, S.; Farioli, F.; Zamagni, A. Progress in sustainability science: Lessons learnt from current methodologies for sustainability assessment: Part. *Int. J. Life Cycle Assess.* **2013**, *18*, 1653–1672. [CrossRef]
99. Sala, S.; Ciuffo, B.; Nijkamp, P. A systemic framework for sustainability assessment. *Ecol. Econ.* **2015**, *119*, 314–325. [CrossRef]
100. Gasparatos, A.; El-Haram, M.; Horner, M. A critical review of reductionist approaches for assessing the progress towards sustainability. *Environ. Impact Assess. Rev.* **2008**, *28*, 286–311. [CrossRef]
101. Reap, J.; Roman, F.; Duncan, S.; Bras, B. A survey of unresolved problems in life cycle assessment. Part 2: Impact assessment and interpretation. *Int. J. Life Cycle Assess.* **2008**, *13*, 374–388. [CrossRef]
102. Reap, J.; Roman, F.; Duncan, S.; Bras, B. A survey of unresolved problems in life cycle assessment. Part 1: Goal and scope and inventory analysis. *Int. J. Life Cycle Assess.* **2008**, *13*, 290–300. [CrossRef]
103. Arvidsson, R.; Tillman, A.; Sandén, B.A.; Janssen, M.; Nordelöf, A.; Kushnir, D.; Molander, S. Environmental Assessment of Emerging Technologies: Recommendations for Prospective LCA. *J. Ind. Ecol.* **2018**, *22*, 1286–1294. [CrossRef]
104. Waas, T.; Hugé, J.; Block, T.; Wright, T.; Benitez-Capistros, F.; Verbruggen, A. Sustainability Assessment and Indicators: Tools in a Decision-Making Strategy for Sustainable Development. *Sustainability* **2014**, *6*, 5512–5534. [CrossRef]
105. Beltran, A.M.; Cox, B.; Mutel, C.; Van Vuuren, D.P.; Vivanco, D.F.; Deetman, S.; Edelenbosch, O.Y.; Guinée, J.; Tukker, A. When the Background Matters: Using Scenarios from Integrated Assessment Models in Prospective Life Cycle Assessment. *J. Ind. Ecol.* **2020**, *24*, 64–79. [CrossRef]
106. Van Der Sluijs, J.P.; Petersen, A.C.; Janssen, P.H.M.; Risbey, J.S.; Ravetz, J.R. Exploring the quality of evidence for complex and contested policy decisions. *Environ. Res. Lett.* **2008**, *3*. [CrossRef]
107. Matthews, N.E.; Stamford, L.; Shapira, P. Aligning sustainability assessment with responsible research and innovation: Towards a framework for Constructive Sustainability Assessment. *Sustain. Prod. Consum.* **2019**, *20*, 58–73. [CrossRef] [PubMed]
108. Weidema, B.P.; Ekvall, T.; Pesonen, H.-L.; Rebitzer, G.; Sonnemann, G.W.; Spielmann, M. *Scenarios in Life-Cycle Assessment*; Society of Environmental Toxicology and Chemistry (SETAC): Pensacola FL, USA, 2004; ISBN 1-880611-57-0.
109. Reed, M.S. Stakeholder participation for environmental management: A literature review. *Biol. Conserv.* **2008**, *141*, 2417–2431. [CrossRef]
110. Reed, M.; Fraser, E.D.G.; Morse, S.; Dougill, A.J. Integrating Methods for Developing Sustainability Indicators to Facilitate Learning and Action. *Ecol. Soc.* **2005**, *10*, 10. [CrossRef]
111. McCabe, A.; Halog, A. Exploring the potential of participatory systems thinking techniques in progressing SLCA. *Int. J. Life Cycle Assess.* **2016**, *23*, 739–750. [CrossRef]
112. Connelly, S. Mapping Sustainable Development as a Contested Concept. *Local Environ.* **2007**, *12*, 259–278. [CrossRef]
113. Scolobig, A.; Lilliestam, J. Comparing Approaches for the Integration of Stakeholder Perspectives in Environmental Decision Making. *Resources* **2016**, *5*, 37. [CrossRef]
114. Astleithner, F.; Hamedinger, A. The Analysis of Sustainability Indicators as Socially Constructed Policy Instruments: Benefits and challenges of 'interactive research'. *Local Environ.* **2003**, *8*, 627–640. [CrossRef]
115. Mathur, V.N.; Price, A.D.; Austin, S.A. Conceptualizing stakeholder engagement in the context of sustainability and its assessment. *Constr. Manag. Econ.* **2008**, *26*, 601–609. [CrossRef]
116. Seidel, C. The application of life cycle assessment to public policy development. *Int. J. Life Cycle Assess.* **2016**, *21*, 337–348. [CrossRef]
117. Holden, M. Sustainability indicator systems within urban governance: Usability analysis of sustainability indicator systems as boundary objects. *Ecol. Indic.* **2013**, *32*, 89–96. [CrossRef]
118. Holman, N. Incorporating local sustainability indicators into structures of local governance: A review of the literature. *Local Environ.* **2009**, *14*, 365–375. [CrossRef]
119. Hezri, A.A.; Dovers, S.R. Sustainability indicators, policy and governance: Issues for ecological economics. *Ecol. Econ.* **2006**, *60*, 86–99. [CrossRef]
120. Freidberg, S. From behind the curtain: Talking about values in LCA. *Int. J. Life Cycle Assess.* **2018**, *23*, 1410–1414. [CrossRef]

sustainability

MDPI

Article

HoReCa Food Waste and Sustainable Development Goals—A Systemic View

Aleksander Buczacki [1,*], Bartłomiej Gładysz [1] and Erika Palmer [2]

[1] Faculty of Production Engineering, Warsaw University of Technology, 02-524 Warszawa, Poland; bartlomiej.gladysz@pw.edu.pl
[2] Sintef Ocean, N-7010 Trondheim, Norway; erika.palmer@sintef.no
* Correspondence: aleksander.buczacki@pw.edu.pl; Tel.: +48-22-234-8126

Abstract: A significant share of food waste originates in the food services domain and HoReCa sector. Organizational improvements leading to the decrease of food waste and related costs in HoReCa are needed to make progress in this issue. A systems engineering approach was applied to examine the links between food waste generated in the HoReCa industry and the Sustainable Development Goals (SDGs). A literature review discovered two dimensions of actions leading to decreasing food waste in HoReCa; i.e., actions triggered by companies and by authorities (e.g., governmental policies). Additionally, customers and society were also considered. A framework is proposed to explicitly illustrate the dependencies of different micro actions devoted to food waste reduction in HoReCa in support of the SDGs. The other dimension of this framework is macro policies and their impact on SDGs. To increase food waste reduction awareness and collaboration, stakeholders on both the macro (launched by authorities for the whole sector) and micro (initiated by single organizations on their own) levels must work together. The results of this research will be useful in coordinating the efforts of all (consumers, HoReCa companies and suppliers, policymakers and administrations on different levels) involved in the supply chain of food production and consumption.

Keywords: food waste; food services; HoReCa; Sustainable Development Goal; regional policy

Citation: Buczacki, A.; Gładysz, B.; Palmer, E. HoReCa Food Waste and Sustainable Development Goals—A Systemic View. *Sustainability* **2021**, *13*, 5510. https://doi.org/10.3390/su13105510

Academic Editor: Michael Carolan

Received: 13 April 2021
Accepted: 13 May 2021
Published: 14 May 2021

1. Introduction

Sluggish progress towards the targets of the 17 Sustainable Development Goals (SDGs) suggests that there is an urgent need to consider new approaches that clarify their interrelationships while also accounting for both their complexity and their sometimes mutually reinforcing or conflicting objectives. Attempting to achieve them in a linear succession is impossible, but pursuing them simultaneously is impractical [1].

Researchers from developed countries (USA and Europe) have been interested in food waste (FW) reduction, also recycling and prevention, since the 1910s [2,3]. In the past, the main focus was on waste disposal, but from the early 1980s, the awareness of the importance of FW prevention and reduction arose again [4,5]. Food is wasted along the entire food chain, starting from farms, plantations and fisheries, through processing facilities, transportation, distribution sites, retail establishments, restaurants and homes [6,7]. In the supply chain, from harvesting to the processing stage, 13.8% of food is wasted [8]. A significant share of food waste is generated in the food services industry, including restaurants, bars, bistros, fast food chains, catering, etc. It is estimated that about 7% of food is wasted in US restaurants before the consumer is served [9], and as determined for 2012, about 12% of food is wasted in the food service industry in the European Union [10]. For effective FW management in HoReCa (hotels, restaurants, catering), consumers' habits should be considered along with additional stakeholder activities [11,12]. Typical sources of FW are oversized portions [13], inflexibility of chain store management, extensive menu choices [9], and meals served mistakenly or delayed. According to research results conducted in selected European countries, on average 20% of meals are wasted in the hospitality sector [14].

Sustainable Development Goal 12 has a target that addresses these statistics, namely, 12.3, which states, "By 2030, halve per capita global food waste at retail and consumer levels and reduce food loses along production and supply chains, including post-harvest losses [15]."

There are a number of studies focused on companies' operational efficiency from the HoReCa sector that consider food waste on the micro (company) level, also taking into account the macro level, but without an SDG context. However, there is research focused on food waste on regional, country or event global (macro) levels that considers different SDGs and their indicators [15]. There is currently a lack of research with an integrative view considering both micro (organization) and macro (policymakers and economies) levels, and the SDG implementation perspective. Due to this, the main objective of this paper is a proposal for food waste reduction practice integration on micro and macro levels in the case of SDG objectives management using a systems engineering approach. In using a systems engineering approach, we apply systems thinking to food waste in HoReCa in an SDG context. In addition to developing a proposal, we place the results of this study into a larger discussion of how using a systems engineering approach addresses criticism of SDG targets in general using HoReCa as an example.

We begin this paper with an explanation of how we conducted the systematic literature review, and then we present the results in Section 3 in a systems engineering context. We conclude this study with a discussion of how a systems engineering approach can foster progress towards SDG targets. This paper provides an overview of dependencies between different SDGs from a food waste reduction point of view. It also provides an introduction to available practices for HoReCa businesses, with consideration of their own actions as well as authority-driven programs. The goal is, therefore, twofold, i.e.,

- To develop a conceptual causal loop diagram for food waste in the HoReCa sector;
- To develop a conceptual causal loop diagram for SDGs' support from HoReCa best practices on food waste.

2. Materials and Methods

For this research, the following systematic literature review research methodology was implemented. First, the following queries (for article title, abstract and keywords) in WoS and Scopus databases were performed:

(1) "Food waste" AND (HoReCa OR Hospitality OR "food service")
(2) "Food waste" AND "Sustainable Development Goals"

Hospitality and the food sector were included in the query, as the term "HoReCa" is not often used in publications on hotel, restaurant and catering operations. In the next step, only papers in WoS in English without date limits, related to environmental sciences, environmental engineering, green sustainable science technology, environmental studies, nutrition dietetics, management, business, economics, behavioral sciences and operations research management science were considered. In Scopus, only papers in English without date limits, related to environmental sciences, business, management and accounting, social sciences, economics, econometrics and finance and decision sciences were considered. Through abstract analysis, papers related to technical questions were excluded.

Only papers from 2015 to the present in English have been taken into consideration because the SDGs were accepted by the UN in 2015. The majority of the papers for the Food Waste AND SDG query focused on the circular bioeconomy and effective energy management, which are not relevant for this research. The second group of papers focused on technical questions. Some of the papers focused on recycling in general, not related to the HoReCa sector and FW reduction practices, so such papers were also not taken into consideration.

Finally, only papers related to food waste and HoReCa and related with HoReCa (hospitality and food service sectors) have been taken into further analysis (Table 1).

Table 1. Number of papers analyzed in particular steps.

Sample Characteristics	WoS	Scopus	WoS + Scopus with No Repetitions
"Food waste" AND (HoReCa OR Hospitality OR "Food service")	145	160	N/A
Selected in first step after refinement	103	105	N/A
Selected in second step after abstracts analysis	19	18	20
Food waste AND Sustainable Development Goal (FW AND SDG)	262	382	N/A
FW AND SDG from 2015 in English	232	310	N/A
FW AND SDG after abstracts analysis	26	27	28

Selected papers have been analyzed according to the level of food waste practices:

- Company level (micro)—closely related to companies' operational activities and customer behavior.
- Policymakers at regional, national or global level (macro)—related to polices and regional, national or global programs.

Best practices on the micro level, and in particular related to restaurants, have been identified. Preliminary impact of the identified best practices has been assessed.

Concurrently, best practices implemented by local, regional and national authorities as well as by international organizations and related to food waste appropriate SDGs have been identified. Later, relationships were analyzed separately for best practices on the macro level and the SDG focus. Finally, based on the analysis results, we developed the integrated framework. Figure 1 illustrates the research methodology.

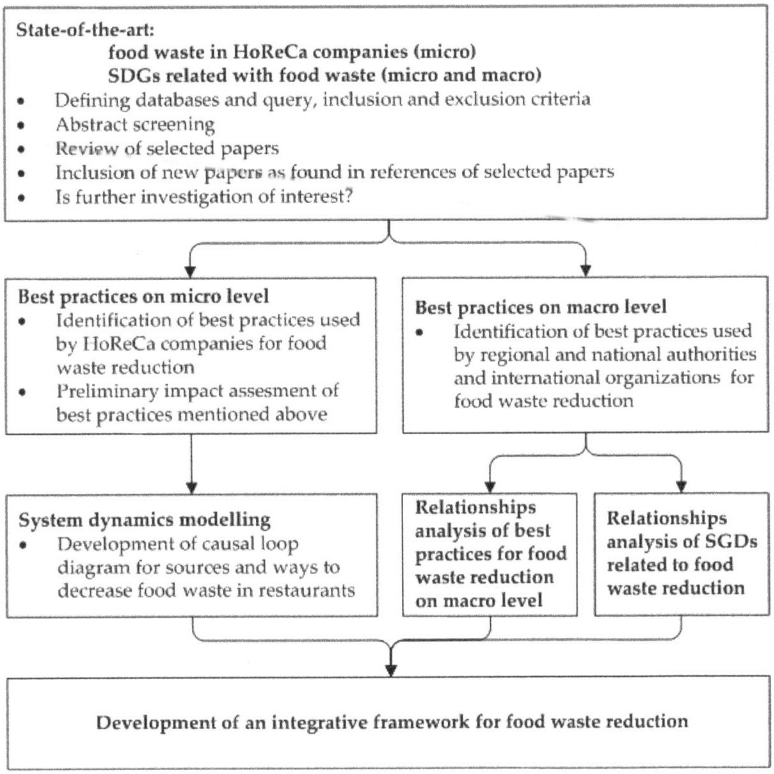

Figure 1. Research methodology. Source: authors' elaboration.

3. Results

3.1. Queries Review Results

The tables below present the literature review results of two queries:

1. FW AND HORECA (Table 2), with year of publication and whether it concerns good practices addressed on the micro and macro level entities of both.
2. FW AND SDGs (Table 3), with year of publication and which SDGs the publication concerns. Additionally, the tables indicate whether the best practices described in the publication related to the micro and macro levels and entities of both.

Table 2. Summary of papers on food waste (FW) and hotels, restaurants and catering (HoReCa).

Ref.	Year	Level		Summary
		Micro	Macro	
[16]	2021	X		Analysis of responses from 206 managers of Brazilian food service companies. Findings focused on FW reduction.
[17]	2021		X	Presentation of sociodemographic and food consumption-related factors generated by tourists in the HoReCa sector. Insights into achieving sustainability objectives in national and international tourism sectors.
[18]	2020	X		Description of best practices implemented by 3 HoReCa companies based on lean management approach and its impact on FW reduction.
[19]	2020	X	X	Description of multi-stakeholder partnership program "Food waste Challenge", involving 172 restaurants. Program focuses on a range of behavioral interventions.
[20]	2020	X	X	Paper focuses on the lifecycle assessment of 15 tapas meals, as well as analysis of nutritional quality of the meals and energy efficiency at the restaurant.
[21]	2020		X	Analysis of social, economic and environmental impact of food waste reduction activities.
[22]	2020		X	Analysis of restaurant food waste management practices in the UK and the Netherlands in case of FW management.
[23]	2020	X		Findings for food waste measurement and waste reduction strategies in the different stages (pre-kitchen, in-kitchen and post-kitchen). Results based on the survey of almost 500 HoReCa managers.
[24]	2020	X		Results of semi-structured interviews with 32 hotel employees and managers. Proposition of possible strategies for reducing FW addressed for restaurants and consumers.
[25]	2019	X		Evaluation of interventions in two university canteens on their effectiveness to reduce visitors' plate waste.
[14]	2019	X	X	Empirical evidence for FW reduction in HoReCa sector in Malaysia. Proposition of FW prevention strategies related to companies' operations activities and consumers' social practices.
[26]	2019	X	X	Analysis of direct-weighing data (from 164 restaurants) and its ecological footprint rather on macro-level.
[27]	2019	X	X	A case study of the restaurant activity in Bulgaria. Proposition of training programs for restaurateurs and public authorities.
[28]	2019		X	Presentation of research results for the quantification of food waste based on the data hospitality sector (1189 kitchens) in Sweden, Norway, Finland and Germany. Proposition of statistical measures for food waste tracking
[29]	2018		X	Insights of food waste generation factors in HoReCa sector at the national level.
[30]	2018	X		Identification of waste management initiatives in the food service sector and evaluation of management practices for waste reduction by managers from food service sector companies.
[31]	2017	X		Analysis based on managerial opinions of the role of menu design in shaping more responsible consumer choice and its impact on FW prevention.

Table 2. *Cont.*

Ref.	Year	Level		Summary
		Micro	Macro	
[32]	2017	X	X	Analysis based on 315 questionnaires from Italian HoReCa SMEs operating in the tourism sector. Presenting general strategies for both levels micro (company) and macro (regional authorities and commercial chambers).
[33]	2016	X		Research results of the management staff of 45 restaurants/hotels. Proposition of strategies that involve the cooperation of the restaurant/hotel staff and the guests.
[34]	2015	X		Presentation of a case study and implications and impact on prevention.

Table 3. Summary of papers on FW and sustainable development goals.

Ref.	Year	Level		SDGs	Summary
		Micro	Macro		
[35]	2021		X	12.3	Review of the food waste scenario, adverse effects, food waste policies and regulations in Bangladesh on national and municipalities levels.
[36]	2021	X	X	12.3	Statistical analysis of 46 food waste items generated by households in the Buk-gu province of Daegu, South Korea.
[37]	2021	X		12.3	Presentation of case study: primary school (in Valencia) pupils' behavior change after intervention focused on food waste reduction (knowledge, awareness, attitudes).
[38]	2020		X	12.3	Presentation of additional value of citizen science as a trigger for policy making on behavior change.
[39]	2020		X	1, 2, 6, 12, 13	Estimation of carbon and water footprint of food. Comparison of the values of environmental footprint in the case of rational consumption of meat. Presentation of evaluation-obtained results by chemical engineering students.
[19]	2020	X	X	12, 17	Description of multi-stakeholder partnership program "Food waste Challenge" involving 172 restaurants. Program focuses on a range of behavioral interventions.
[40]	2020	X		12.3	Study of Covid-19 pandemic impact on food waste generation, financial costs and nutritional losses in Italian households.
[41]	2020	X		12.8	Study of 19 passengers from 21 full-service flights.
[42]	2020		X	1.2, 1.3, 3.4, 4.7, 7.2, 8.2, 8.5, 9.4, 11.6, 12.3, 12.5, 13.3, 17.16, 17.17	Food waste policy analysis provided by 40 cities across 16 European countries. Links between different types of policies provided and their impact on selected SDGs.
[21]	2020		X	12	Analysis of social, economic and environmental impact of food waste reduction activities.
[43]	2020		X	2, 3, 8	Presentation of the interrelations between SDGs, food access and waste also in the case of COVID-19 pandemic conditions.
[44]	2020	X	X		Presentation of quantification methodology of food waste in Swedish hospitals. Different types of waste defined (serving waste, plate waste, kitchen waste).
[45]	2020		X	12.3	Results of research (in 2016 and 2019) of 165 Hungarian households based on FUSIONS methodology.
[46]	2020	X	X	12.3	Presentation of research results on the application of different types of awareness techniques (passive approach (handouts), community engagement approach and gamification for 501 households.

Table 3. *Cont.*

Ref.	Year	Level		SDGs	Summary
		Micro	Macro		
[47]	2019	X	X	12.3	Presentation of research using different approaches for self-reporting—passive (offline), proactive (online)—and evaluation of both approaches in case of food waste reduction.
[48]	2019		X	12.3	Results of supply chain analysis, including production, consumption and utilization per different food groups according to the food waste reduction.
[49]	2019		X	12.3	Results of questionnaire-based research in the UK. Based on the proposition of categorization of what is considered edible in the case of the quantification of food waste.
[28]	2019		X	12.3	Presentation of research results for the quantification of food waste based on the data hospitality sector (1189 kitchens) in Sweden, Norway, Finland and Germany. Proposition of statistical measures for food waste tracking.
[50]	2019	X		12.3	Results of the analysis of 411 individuals from central Italy. Identification of support programs related to FW awareness, business investment in innovations and digital solutions focused on FW reduction.
[51]	2019		X	12.3	A footprint analysis of the food loss at the stage of vegetable production in Japan. Proposition of actions helping farmers to make a crop production and distribution plan.
[52]	2019		X	2, 12.3	Proposition of the actions of food and nutrition security for supporting developing countries.
[53]	2019	X	X		Presentation of study results of 680 Danish canteens related to nutrition and service management focused on food waste reduction.
[54]	2019	X	X	12.3	Review of interventions focused on food waste reduction in the hospitality sector and society (relating to nutrition behaviors).
[55]	2018		X	12.3	Brief description of UK program designed to prevent food waste generation in the hospitality and food service sector launched in 2017.
[56]	2018		X	12.3	Review of studies on food waste generation at the global and European scales.
[30]	2018	X			Identification of waste management initiatives in the food service sector and evaluation of management practices for waste reduction by managers from food service sector companies.
[57]	2018	X	X	12	Study of middle school students from 11 Polish schools. Analysis of food, nutrient, and energy waste and its impact on households' food waste and nutrition and energy losses.
[58]	2015	X	X		Presentation of 17 interview results with 17 Swedish food retailer representatives.

3.2. Best Practices on the Micro Level

As mentioned, practices on the micro level are related to the ways HoReCa companies operate and consumer behavior. Analysis of the HoReCa companies may include three phases: pre-kitchen (FW in warehouse, storage before meal preparation), in-kitchen (FW during meal preparation) and post-kitchen (FW on plate). Food waste generated in pre-kitchen and in-kitchen phases largely depends on company business practice and supplier engagement. Food waste generated in the post-kitchen phase depends mainly on consumer behavior.

According to Vizzoto et al. [23], companies should focus on constantly revising the dishes offered in the menu, reduction of overcooking, creative reproduction, donations to staff/charity organizations, offering options of ordering smaller portions (e.g., for kids) and marketing actions. In general for waste reduction, many companies use the following types of actions [24,59]: measuring, engaging staff, reducing overproduction, rethinking inventory and purchasing practices, and repurposing excess food. For food waste reduction, there could also be the utilization of management practices, which are common in manufacturing companies. Gladysz at al. [18] examined the application of lean management practices and found the following to be most useful: 5S (technique for station organization), new layout design, Kanban (technique for process flow management), Gemba walk (technique for process analysis), TWI (training within industry) job instructions, standardization, visual management, personnel motivational system, matrix of competences and a suggestion system. To have a complete picture of the best practices in HoReCa, companies focused on food waste reduction should also account for the impact of suppliers and consumer behavior. According to da Rosa et al. [16], FW reduction activities should be focused on innovations in the planning of menus, purchases and process of food preparation. Local suppliers usually can react more flexibly and respond quicker to changes in demand. Especially in the tourism sector, there is an additional advantage of cooperation with local suppliers [32]. Better understanding of customer needs also has an impact on food waste reduction; e.g., extension of lunch breaks from 20 to 30 min in middle schools resulted in a reduction of plate leftovers [57].

Secondi et al. [50] suggested that the use and implementation of smart applications and digital solutions can support additional studies on out-of-home food waste from a multiple stakeholder perspective, to better understand the amount of FW generated in this part of the food supply chain and multi-stakeholder collaboration along the entire food supply chain [60].

After analysis of the best practices, the authors considered the value of capturing the dynamics surrounding food waste in the HoReCa sector in a causal loop diagram (CLD). Analysis of three HoReCa case studies conducted in Poland provided the rich details used to derive this diagram [17]. A causal loop diagram is a useful visualization of how different elements are interrelated. The diagram consists of a set of nodes and edges. Nodes represent the elements and edges are the links that represent a connection or a relation between them. Thus, if you consider the diagram in Figure 2, FoodWaste is placed at the center of the diagram. The research has indicated that the source of this waste comes from OutdatedFood and PlateLeftovers, which in turn are affected by the SizeOfMeals. Additional information about creating CLDs is abundantly available [61,62], and this diagram was generated using the AnyLogic8.2 software.

TotalDistancePerMeal is the total travel distance from stock of ingredients to customer plate. The greater Space, the greater this distance. Space in this model is used to mean overall space, including storage of ingredients, storage of leftovers, storage of meals to be served and meals in progress, but it also means space in the guestroom. Sales is mostly related to difficulties in planning due to the uncertainty of the number of guests. Those issues lead to fluctuations in other entities in the CLD, like IngredientsStocks that are linked with NumberOfMeals. The more complex is the meal (NumberOfOperationsPerMeal and OperationsCycleTimes), the longer the lead time from ordering to serving the meal to a customer (ServingLeadTime). The lead time is also growing with an increase in total distance (transportation) per meal. On the other hand, lead time is also growing if batches (CookingBatches) are increasing. The greater the batches, the bigger the stocks of ingredients are needed. From this point, one can see that increase of stock directly leads to outdated food (OutdatedFood), which is a kind of food waste itself. In the bottom part of CLD, the loops for leftovers are represented. The greater the Sales, the more EdibleLeftovers. Actions that should directly address EdibleLeftovers, and therefore, the more EdibleLeftovers, the greater need for these actions are transforming Edible into ReprocessedLeftovers, delivering DonationsToCharity, and serving Staffmeals from EdibleLeftovers. All those actions lead to a final decrease in the total FoodWaste. Additionally, it is important to decrease FoodWaste through customer behavior and encourage them to take PlateTakeovers (LeftoversTakenByCustomer decreasing total FoodWaste). Importantly, PlateLeftovers themselves increase with the SizeOfMeals. On the contrary to actions decreasing total FoodWaste (action leading to an increase of ReprocesedLeftovers, DonationsToCharity,

StaffMeals, LeftoversTakenByCustomer), there are factors that increase total FoodWaste, i.e., NonEdibleLeftovers and PlateLeftovers.

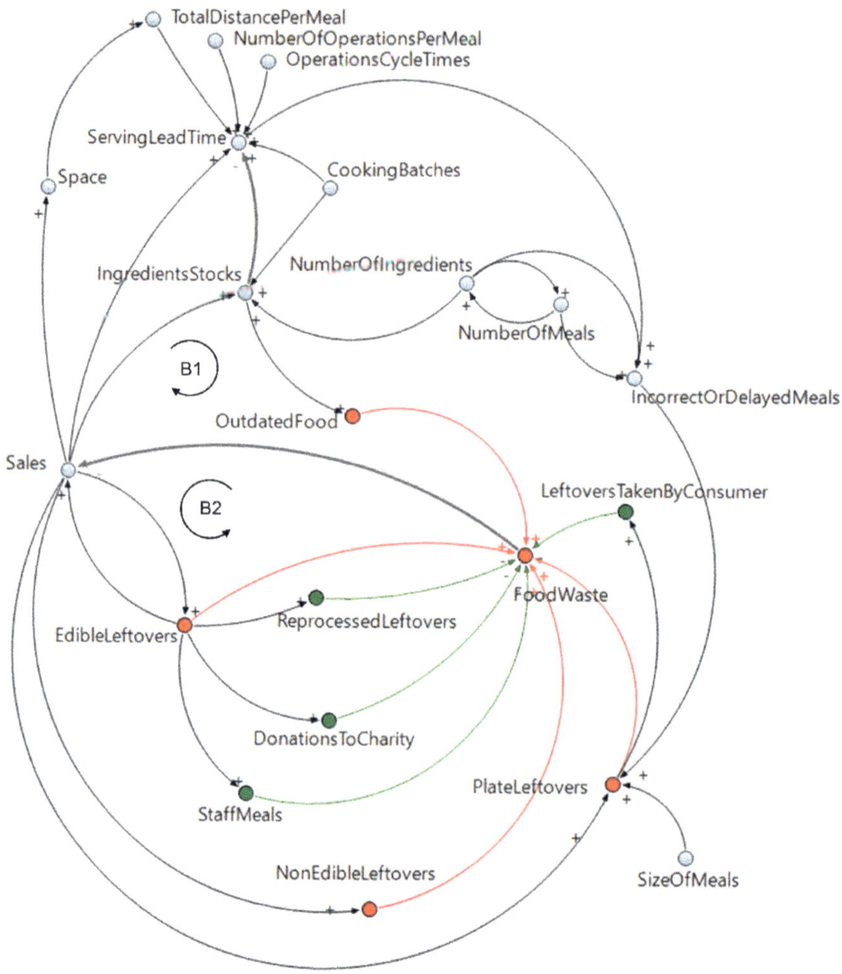

Figure 2. Conceptual causal loop diagram for food waste in HoReCa sector (drawn in AnyLogic8.2): Red points and arrows indicate a positive reinforcing effect, i.e., increased food waste; green points and arrows indicate a negative reinforcing effect, i.e., reduced food waste; other points and arrows illustrate contributing factors that lead to the respective effects. B1—Balancing Loop 1. B2—Balancing Loop 2.

The CLD includes many loops, most of which are balancing loops, because we focus on the analysis activities that will reduce a FoodWaste. The two most important loops for the purposes of this paper are from the HoReCa company's operations point of view.

Balancing Loop 1 (B1). Growing Sales increases IngredientsStocks, which creates the risk of increasing OutdatedFood. This then in turn will increase FoodWaste. In contrast, the increased FoodWaste consequently lowers the Sales volume. For this reason, there is a balancing loop. The main conclusion of this analysis is that in the case of planning to increase sales, more precise IngredientsStocks control mechanisms should be implemented first. Here, an implementation of selected lean tools (e.g., kanban) can be helpful.

Balancing Loop 2 (B2). Growing Sales increases EdibleLeftovers, which contributes to increasing FoodWaste. This then in turn will decrease Sales. At the same time, in order to reduce EdibleLeftovers, they can be used as ReprocessedLeftovers, DonationsToCharity or StaffMeals, which will reduce FoodWaste. An important conclusion to be drawn from the analysis of such a loop is that by implementing ReprocessedLeftovers, DonationsToCharity or StaffMeals, we are not able to eliminate this type of FoodWaste—we are only able to reduce it. For this reason, from the company's point of view, it may be more important to analyze the reasons for the creation of EdibleLeftovers (e.g., on the basis of the Ishikawa chart) than to introduce reduction mechanisms for EdibleLeftovers.

The diagram presents a model that could be of interest to HoReCa businesses considering ways that food waste reduction creates a positive overall impact on sales and eventual profitability of their enterprise. Combining the knowledge of interrelated factors with the management practices can help a business set reasonable goals, such as space layouts and simplified menus to improve their operations. The CLD could be used to define a roadmap for program improvements. In applying the model, practitioners could see clear relations, analyze feedback loops, plan and coordinate their actions accordingly.

3.3. Best Practices on the Macro Level

Implementation of the 2030 Agenda relies on actions that take place mainly at the national and sub-national level [63]. Any analyses of SDG interdependencies that will influence policymaking need to take a macro perspective. The complexity that derives from these interdependencies and the aforementioned urgency to achieve change requires a trans-disciplinary process that combines research results with policymaking. Eventually, policymakers need a rubric for thinking systematically about the many interactions to support their identification of the stakeholders of a proposition; i.e., which groups will be allies and which ones require negotiation. To make coherent policies and strategies—beyond simply synergies and trade-offs—they need up-to-date empirical knowledge on how the goals and interventions of one sector affect another positively or negatively [64].

To achieve an effective implementation, executive acts, policies and strategies for achieving individual goals adopted at the international level are necessary [65]. Individual programs will then be implemented at the national level by involving regional and/or local authorities. Especially on national and international levels, the SDG perspective could be taken into account as an indirect support for FW reduction initiatives related to resource efficiency and circular economy, e.g., [66–68] [NDCs, Farm to Fork Strategy, Circular Economy Package]. Food waste reduction policy may be connected with environmental policy statements and environmental targets such as energy and transport efficiency and recycling of waste. Additionally, guidelines for consumers can encourage climate-smarter food choices, such as meat reduction or promotion of local food or considering the greenhouse gas emissions for food production (e.g., reduction of meat consumption and production). However, it is not just the environmental pillar that should be taken into account by stakeholders, but also the economic and social pillars.

Activities at the macro level are mainly carried out and coordinated alone or in alliances by public authorities, non-governmental organizations and other types of organizations, e.g., schools, universities, etc. Such activities can be divided into two areas: soft and hard (Figure 3).

Soft activities focus on increasing awareness among society members and representatives of various organizational entities: enterprises, business environment institutions, academia and secondary schools, e.g., promotion of good nutrition practices in secondary schools [57]. These activities can be divided into passive and active. Passive mainly concerns passive influence on recipients, e.g., information campaigns: leaflets, actions in multimedia and mass media. Active requires active action, commitment, implementation of desired behaviors, e.g., implementation of biodegradable dishes or food collection [19]. In practice, much better results or even a synergy result will be obtained through a mix of different policies, e.g., passive awareness, gamification or passive awareness and community engagement provide a synergy effect of food waste reduction [46]. Soft actions should be initiated at the international and/or national levels, while regional and/or local authorities should be responsible for their implementation, as they have direct contact with society (people).

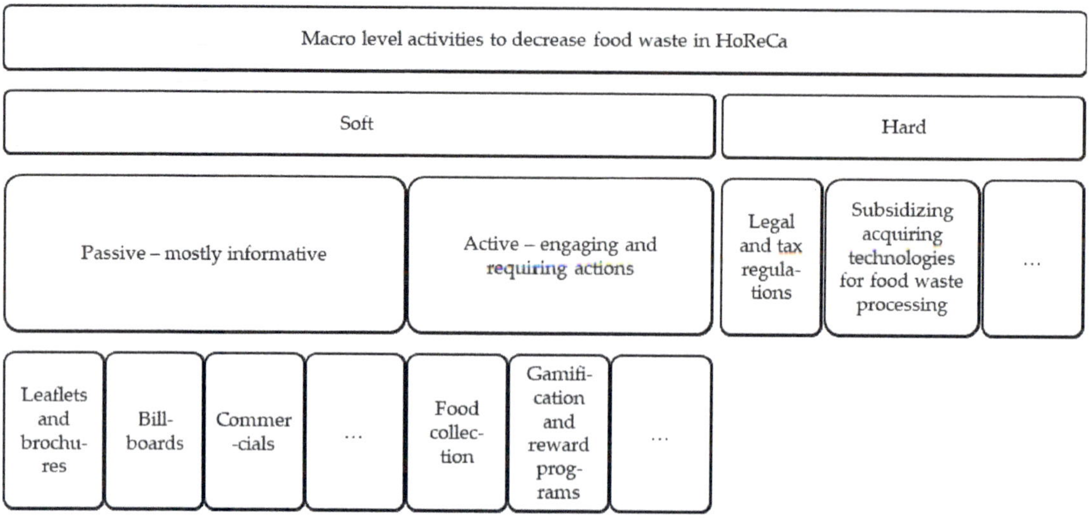

Figure 3. Types of actions for decrease of FW on macro level. Legend: " ... "—hybrid or not defined activities.

Hard activities can be divided into: legal regulations, e.g., requiring the use of selected solutions in business and agricultural activities or a ban on the use of selected solutions or behaviors; tax regulations, e.g., introducing tax breaks for enterprises, including farms or VAT reductions; dedicated support programs, e.g., financial support for enterprises for the purchase of clean technologies. Most of the hard activities should be carried out by government administration.

Often authorities are implementing sets of soft and hard activities [29]. Activities dedicated to FW reduction usually address society and/ or people; companies, including HoReCa companies; science, research and technology development organizations; schools, including primary, secondary and high schools; other institutions, including chambers of commerce and public authorities.

3.4. Analysis of the SDGs Related to Food Waste

There are a number of publications in which the relations between individual SDGs in the context of FW were analyzed [39,42,43]. Pradhan et al. [69] analyzed the SDG interactions to identify the synergies and trade-offs between them. They classified a significant positive correlation between a pair of SDG indicators as a synergy, i.e., progress in one goal promotes progress in another. A significant negative correlation was classified as a trade-off where progress in one goal hinders progress in another. Most trade-offs relate to a non-sustainable development paradigm that focuses on economic growth to generate human welfare at the expense of the environment and natural resources. They uncovered some global patterns among the SDGs based on positive and negative correlations between indicator pairs. For example, in their analysis, SDG 1 (No poverty) has a synergistic relationship with most of the other goals, whereas SDG 12 (Responsible consumption and production) is the goal most commonly associated with trade-offs. Sustainable Development Goal 12 is identified as conflicting with most other SDGs and is thereby non-supportive of sustainable development. They found that SDG 12 has negative correlations with 10 goals (SDGs 1–7,

9, 10, 17). Eventually, attainment of Agenda 2030 will greatly depend on whether synergies can be leveraged, and trade-offs identified and tackled.

As already indicated, Target 12.3 explicitly address the reduction of global food waste at retail and consumer levels and FW along production and supply chains. Malefors et al. [28] make a strong case for more robust measurements of current food waste statistics, without which it is not possible to ascertain that food waste has been halved. Additional targets that reinforce or are reinforced by 12.3 are given in (Table 4). However, this should not be considered as an exhaustive list of possible synergies or trade-offs.

Table 4. Relationships between Target 12.3 and additional targets of SDGs from a HoReCa perspective.

Interdependent SDG Targets	Rationale—Synergy, Trade-off or Both with Relation to Target 12.3
4.7—By 2030, ensure that all learners acquire the knowledge and skills needed to promote sustainable development, including, among others, through education for sustainable development and sustainable lifestyles. 12.8—By 2030, ensure that people everywhere have the relevant information and awareness for sustainable development and lifestyles in harmony with nature. 13.3— Improve education, awareness-raising and human and institutional capacity on climate change mitigation, adaptation, impact reduction and early warning.	Synergy: addressing consumer and retail food waste will require a re-education and increased awareness of the importance of food resources for both local and global sustainability.
8.2—Achieve higher levels of economic productivity through diversification, technological upgrading and innovation, including through a focus on high-value added and labor-intensive sectors. 8.4—Improve progressively, through 2030, global resource efficiency in consumption and production and endeavor to decouple economic growth from environmental degradation, in accordance with the 10-Year Framework of Programs on Sustainable Consumption and Production, with developed countries taking the lead. 12.5—By 2030, substantially reduce waste generation through prevention, reduction, recycling and reuse.	Both: food waste reduction will require the food service industries to look at packaging, transportation, and procedural innovations and improvements across the entire food value chain.
2.4—By 2030, ensure sustainable food production systems and implement resilient agricultural practices that increase productivity and production, that help maintain ecosystems, that strengthen capacity for adaptation to climate change, extreme weather, drought, flooding and other disasters and that progressively improve land and soil quality.	Both: if this target is addressed, the availability of food will be secured, but waste should still be reduced to improve availability to the less advantaged in society.
11.6—By 2030, reduce the adverse per capita environmental impact of cities, including by paying special attention to air quality and municipal and other waste management.	Synergy: addressing this target will not reduce food waste, but it could help ensure a proper handling of food and other organic wastes.

These related targets reflect the observed asymmetry regarding FW within the food value chain. Developed countries have been found to be inclined to waste food later in the supply chain, whereas food waste in developing countries occurs early in the food chain [70]. In addition, there is also the problem of the severe imbalance between surplus food suitable for human consumption going to waste in the developed countries while people living in poverty (anywhere) or famine struck regions are unable to provide sufficient food. As an example, the topic of education broaches both the need for good research to provide the scientific foundations for sustainable agriculture or innovations in biodegradable packaging [71]. In addition, basic education worldwide is needed to increase the awareness of FW as a problem and provide advice on how to avoid FW in the home. HoReCa professionals will also benefit from sharing workable approaches to FW, such as those proposed by Secondi et al. [50] and based on research on food surplus redistribution reported by Galli et al. [72].

3.5. Framework

In general, all programs/activities are designed and tend to have a positive impact on FW reduction. However, additional research is needed to evaluate dependencies and impacts of specific programs and actions on FW reduction and other specific SDGs indirectly related to 12.3.

The framework (Figure 4) shows the traceability of best practices on the micro level for SDGs according to the food waste reduction. The crucial role in this framework is to have public authorities, especially for national and regional levels. Programs launched by public authorities are usually defined as a set of actions that consist of soft and hard program elements.

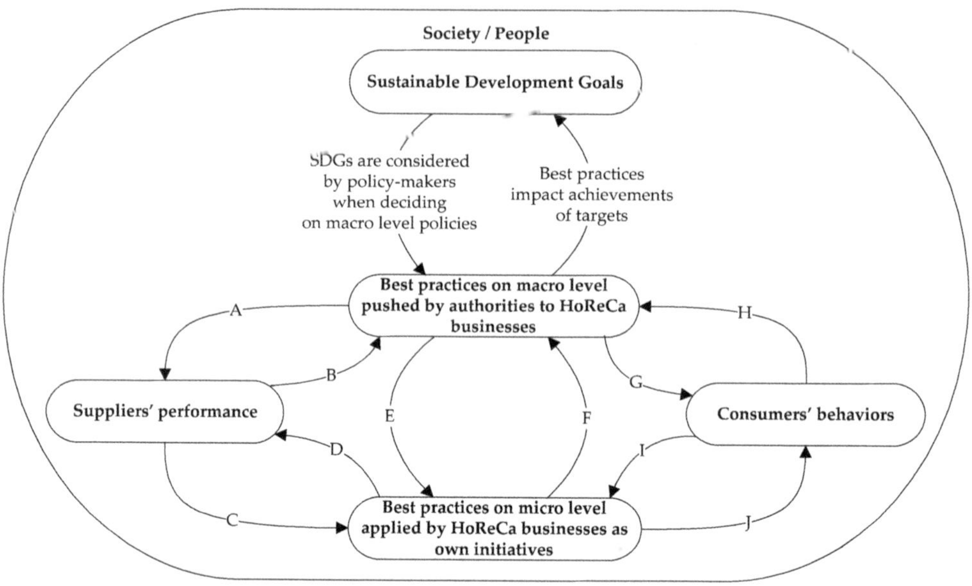

Figure 4. Conceptual causal loop diagram for SDG support from HoReCa best practices on food waste.

Legend:

A Programs launched by public authorities (regional, national and international levels) addressed to HoReCa suppliers have indirect influence on FW reduction by HoReCa sector.

B Suppliers' performance in terms of FW is considered by authorities when deciding on best practices on macro level.

C Suppliers' performance in term of FW drives their own decisions on best practices on the micro level.

D Available best practices on micro level and access to information about them affects suppliers' performance in terms of FW.

E Programs launched by authorities serve as guidelines for suppliers and consumers when thinking about best practices on the micro level.

F Available best practices on the micro level and access to information about serving for a bottom-up formulation of best practices on macro level by authorities.

G Programs launched by public authorities (regional, national and international levels) addressed to consumers, mainly related to awareness have indirect influence on FW reduction by HoReCa sector.

H Consumer behavior is considered by authorities when deciding on macro level policies.

I Consumer behavior is a driver for formulation of best practices (and worst as well) on micro level.

J Consumer' behavior is impacted by best available practices on micro level and information about them.

4. Discussion—Systems Engineering for SDGs in HoReCa

The SDGs are a network of interconnected goals used as a reference for the international community in working towards sustainable development [69,73,74]. The interdependencies of the SDGs in HoReCa described above are presented in Table 4, but questions raised by this analysis are: why do these relationships between SDGs in HoReCa matter and what is the relevance of a systems engineering approach? In this study, we conducted a systematic literature review and used a systems engineering approach by applying systems thinking to these results in the context of the SDGs. The SDGs act as a large system made up of smaller systems [75], i.e., a system of systems. Recent critiques of the SDGs highlight the importance of these interdependencies. Spangenberg [76] found them to be weak on agency, while requiring little from governments and nothing from business or consumers. The SDGs focus on the state and impact, ignoring conflict and stakeholder knowledge of the issues [77,78], as well as the pressures and drivers that counteract each other in competing impact categories in the same or different geographic contexts [76].

These critiques show areas where systems engineering can contribute. Systems engineering is quite advanced in methods for stakeholder inclusivity, and including stakeholders in actions directed towards implementing these goals, will help address the criticisms listed by Spangenberg [76] and others. The systems engineering community, and by this we specifically call attention to the progress and initiatives of the International Council for Systems Engineering (INCOSE), is actively venturing into the social domain, where the technical and social systems are being addressed as one [79]. This inclusivity of not only stakeholders, but also non-traditional engineering domains, places systems engineering practices in a key position to develop an inclusive, representative approach that addresses these goals as a holistic system for HoReCa enterprises.

Meeting or attempting to find ways to implement the SDGs in HoReCa lends itself naturally towards systems engineering because the SDGs are easily framed in a systems theoretical context [80], and capacity building to meet SDGs has been argued to require systems thinking [81]. This is illustrated with Table 4 and Figure 4 in the context of HoReCa. System archetypes [82,83] can be helpful to gain insight into patterns of behavior, especially in contexts of interconnected and competing goals [84]. In addition, SDGs intersect many societal domains, and a systems approach to investigating persistent social and environmental problems has been shown to help policymakers tackle the complexity associated with many intersected societal domains [85].

5. Conclusions

Food waste reduction should be analyzed holistically. It should be noted that there are many interdependencies between individual SDGs. These interdependencies should also be taken into account when policies, strategies and programs at the international, national, interregional, regional and local levels are defined. It requires a systemic approach not only with regard to the impact of one or more policies, but also considering the relationships between SDGs (also on sub-goal levels) which these policies address.

Author Contributions: Conceptualization, A.B. and B.G.; methodology, B.G., A.B. and E.P.; software, B.G. and A.B.; validation, A.B., B.G. and E.P.; formal analysis, A.B., B.G. and E.P.; investigation, A.B. and B.G.; resources, A.B. and B.G.; data curation, A.B. and B.G.; writing—original draft preparation, A.B. and B.G.; writing—review and editing, A.B., B.G. and E.P.; visualization, B.G. and A.B.; supervision, A.B. and B.G.; project administration, A.B. and B.G.; funding acquisition, A.B. and B.G. All authors have read and agreed to the published version of the manuscript.

Funding: The APC was covered from internal Warsaw University of Technology grant for management and quality sciences.

Institutional Review Board Statement: Not applicable.

Informed Consent Statement: Not applicable.

Data Availability Statement: Data available on request due privacy restrictions.

Acknowledgments: Authors would like to thank Cecilia Haskins from Norwegian University of Science and Technology for substantial support during paper preparation.

Conflicts of Interest: The authors declare no conflict of interest.

References

1. Fu, B.; Wang, S.; Zhang, J.; Hou, Z.; Li, J. Unravelling the Complexity in Achieving the 17 Sustainable-Development Goals. *Natl. Sci. Rev.* **2019**, *6*, 386–388. [CrossRef]
2. Frank, A. Über Verwertung Der Küchenabfälle Für Gewinnung von Trockenfutter Und von Fett Durch Ausnutzung Der Abgehenden Hitze Der Gasanstalten, Sowie Anderer Industrieller Betriebe. *J. Prakt. Chem.* **1915**, *92*, 419–424. [CrossRef]
3. Brown, L.P. Food Wastes—Some Causes and Remedies. *J. Frankl. Inst.* **1918**, *185*, 585–610. [CrossRef]
4. Lau, W.; Cipani, E. Reducing Student Food Waste in a Cafeteria-Style Dining Setting through Contingency Management. *Child Youth Care Forum* **1983**, *12*, 301–310. [CrossRef]
5. Youngs, A.J.; Nobis, G.; Town, P. Food Waste from Hotels and Restaurants in the U.K. *Waste Manag. Res.* **1983**, *1*, 295–308. [CrossRef]
6. Gunders, D.; Bloom, J. *Wasted: How America Is Losing up to 40 Percent of Its Food from Farm to Fork to Landfill*; Natural Resources Defense Council: New York, NY, USA, 2017.
7. Stenmarck, Â.; Jensen, C.; Quested, T.; Moates, G.; Buksti, M.; Cseh, B.; Juul, S.; Parry, A.; Politano, A.; Redlingshofer, B. *Estimates of European Food Waste Levels*; IVL Swedish Environmental Research Institute: Stockholm, Sweden, 2016.
8. UN. *The Sustainable Development Goals Report 2020*; United Nations: New York, NY, USA, 2020; ISBN 978-92-1-004960-3. Available online: www.un-ilibrary.org (accessed on 10 April 2021).
9. Bloom, J. *American Wasteland: How America Throws Away Nearly Half of Its Food*; Da Capo Books: Cambridge, MA, USA, 2010.
10. Benton, D. Portion Size: What We Know and What We Need to Know. *Crit. Rev. Food Sci. Nutr.* **2015**, *55*, 988–1004. [CrossRef] [PubMed]
11. Stöckli, S.; Dorn, M.; Liechti, S. Normative Prompts Reduce Consumer Food Waste in Restaurants. *Waste Manag.* **2018**, *77*, 532–536. [CrossRef] [PubMed]
12. Lasek, A.; Cercone, N.; Saunders, J. Restaurant Sales and Customer Demand Forecasting: Literature Survey and Categorization of Methods. In *Smart City 360°, Proceedings of the First EAI International Summit, Smart City 360°, Toronto, ON, Canada, 13–16 October 2015*; Leon-Garcia, A., Lenort, R., Holman, D., Staš, D., Krutilova, V., Wicher, P., Cagáňová, D., Špirková, D., Golej, J., Nguyen, K., Eds.; Springer International Publishing: Cham, Switzerland, 2016; pp. 479–491.
13. Sirieix, L.; Lála, J.; Kocmanová, K. Understanding the Antecedents of Consumers' Attitudes towards Doggy Bags in Restaurants: Concern about Food Waste, Culture, Norms and Emotions. *J. Retail. Consum. Serv.* **2017**, *34*, 153–158. [CrossRef]
14. Papargyropoulou, E.; Steinberger, J.K.; Wright, N.; Lozano, R.; Padfield, R.; Ujang, Z. Patterns and Causes of Food Waste in the Hospitality and Food Service Sector: Food Waste Prevention Insights from Malaysia. *Sustainability* **2019**, *11*, 6016. [CrossRef]
15. UN. *Global Indicator Framework for the Sustainable Development Goalsand Targets of the 2030 Agenda for Sustainable Development 2020*; United Nations: New York, NY, USA, 2020.
16. Da Rosa, F.S.; Lunkes, R.J.; Spigarelli, F.; Compagnucci, L. Environmental Innovation and the Food, Energy and Water Nexus in the Food Service Industry. *Resour. Conserv. Recycl.* **2021**, *166*, 105350. [CrossRef]
17. Wang, L.-E.; Filimonau, V.; Li, Y. Exploring the Patterns of Food Waste Generation by Tourists in a Popular Destination. *J. Clean. Prod.* **2021**, *279*, 123890. [CrossRef]
18. Gładysz, B.; Buczacki, A.; Haskins, C. Lean Management Approach to Reduce Waste in Horeca Food Services. *Resources* **2020**, *9*, 144. [CrossRef]
19. De Visser-Amundson, A. A Multi-Stakeholder Partnership to Fight Food Waste in the Hospitality Industry: A Contribution to the United Nations Sustainable Development Goals 12 and 17. *J. Sustain. Tour.* **2020**. [CrossRef]
20. Batlle-Bayer, L.; Bala, A.; Roca, M.; Lemaire, E.; Aldaco, R.; Fullana-i-Palmer, P. Nutritional and Environmental Co-Benefits of Shifting to "Planetary Health" Spanish Tapas. *J. Clean. Prod.* **2020**, *271*, 122561. [CrossRef]
21. Chinie, A.-C. Challenges for Reducing Food Waste. *Proc. Int. Conf. Bus. Excell.* **2020**, *14*, 819–828. [CrossRef]
22. Filimonau, V.; Todorova, E.; Mzembe, A.; Sauer, L.; Yankholmes, A. A Comparative Study of Food Waste Management in Full Service Restaurants of the United Kingdom and the Netherlands. *J. Clean. Prod.* **2020**, *258*, 120775. [CrossRef]
23. Vizzoto, F.; Tessitore, S.; Iraldo, F.; Testa, F. Passively Concerned: Horeca Managers' Recognition of the Importance of Food Waste Hardly Leads to the Adoption of More Strategies to Reduce It. *Waste Manag.* **2020**, *107*, 266–275. [CrossRef] [PubMed]
24. Okumus, B. How Do Hotels Manage Food Waste? Evidence from Hotels in Orlando, Florida. *J. Hosp. Mark. Manag.* **2020**, *29*, 291–309. [CrossRef]
25. Visschers, V.H.M.; Gundlach, D.; Beretta, C. Smaller Servings vs. Information Provision: Results of Two Interventions to Reduce Plate Waste in Two University Canteens. *Waste Manag.* **2020**, *103*, 323–333. [CrossRef]
26. Li, Y.; Wang, L.; Cheng, S. Spatiotemporal Variability in Urban HORECA Food Consumption and Its Ecological Footprint in China. *Sci. Total Environ.* **2019**, *687*, 1232–1244. [CrossRef] [PubMed]
27. Filimonau, V.; Fidan, H.; Alexieva, I.; Dragoev, S.; Marinova, D.D. Restaurant Food Waste and the Determinants of Its Effective Management in Bulgaria: An Exploratory Case Study of Restaurants in Plovdiv. *Tour. Manag. Perspect.* **2019**, *32*, 100577. [CrossRef]

28. Malefors, C.; Callewaert, P.; Hansson, P.-A.; Hartikainen, H.; Pietiläinen, O.; Strid, I.; Strotmann, C.; Eriksson, M. Towards a Baseline for Food-Waste Quantification in the Hospitality Sector—Quantities and Data Processing Criteria. *Sustainability* **2019**, *11*, 3541. [CrossRef]
29. Chalak, A.; Abou-Daher, C.; Abiad, M.G. Generation of Food Waste in the Hospitality and Food Retail and Wholesale Sectors: Lessons from Developed Economies. *Food Secur.* **2018**, *10*, 1279–1290. [CrossRef]
30. Martin-Rios, C.; Demen-Meier, C.; Gössling, S.; Cornuz, C. Food Waste Management Innovations in the Foodservice Industry. *Waste Manag.* **2018**, *79*, 196–206. [CrossRef]
31. Filimonau, V.; Krivcova, M. Restaurant Menu Design and More Responsible Consumer Food Choice: An Exploratory Study of Managerial Perceptions. *J. Clean. Prod.* **2017**, *143*, 516–527. [CrossRef]
32. Iraldo, F.; Testa, F.; Lanzini, P.; Battaglia, M. Greening Competitiveness for Hotels and Restaurants. *J. Small Bus. Enterp. Dev.* **2017**, *24*, 607–628. [CrossRef]
33. Pirani, S.I.; Arafat, H.A. Reduction of Food Waste Generation in the Hospitality Industry. *J. Clean. Prod.* **2016**, *132*, 129–145. [CrossRef]
34. Falasconi, L.; Vittuari, M.; Politano, A.; Segrè, A. Food Waste in School Catering: An Italian Case Study. *Sustainability* **2015**, *7*, 14745–14760. [CrossRef]
35. Ananno, A.A.; Masud, M.H.; Chowdhury, S.A.; Dabnichki, P.; Ahmed, N.; Arefin, A.M.E. Sustainable Food Waste Management Model for Bangladesh. *Sustain. Prod. Consum.* **2021**, *27*, 35–51. [CrossRef]
36. Adelodun, B.; Kim, S.H.; Choi, K.-S. Assessment of Food Waste Generation and Composition among Korean Households Using Novel Sampling and Statistical Approaches. *Waste Manag.* **2021**, *122*, 71–80. [CrossRef]
37. Antón-Peset, A.; Fernandez-Zamudio, M.-A.; Pina, T. Promoting Food Waste Reduction at Primary Schools. A Case Study. *Sustainability* **2021**, *13*, 600. [CrossRef]
38. Pateman, R.M.; de Bruin, A.; Piirsalu, E.; Reynolds, C.; Stokeld, E.; West, S.E. Citizen Science for Quantifying and Reducing Food Loss and Food Waste. *Front. Sustain. Food Syst.* **2020**, *4*, 589089. [CrossRef]
39. Feijoo, G.; Moreira, M.T. Fostering Environmental Awareness towards Responsible Food Consumption and Reduced Food Waste in Chemical Engineering Students. *Educ. Chem. Eng.* **2020**, *33*, 27–35. [CrossRef]
40. Amicarelli, V.; Bux, C. Food Waste in Italian Households during the Covid-19 Pandemic: A Self-Reporting Approach. *Food Secur.* **2021**, *13*, 25–37. [CrossRef] [PubMed]
41. You, F.; Bhamra, T.; Lilley, D. Why Is Airline Food Always Dreadful? Analysis of Factors Influencing Passengers' Food Wasting Behaviour. *Sustainability* **2020**, *12*, 8571. [CrossRef]
42. Fattibene, D.; Recanati, F.; Dembska, K.; Antonelli, M. Urban Food Waste: A Framework to Analyse Policies and Initiatives. *Resources* **2020**, *9*, 99. [CrossRef]
43. Fleetwood, J. Social Justice, Food Loss, and the Sustainable Development Goals in the Era of COVID-19. *Sustainability* **2020**, *12*, 5027. [CrossRef]
44. Eriksson, M.; Malefors, C.; Bergström, P.; Eriksson, E.; Osowski, C.P. Quantities and Quantification Methodologies of Food Waste in Swedish Hospitals. *Sustainability* **2020**, *12*, 3116. [CrossRef]
45. Kasza, G.; Dorkó, A.; Kunszabó, A.; Szakos, D. Quantification of Household Food Waste in Hungary: A Replication Study Using the FUSIONS Methodology. *Sustainability* **2020**, *12*, 3069. [CrossRef]
46. Soma, T.; Li, B.; Maclaren, V. Food Waste Reduction: A Test of Three Consumer Awareness Interventions. *Sustainability* **2020**, *12*, 907. [CrossRef]
47. Leverenz, D.; Moussawel, S.; Maurer, C.; Hafner, G.; Schneider, F.; Schmidt, T.; Kranert, M. Quantifying the Prevention Potential of Avoidable Food Waste in Households Using a Self-Reporting Approach. *Resour. Conserv. Recycl.* **2019**, *150*, 104417. [CrossRef]
48. Caldeira, C.; De Laurentiis, V.; Corrado, S.; van Holsteijn, F.; Sala, S. Quantification of Food Waste per Product Group along the Food Supply Chain in the European Union: A Mass Flow Analysis. *Resour. Conserv. Recycl.* **2019**, *149*, 479–488. [CrossRef] [PubMed]
49. Nicholes, M.J.; Quested, T.E.; Reynolds, C.; Gillick, S.; Parry, A.D. Surely You Don't Eat Parsnip Skins? Categorising the Edibility of Food Waste. *Resour. Conserv. Recycl.* **2019**, *147*, 179–188. [CrossRef]
50. Secondi, L.; Principato, L.; Mattia, G. Can Digital Solutions Help in the Minimization of Out-of-Home Waste? An Analysis from the Client and Business Perspective. *Br. Food J.* **2019**, *122*, 1341–1359. [CrossRef]
51. Wakiyama, T.; Lenzen, M.; Faturay, F.; Geschke, A.; Malik, A.; Fry, J.; Nansai, K. Responsibility for Food Loss from a Regional Supply-Chain Perspective. *Resour. Conserv. Recycl.* **2019**, *146*, 373–383. [CrossRef]
52. Pollard, C.M.; Booth, S. Food Insecurity and Hunger in Rich Countries—It Is Time for Action against Inequality. *Int. J. Environ. Res. Public Health* **2019**, *16*, 1804. [CrossRef]
53. Lassen, A.D.; Christensen, L.M.; Spooner, M.P.; Trolle, E. Characteristics of Canteens at Elementary Schools, Upper Secondary Schools and Workplaces That Comply with Food Service Guidelines and Have a Greater Focus on Food Waste. *Int. J. Environ. Res. Public Health* **2019**, *16*, 1115. [CrossRef]
54. Reynolds, C.; Goucher, L.; Quested, T.; Bromley, S.; Gillick, S.; Wells, V.K.; Evans, D.; Koh, L.; Carlsson Kanyama, A.; Katzeff, C.; et al. Review: Consumption-Stage Food Waste Reduction Interventions—What Works and How to Design Better Interventions. *Food Policy* **2019**, *83*, 7–27. [CrossRef]

55. Cooper, J. Briefing: Food Waste—Next Steps for Food Processors and Manufacturers. *Proc. Inst. Civ. Eng. Waste Resour. Manag.* **2018**, *171*, 91–93. [CrossRef]
56. Corrado, S.; Sala, S. Food Waste Accounting along Global and European Food Supply Chains: State of the Art and Outlook. *Waste Manag.* **2018**, *79*, 120–131. [CrossRef]
57. Kowalewska, M.T.; Kołłajtis-Dołowy, A. Food, Nutrient, and Energy Waste among School Students. *Br. Food J.* **2018**, *120*, 1807–1831. [CrossRef]
58. Tjarnemo, H.; Sodahl, L. Swedish Food Retailers Promoting Climate Smarter Food Choices-Trapped between Visions and Reality? *J. Retail. Consum. Serv.* **2015**, *24*, 130–139. [CrossRef]
59. Clowes, A.; Hanson, C.; Swanell, R. The Business Case for Reducing Food Loss and Waste: Restaurants. Champions 12.3.2019. Available online: www.champions123.org (accessed on 10 April 2021).
60. De Steur, H.; Wesana, J.; Dora, M.K.; Pearce, D.; Gellynck, X. Applying Value Stream Mapping to Reduce Food Losses and Wastes in Supply Chains: A Systematic Review. *Waste Manag.* **2016**, *58*, 359–368. [CrossRef] [PubMed]
61. Morecroft, J. *Strategic Modelling and Business*; John Wiley &Sons: Chichester, UK; Hoboken, NJ, USA, 2007; ISBN 978-0-470-01286-4.
62. Sterman, J.D. System Dynamics Modeling: Tools for Learning in a Complex World. *Calif. Manag. Rev.* **2001**, *43*, 8–25. [CrossRef]
63. Breuer, A.; Janetschek, H.; Malerba, D. Translating Sustainable Development Goal (SDG) Interdependencies into Policy Advice. *Sustainability* **2019**, *11*, 2092. [CrossRef]
64. Nilsson, M.; Griggs, D.; Visbeck, M. Policy: Map the Interactions between Sustainable Development Goals. *Nat. News* **2016**, *534*, 320. [CrossRef]
65. EC. Closing the Loop—An EU Action Plan for the Circular Economy COM/2015/0614 Final—European Environment Agency. Available online: https://www.eea.europa.eu/policy-documents/com-2015-0614-final (accessed on 8 April 2021).
66. EC. Report from the Commission to the European Parliament, the Council COM/2019/190 Final—The European Economic and Social Committee and the Committee of the Regions on the Implementation of the Circular Economy Action Plan. Available online: https://eur-lex.europa.eu/legal-content/pl/TXT/?uri=CELEX:52019DC0190 (accessed on 10 April 2021).
67. EC. Farm to Fork Strategy. For a Fair, Healthy and Environmentally-Friendly Food System. Available online: https://ec.europa.eu/food/farm2fork_en (accessed on 10 April 2021).
68. UN. Nationally Determined Contributions under the Paris Agreement. Synthesis Report by the Secretariat, UN. Available online: https://unfccc.int/process-and-meetings/the-paris-agreement/nationally-determined-contributions-ndcs/nationally-determined-contributions-ndcs (accessed on 10 April 2021).
69. Pradhan, P.; Costa, L.; Rybski, D.; Lucht, W.; Kropp, J.P. A Systematic Study of Sustainable Development Goal (SDG) Interactions. *Earth's Future* **2017**, *5*, 1169–1179. [CrossRef]
70. Gustavsson, J.; Cederberg, C.; Sonesson, U.; Van Otterdijk, R.; Meybeck, A. *FAO Global Food Losses and Food Waste—Extent, Causes and Prevention*; Technical Report; FAO: Rome, Italy; Swedish Institute for Food and Biotechnology (SIK): Göteborg, Sweden, 2011.
71. Bergman, Z.; Bergman, M.M.; Fernandes, K.; Grossrieder, D.; Schneider, L. The Contribution of UNESCO Chairs toward Achieving the UN Sustainable Development Goals. *Sustainability* **2018**, *10*, 4471. [CrossRef]
72. Galli, F.; Cavicchi, A.; Brunori, G. Food Waste Reduction and Food Poverty Alleviation: A System Dynamics Conceptual Model. *Agric. Hum. Values* **2019**, *36*, 289–300. [CrossRef]
73. Blanc, D.L. Towards Integration at Last? The Sustainable Development Goals as a Network of Targets. *Sustain. Dev.* **2015**, *23*, 176–187. [CrossRef]
74. UN. *Resolution Adopted by the General Assembly on 25 September 2015. Transforming Our World: The 2030 Agenda for Sustainable Development 2015*; United Nations: New York, NY, USA, 2015.
75. Palmer, E.; Burton, R.; Haskins, C. A Systems Engineering Framework for Bioeconomic Transitions in a Sustainable Development Goal Context. *Sustainability* **2020**, *12*, 6650. [CrossRef]
76. Spangenberg, J.H. Hot Air or Comprehensive Progress? A Critical Assessment of the SDGs. *Sustain. Dev.* **2017**, *25*, 311–321. [CrossRef]
77. Wong, R.; van der Heijden, J. Avoidance of Conflicts and Trade-Offs: A Challenge for the Policy Integration of the United Nations Sustainable Development Goals. *Sustain. Dev.* **2019**, *27*, 838–845. [CrossRef]
78. Sanz-Hernández, A.; Esteban, E.; Garrido, P. Transition to a Bioeconomy: Perspectives from Social Sciences. *J. Clean. Prod.* **2019**, *224*, 107–119. [CrossRef]
79. Palmer, E.; Rhodes, D.; Watson, M.; Haskins, C.; Olaya, C.; Presland, I.; Fossum, K. *Putting the Social in Systems Engineering: An Overview and Conceptual Development*; INCOSE: San Diego, CA, USA, 2021; in press.
80. Skene, K.R.; Malcolm, J. Using the SDGs to Nurture Connectivity and Promote Change. *Des. J.* **2019**, *22*, 1629–1646. [CrossRef]
81. Stafford-Smith, M.; Griggs, D.; Gaffney, O.; Ullah, F.; Reyers, B.; Kanie, N.; Stigson, B.; Shrivastava, P.; Leach, M.; O'Connell, D. Integration: The Key to Implementing the Sustainable Development Goals. *Sustain. Sci.* **2017**, *12*, 911–919. [CrossRef] [PubMed]
82. Senge, P.M. *The Fifth Discipline Fieldbook: Strategies and Tools for Building a Learning Organization*; Currency, Doubleday: New York, NY, USA, 1994; ISBN 978-0-385-47256-2.
83. Braun, W. The System Archetypes. *System* **2002**, *1*, 1–26.
84. Wolstenholme, E. Using Generic System Archetypes to Support Thinking and Modelling. *Syst. Dyn. Rev.* **2004**, *20*, 341–356. [CrossRef]
85. Probst, G.; Bassi, A.M. *Tackling Complexity*; Routledge: London, UK, 2014.

sustainability

MDPI

Article

Developing a Participatory Planning Support System for Sustainable Regional Planning—A Problem Structuring Case Study

Dina Margrethe Aspen [1,*] and Andreas Amundsen [2]

1 Department of International Business, Norwegian University of Science and Technology, 6009 Ålesund, Norway
2 United Future Lab Norway, 6009 Ålesund, Norway; Andreas.Amundsen@alesund.kommune.no
* Correspondence: dina.aspen@ntnu.no

Abstract: In this paper, we report on the application of systems engineering in initiating the synthesis of a participatory planning support system (PSS) for sustainable regional planning. The systems engineering SPADE approach is applied in a model-based fashion to define and link sustainable development goals (SDGs) to regional and urban planning policies in a co-creative multi-stakeholder environment. The approach is demonstrated through a case study from the interregional climate, land-use, and transportation planning process (PAKT) in the Ålesund region in Norway. The work was performed using focus groups with planning stakeholders over a series of workshops to analyze, design, verify and validate the problem structure. Our study shows that the approach is useful for integrating and operationalizing the SDGs in a planning context. The methodology also brings clarity and structure to planning problems and provides a pedagogical frame to engage stakeholders in co-creative PSS synthesis. Further research is necessary to explore how structured elements may be exploited in PSS synthesis.

Keywords: regional and urban planning; sustainable development goals; planning support systems; systems engineering

Citation: Aspen, D.M.; Amundsen, A. Developing a Participatory Planning Support System for Sustainable Regional Planning—A Problem Structuring Case Study. *Sustainability* **2021**, *13*, 5723. https://doi.org/10.3390/su13105723

Academic Editor: Åsa Gren

Received: 31 March 2021
Accepted: 17 May 2021
Published: 20 May 2021

1. Introduction

Integrating the Sustainable Development Goals (SDGs) in regional and urban planning is critical for goal attainment. The Organisation for Economic Co-operation and Development (OECD) estimates that 105 out of the 160 SDG targets will not be achieved without engagement at the sub-national level [1], and all SDGs have targets related to responsibilities at the local and regional level. This integration process is often referred to as "localizing", and urban and regional planning are considered critical processes in which this integration must occur [2]. Literature is abundant with case reports from various localizing efforts, where municipal and regional authorities have attempted to use the goals in creating visions, strategies, and plans (see, e.g., [3,4]). These efforts show that successful integration must be performed systematically [5], building on a coherent and holistic approach [6] tailored to the local context and plans [7].

While there currently exists a wide range of planning support systems (PSSs) [8–10], their common objective is to assist planning practitioners and other stakeholders in accessing, assessing, and communicating information and knowledge in the planning process. Following the communicative and collaborative development of planning practices, participatory PSS to facilitate stakeholder engagement and interaction have also been developed (see, e.g., [11]). Planning support system functionality may vary depending on the selected user and planning tasks but typically include capabilities to acquire, store, analyze and visualize data [12]. This is achieved by integrating various computational tools that include, but are not necessarily limited to, geographical information and spatial modeling systems [13]. Although land-use models are necessary for planning decision support, aspatial models and information are also considered valuable and essential [14].

This study aims to demonstrate how systems engineering may be utilized to organize and facilitate the co-creative development of PSSs for sustainable regional and urban planning. The systems engineering SPADE (stakeholders, problems, alternatives, decisions, evaluation) methodology is applied in a model-based fashion to develop a problem structure for urban planning utilizing the SDGs. The article reports on an application of this approach in the climate, land-use, and transportation planning process (PAKT) in the Ålesund region, where a series of stakeholder involvement workshops were held to link the sustainable development goals to planning decisions.

In the following section, we elaborate on contemporary developments and challenges in PSS development before study methods are introduced in Section 3. Results from the problem structuring exercise are presented in Section 4 and further discussed in Section 5. Concluding remarks and further research are presented in Section 6.

2. Background

2.1. Planning Support in Ill-Structured Contexts

Although multiple PSSs have been developed to aid sustainable regional and urban development, land-use and transport practitioners still largely rely on conventional methods, tools, and techniques in their everyday practice. Several scholars have explored this mismatch between the supply and demand of PSSs, often referred to as the implementation gap in PSS science [15,16]. In surveying planning practitioners and PSS users in the Netherlands, Te Brömmelstroet [17] found that "soft issues" such as lack of transparency, poor communication value, and low user-friendliness were considered primary bottlenecks for tool implementation. These findings were echoed in a study performed by Vonk and Geertman [18]. They identified challenges related to insufficient instrument quality and insufficient diffusion in planning practice, and low user acceptance. As a result of these shortcomings, PSSs often fail to support planning practitioners and other stakeholders in strategy-making processes [15].

The well-structured environment offered by PSSs is in stark contrast to the complex and dynamic planning and problem-solving processes at the strategic level [15]. According to Pidd [19], problems may be classified according to the level of stakeholder agreement on what the problem is and how it may be solved. Where there exists consensus on both, problems are merely puzzles; identifying a course of action is a matter of identifying the best option within the given context. At the other extreme, we find messes where stakeholders disagree on the problem and how these (different) problems should be solved. In between, we have problems where a unified problem definition is achievable, but work is required to formulate the problem and its potential solutions. Rittel and Webber [20] argue that planning problems are wicked in that it is impossible to achieve a definite answer to what the problem is and how it may be solved. A key property of wicked problems is that they are never definitely formulated. The problem and solutions emerge gradually among stakeholders through a continuous process of judgment, argument, and negotiation [20].

In order to offer a decision aid in these ill-structured contexts, PSS developers need to approach planning problems more holistically. In their summary of planning support science advancements and challenges, Geertman and Stillwell [16] assert that development in instrumentation needs to be supplemented with the continued research on the dimensions of application and governance to move beyond the former technology-driven approach in PSS. The application aspect pertains to the object-oriented goal of planning support and covers the content of planning and the objectives to be achieved. Governance aspects, in turn, relate to planning practices and processes by which these objectives are achieved [16]. This paper targets both the application and governance dimensions of planning explicitly using systems engineering to determine what sustainable urban development is and how it may be structured within a strategic planning context. This approach permits specifying the problems a PSS must solve in an open, structured manner, outside the realm of technology and its current limitations.

2.2. Sustainable Urban Planning at the Strategic Level

The planning context explored in this study is the establishment of a regional land-use and transportation master plan. Planning tasks at this level requires gathering and metabolizing information from various repositories to address multiple interrelated themes. The planning proposal should present key trends and development patterns in the region and identify preferable strategies and policies to support sustainable growth. This entails specifying guidelines and principles for land-use change and the development of transportation infrastructure and mobility services. In scoping and preparing the proposal, planners need to engage a wide range of planning stakeholders and facilitate their expression of opinions, perspectives, and expertise.

The case study presented in this paper follows a regional master plan process. In 2019, Ålesund, Giske, and Sula municipalities, commonly referred to as the Ålesund region, decided to jointly develop an inter-municipal master plan for climate, land-use, and transportation planning processes (abbreviated PAKT). The municipalities are located on the west coast of Norway and cover an area of 731 km^2. The municipalities are interlinked as their 84,000 residents live, work and utilize services across municipal borders. An important objective in PAKT is to identify strategies to achieve the SDGs [21]. This is also considered a challenge as planners need to operationalize and link SDGs to their planning strategies in a meaningful and consistent manner.

While PAKT is inter-municipal, it links to the Norwegian planning hierarchy by providing a thematic input to each municipality's social master plans, as shown in Figure 1. The preparation of PAKT follows a four-stage process that takes place over two years. The first step involves developing a planning program that describes the current situation in the planning area pertinent to the theme, including objectives, conflicts, and future scenarios, and the need for further knowledge to support the planning process. This program was developed and adopted by the municipalities in early 2020. Next, the practitioners develop the knowledge basis and planning proposal, which outlines high-level strategies and guidance for further adoption at the municipal level. While the resulting document is not legally binding for the municipalities, it provides guidance for subsequent planning, states how areas should be utilized, and details special considerations and guidance. The draft planning proposal is currently under development and will be announced for public consultation during 2021. Once all interested parties and stakeholders have provided their inputs to the proposal, a revised proposal will be adopted by the municipalities in 2022.

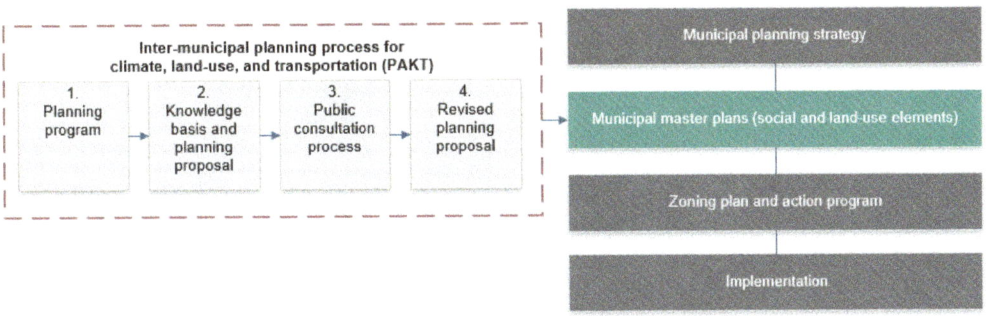

Figure 1. The climate, land-use, and transportation planning (PAKT) process and the municipal planning hierarchy, modified from [21,22].

3. Method

3.1. Methodology: Systems Engineering as an Integrative Framework

Systems engineering may be viewed both as a discipline and a process. As a discipline, it provides a basis for designing and applying systems from a holistic perspective [23]. As a process, it entails a systematic, top-down, iterative approach to system design, development, and deployment [23–25]. While initially developed to handle large-scale complex technical systems, it has evolved to support problem-solving in socio-technical systems [24].

The systems engineering SPADE methodology developed by Haskins [24] is derived from essential systems engineering processes to offer a generic, jargon-free framework for problem-solving. The acronym is built on the steps of the methodology, which encompasses defining and analyzing stakeholders, problems, alternatives, decision-making, and continuous evaluation. The methodology has been applied in multiple problem-solving efforts, with recent examples from applications to support bioeconomy transitions [26] and digital twin development of offshore cranes [27]. In [28], SPADE is used to develop a problem structure for marine emission reduction technology acquisitions. In this article, we deploy the methodology in the same manner, with the objectives of structuring planning decisions and reorienting them according to the SDGs. Table 1 summarizes the steps of SPADE and its application in our case study.

Table 1. The SPADE methodology applied to planning problem structuring, adapted from [24,28].

Step	Inquiry
S	Stakeholders: Who are the key stakeholders to the planning process?
P	Problems: What are the stakeholder sustainability objectives and criteria and how do they link to the SDGs?
A	Alternatives: What are the alternative strategies, policies and measures that may be implemented to achieve the objectives?
D	Decisions: How do alternative courses of action comparatively evaluate towards the stated objectives?
E	Evaluation: Continuous effort in, between and across all steps.

Identifying planning *stakeholders* is the first step of the approach. These are actors who influence and/or are influenced by the decisions at hand, in our case, the decisions and outcomes of the planning process. Stakeholders may be classified in several ways, e.g., considering their influence [29], but a practical approach is to differentiate between primary and secondary stakeholders [30]. In our study, we define a primary stakeholder as an individual or group directly involved in the planning process. Secondary stakeholders are additional individuals or groups whose interests are influenced by the outcomes of the planning process, resulting from interaction among primary stakeholders.

Next, stakeholder *problems* need to be analyzed. In a problem structuring context, this translates to their values, objectives to be achieved, and the partial ways they may be described through criteria [31–33]. If planning objectives are not already based on SDGs, they are reoriented in this step. This also entails identifying and linking objectives and criteria to measurable SDG-based indicators.

The third step requires formulating *alternative* courses of action to achieve the stated objectives. In a planning context, the level of detail in possible alternatives depends on the maturity of the planning process. High-level strategies are usually formulated at the initial stage before more detailed policies and measures are identified. In our approach, we attempt to structure these alternatives in a hierarchical manner. Once stakeholders, problems, and alternatives are identified, a *decision* analytical effort may be initiated. This requires collecting data to understand how alternative courses of action affect SDGs targeted in previous steps. In this article, we provide a linked problem structure to outline the decision step. *Evaluation* is performed within, between, and across all steps.

3.2. *Stakeholder Focus Groups and Workshops*

In order to inform the problem structure and support its verification and validation, a series of stakeholder workshops were held during 2020. The workflow of the stakeholder involvement process in developing the problem structure for PAKT is shown in Figure 2. The process was initiated with scoping workshops held with a wide range of stakeholders representing the municipal, county, and state agencies to provide high-level expectations to the participatory planning support tool and ideation on its potential use in the PAKT planning.

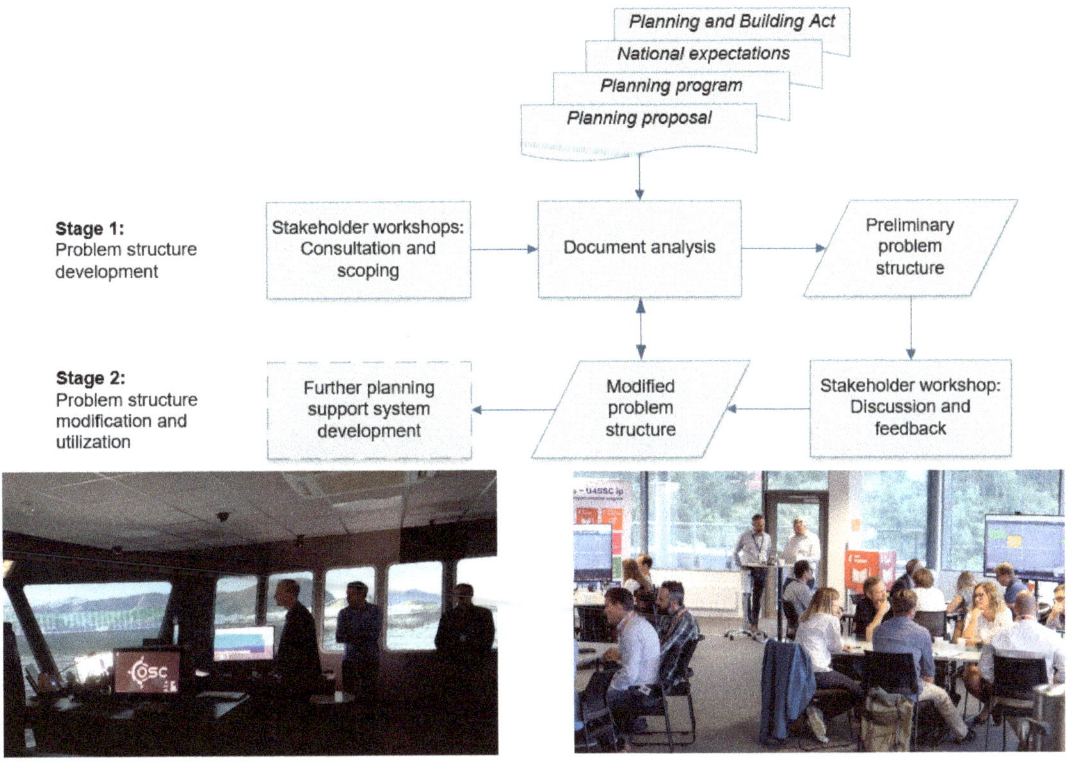

Figure 2. Stakeholder workshops in the PAKT problem structuring process.

The problem structuring process was continued through document analysis, where critical planning documents for the PAKT process were used to provide an initial problem structure. These encompass the Planning and Building Act, [34] which regulates planning activities, the National Expectations Regarding Regional and Municipal Planning [35], and the PAKT planning program [21]. This was presented in another stakeholder workshop for stakeholder feedback and inputs. Based on this exercise, a modified problem structure was developed. This structure was updated as the planning proposal [36] was presented.

4. Results

4.1. *Stakeholders*

The planning process for climate, land-use, and transportation involves many stakeholders representing various interests, viewpoints, and expertise pertinent to the process and decisions. We may divide these stakeholders into three main groups, as shown in Figure 3.

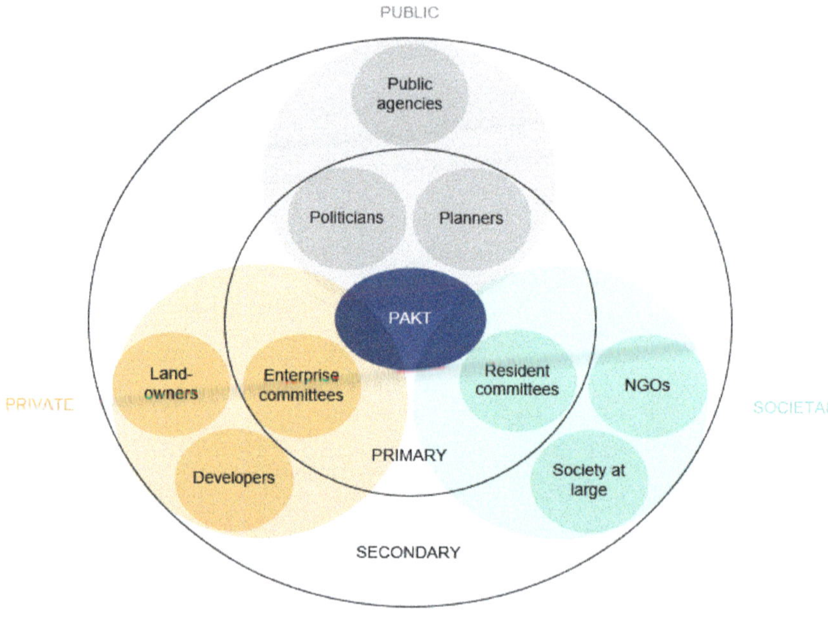

Figure 3. Stakeholders in the PAKT planning program.

Firstly, the public sector encompasses both the administrative and political resources necessary to develop and implement planning strategies and policies. In this group, we find the planners, politicians, and other state agencies that develop, decide, and support the planning processes. Planners and politicians are considered primary public stakeholders as they are directly and continuously engaged in the process as the PAKT strategies mature.

Next, we have stakeholders in the private sector, which influence and are influenced by planning decisions. While land-use policies may strongly influence landowners and developers, their involvement is less direct at the initial strategic level. In PAKT, enterprise committees representing private interests are established and hold regular meetings to provide direct feedback to planning documents. Due to their direct involvement, they are considered primary stakeholders.

Finally, we have the wider societal stakeholders, with residents, the society at large, and others represented through interest organizations. Residents are represented through various committees that provide feedback to PAKT during the planning process. They may therefore be considered primary stakeholders.

4.2. Problems

The objectives of the PAKT program set out the goals that the proposed PAKT plan and subsequent adoption of the plan in spatial and transport planning should strive to achieve. The planning program, draft proposals, and stakeholder feedback resulted in the objectives hierarchy in Figure 4. Eight initial objectives were distilled to four objectives with sub-objectives throughout this process. In the initial planning program, objectives were formulated without explicit links to the SDGs. Through the problem structuring process and subsequent draft plan proposal, SDGs pertinent to each objective were identified and formulated.

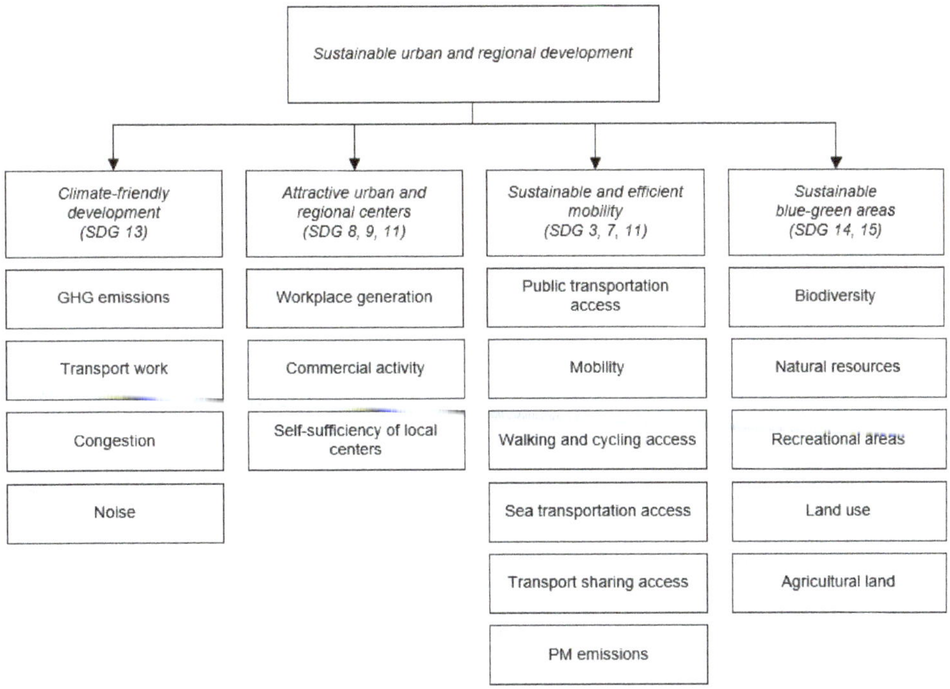

Figure 4. Structured objective hierarchy, derived from the PAKT planning program.

In order to operationalize objectives, they were linked to SDG-based indicators, as displayed in Figure 5. The United 4 Smart Sustainable Cities (U4SSC) indicator set [37] was used as an initial framework for providing indicators as the municipalities had recently performed self-assessment using this framework [38]. This permitted using existing data repositories and analytical models established in the municipalities to support planning decisions. The mapping between indicators and PAKT objectives helped identify which objectives were already covered and which objectives needed new indicators. The exercise showed that most of the mobility-related objectives, and anthropocentric land-use objectives, were well covered. Meanwhile, there was a lack of indicators to evaluate ecological and economic impacts from land use. As a result, it was decided to further develop indicators to accommodate the assessment of these objectives in subsequent work.

4.3. Alternatives

While the decisions following from the PAKT document will continue to be implemented in the period for which it has been adopted, the document itself also decides on some high-level strategies and policies for area development and lower-level measures to be implemented, as shown in Figure 6. These decisions will propagate through the planning hierarchy, ultimately providing the frames for day-to-day decisions made by planners. At the highest level in PAKT, we find the land-use strategies. These provide the rules for land-use in the region over the period, such as where residential and commercial growth will take place and implicitly, the land to be conserved. The PAKT document also provides a set of measures that may be further defined within the high-level land-use change strategies, i.e., area-focused measures concerning site selection for residential and commercial development and location of public services. We may also identify a set of transport-focused measures that aim to support sustainable transport within the region.

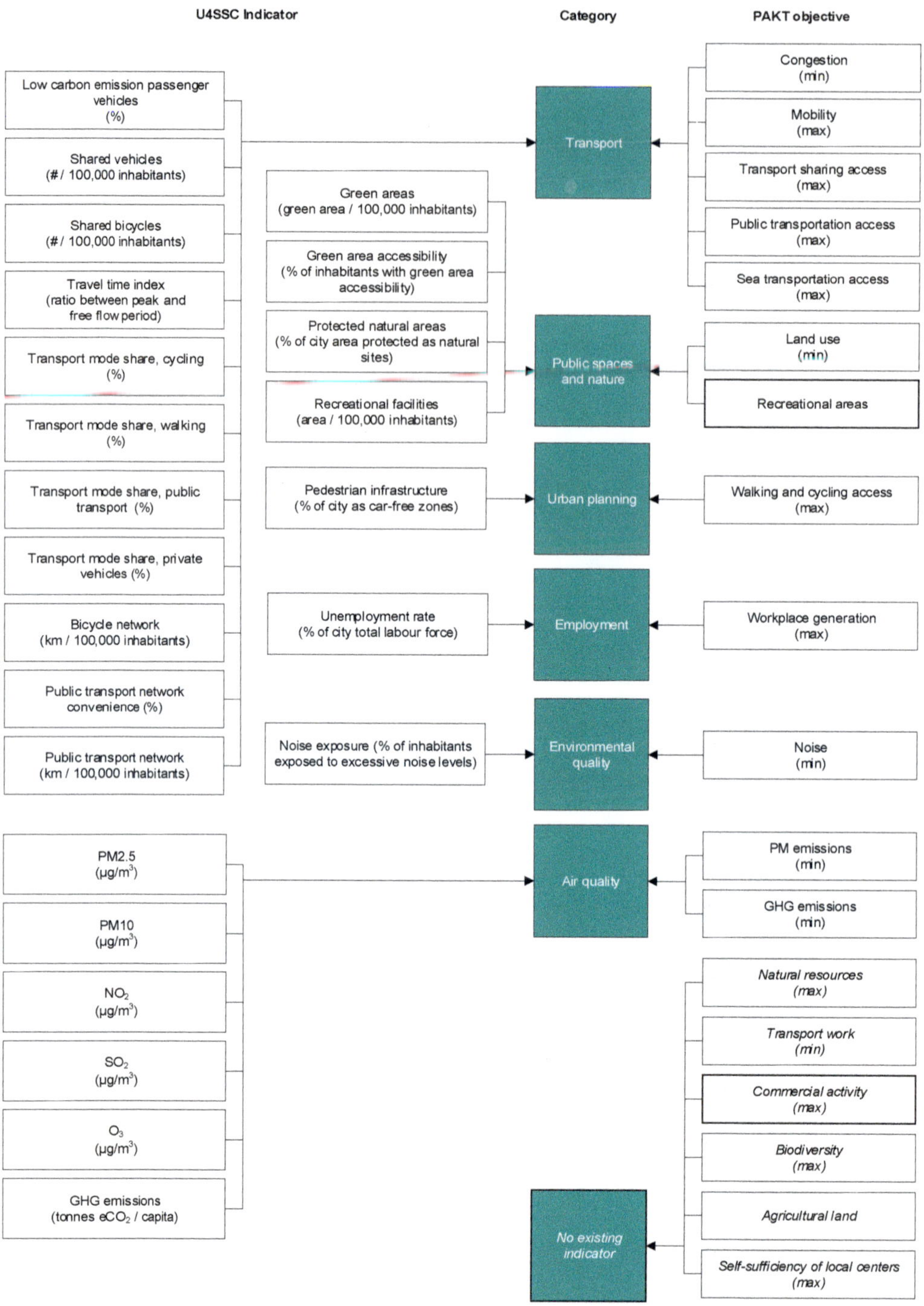

Figure 5. Linking objectives in the PAKT planning program to SDG-based KPIs.

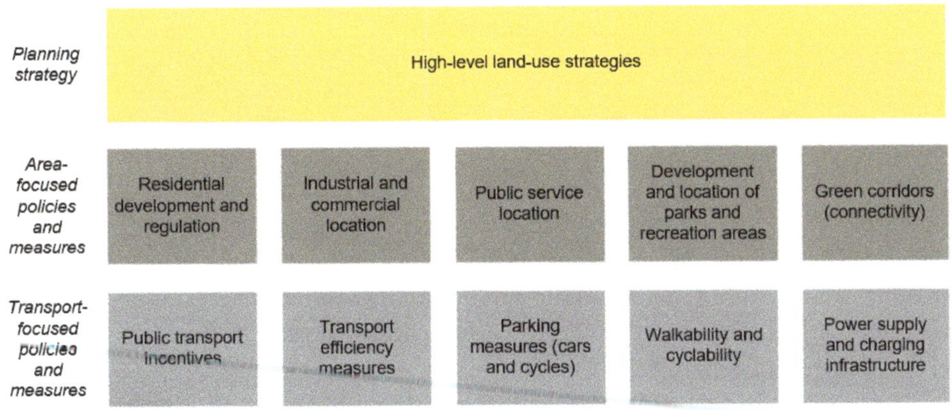

Figure 6. Decision hierarchy derived from the PAKT planning proposal.

During the second stakeholder workshop, high-level land-use strategies were identified as central to all further spatial and transport planning, as shown in Figure 7. These strategies determine the principles for regional development and therefore have a high impact on all the identified objectives. Based on the planning proposal, four main options were identified on a continuum from centralization to decentralization. The first planning strategy is based on monocentric concentration, where land-use change takes place in a few select hubs, e.g., the centers in each of the three municipalities. The policy entails that all residential, commercial and public service development occur in or close to the hubs to provide high-density service centers. Locating residential areas in the vicinity of these areas may also reduce transport work and associated negative impacts. The second strategy is a more polycentric development, where multiple existing hubs are used as a basis for further regional development. This opens up opportunities for multiple types of hubs within each municipality, e.g., regional, district, and local hubs that are established. The third strategy is to perform no changes to existing strategies in PAKT, i.e., to continue the current plans each municipality has adopted. While these plans are not developed for regional efficiency but rather to optimize development within a smaller geographic area, the planners expected this policy alternative as pointing towards a more decentralized type of policy. The fourth and final strategy option is to deregulate spatial plans for the region, essentially permitting all land-use change to take place wherever municipalities, landowners, developers, and other commercial interests see fit. In this option, costs are the primary driver for site selection of residential development as well as the location of public and private services.

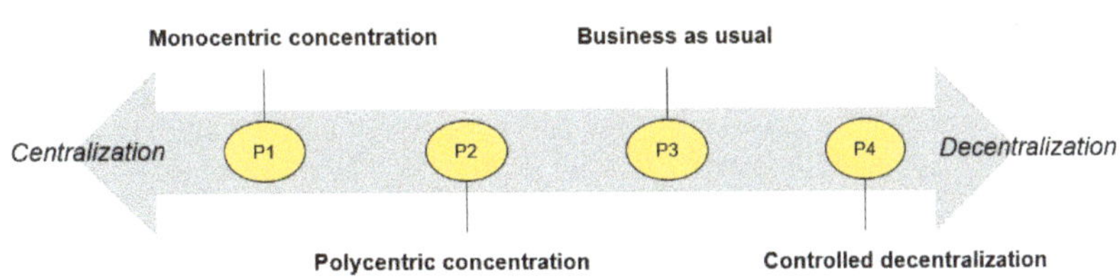

Figure 7. High-level land-use change strategies in the PAKT program and proposal.

4.4. Decisions

As part of planning decision-making, it is also necessary to evaluate how planning strategies and other measures comparatively perform against the stated objectives. This task would require further specifying model parameters and delimiting the decision problem temporally and spatially to perform a policy evaluation. Figure 8 is however useful as it shows the problem structure derived from the SPADE process to support planning decision-making. This structure will form the basis for evaluation tasks to be performed by a PSS to support spatial and transport planning in Ålesund, Giske, and Sula municipality as part of the Smart Plan project. The further design of the planning support tool also warrants systems engineering techniques, as it is necessary to identify user needs and requirements for the tool before defining its function and architecture.

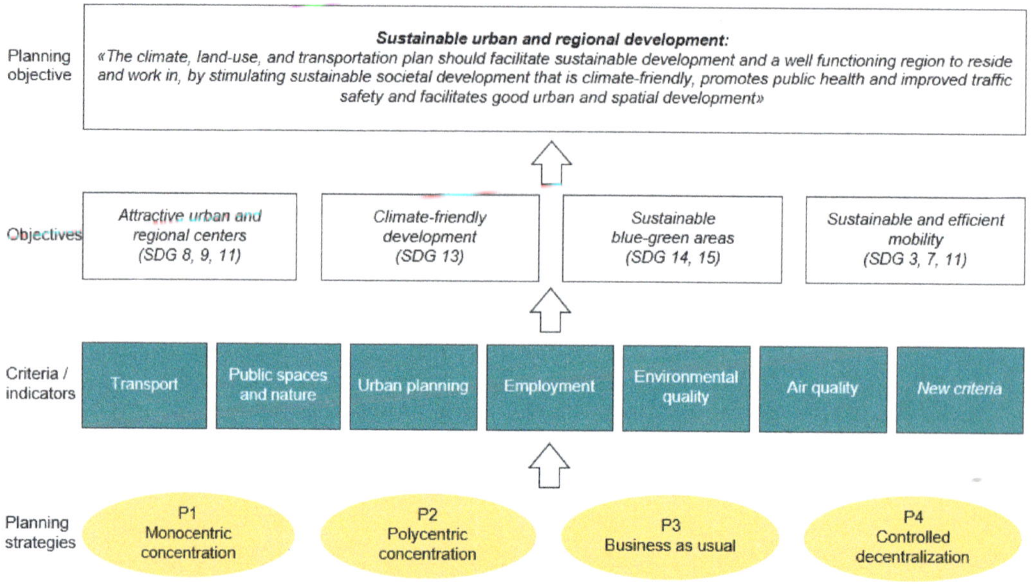

Figure 8. High-level problem structure for PAKT.

4.5. Evaluation

Evaluation is a continuous task performed in and between each step in the systems engineering SPADE method. As part of the stakeholder involvement process, feedback was given for each element of the problem structure during workshops to update and improve it. The within-step evaluation was performed by checking consistency with planning documents and correcting any misinterpretations through stakeholder dialogue. Consistencies across steps were also checked by tracing between models.

5. Discussion

5.1. Workability of Systems Engineering in Planning Problem Structuring

The application of systems engineering to establish a problem structure for sustainable urban planning demonstrates how consistent and traceable representations of an initially complex planning context may be achieved. The approach generates models elicited from and with stakeholders that help clarify the key planning context elements and their interrelationships.

While the literature is abundant with PSS tool descriptions, there is little knowledge about how to scope these tools properly. This requires moving beyond the domain of information and communication technology to consider the dimensions of application and governance [16]. SPADE enables PSS developers to gain a comprehensive understanding of the themes and issues that practitioners and planning stakeholders handle. This may be useful in establishing explicit and justifiable thoughts on tool scoping.

The approach also proved useful in engaging a wide range of planning stakeholders in a joint problem structuring effort. As the approach is jargon-free and intuitive, it presents a pedagogical frame for stakeholder involvement and co-creative problem-solving. The model-based implementation in the case study also proved practical as stakeholders easily could reflect upon and adjust models during and between workshop sessions. As models are sequenced according to the steps of SPADE, navigation between models was easily facilitated during discussions.

5.2. Operationalizing and Integrating the Sustainable Development Goals in Planning

The case study also demonstrates how explicit consideration of SDGs in the planning context may be achieved. While SDG-based planning was stated as an essential objective of the planning process in initial planning documents [21], they were not directly linked to planning objectives and strategies. This linkage was made possible once the problem structure was established. As this reorientation of objectives according to SDGs was made, it was also possible to link objectives to measurable criteria and indicators, which was another stated objective in the planning documents.

The case study utilized and linked an existing SDG indicator framework [37] to planning strategies, which allowed us to use a globally harmonized system. The model traceability rapidly helped identify entities for which new indicators were required. This ability to structure existing information and identify missing elements shows the benefit of creating a consistent model-based problem structure.

Although a specific indicator framework was operationalized in the case study, the systems engineering approach may also be used in instances where new indicators need to be developed. Stakeholders and experts could either be engaged to help synthesize new indicators or help sort and select indicators from a predefined set.

5.3. Theoretical and Practical Limitations

The models generated from the problem structuring exercise reported in this paper provide a basis for further synthesis of a PSS for strategic planning in the case study region. As each planning process and region is unique, the structured outputs are also idiosyncratic to the particular context in which they are developed [15]. The transferability of models between planning contexts may therefore be low. However, models may be practical in inspiring, comparing, and contrasting problem structures across planning contexts.

This limitation of model validity may also apply to the same region over time. As planning is a constantly evolving process, stakeholders' ideas, opinions, and perspectives may change, and new problems may emerge [16]. Therefore, a problem structure should not be considered a permanent representation of the planning context but rather as a snapshot at a specific point in time. SPADE is a cyclical methodology, and the reported procedure may be repeated with regular intervals to update and reorient the structured elements as needed.

Problem structuring and decision support literature highlights the need to create hierarchies with unique and non-overlapping objectives which reflect the fundamental values of stakeholders [31,39]. This is particularly important when alternatives are to be comparatively evaluated using formal models. The objectives and indicators established in Figure 5 have not been scrutinized for this purpose. Further iterations using SPADE could be used to validate the objectives and indicators against this formal requirement. This is a natural step in the continued work of synthesizing a PSS.

The reported application of systems engineering focused on structuring the decision context for planning stakeholders. While this provides a common frame of reference for main planning themes, objectives, and strategies, further tool synthesis must also build on extensive mapping of the practical context of the tool itself. This warrants its own problem structuring that focuses on PSS users and the tasks they need to perform in addressing planning decision problems. This subsequent step in the synthesis process could potentially also be explored from a systems engineering approach, where user needs and requirements are defined to further specify the PSS functions and architecture.

6. Concluding Remarks

In this article, we have demonstrated the use of systems engineering to establish a problem structure to initiate the synthesis of a participatory planning support system (PSS) for sustainable regional planning. The systems engineering SPADE methodology has been applied in a model-based way to define and link sustainable development goals (SDGs) in a practical planning case involving multiple planning stakeholders. The case study has demonstrated that the approach is applicable in generating consistent and traceable representations of an initially ill-structured context and practical for engaging stakeholders in a co-creative problem structuring exercise. The model-based approach to problem structuring also proved practical in integrating and operationalizing the SDGs by linking them to indicators and planning strategies.

While the workability of the approach is demonstrated for the given context, additional applications should be made to further test the applicability and robustness of the approach across regions, planning domains, and planning levels. Mechanisms and procedures for updating problem structures within and across planning cycles would also prove useful to avoid static representations of dynamic planning environments. Further synthesis of PSSs requires additional mapping of user needs and requirements. Exploring the translation of planning problem structures to tool functions and architecture is a critical topic for further research.

Author Contributions: Conceptualization, D.M.A.; methodology, D.M.A.; validation, D.M.A. and A.A.; formal analysis, D.M.A.; writing—original draft preparation, D.M.A. and A.A.; writing—review and editing, D.M.A. and A.A. All authors have read and agreed to the published version of the manuscript.

Funding: This research was funded by The Research Council of Norway and performed as part of the Smart Plan project, grant number 310056 and the SUSTRANS project, grant number 267887.

Institutional Review Board Statement: Not applicable.

Informed Consent Statement: Not applicable.

Data Availability Statement: The data presented in this study are available in the article.

Acknowledgments: The authors wish to thank urban planners, architects, and managers in the Ålesund region for their participation in and inputs to the case study. We also wish to thank project partners and colleagues from the Sustainability Analytics research arena at NTNU, and colleagues at Volda University College and Offshore Simulator Centre for feedback during the process. The authors wish to express their sincere gratitude to three anonymous reviewers for their useful suggestions on how to improve the quality of the paper.

Conflicts of Interest: The authors declare no conflict of interest.

References

1. OECD. *A Territorial Approach to the Sustainable Development Goals*; OECD: Paris, France, 2020.
2. Global Taskforce of Local and Regional Governments; UN Habitat; UNDP. *Roadmap for Localizing the SDGs: Implementation and Monitoring at Subnational Level*. 2016. Available online: https://www.uclg.org/sites/default/files/roadmap_for_localizing_the_sdgs_0.pdf (accessed on 20 January 2021).
3. Asker Municipality. *From Global Goals to Local Action-How We Implement the UN Sustainable Development Goals (SDGs) in Asker Municipality*; Asker Municipality: Asker, Norway, 2020.
4. Lundberg, A.K.; Bardal, K.G.; Vangelsten, B.V.; Brynildsen, M.; Bjørkan, R.; Bjørkan, M.; Richardson, T. *Strekk i Laget: En Kartlegging av Hvordan FNs Bærekraftsmål Implementeres i Regional og Kommunal Planlegging (Mind the Gap-Mapping How UN SDGs Is Implemented in Regional and Municipal Planning)*; Nordlandsforskning-Nordland Research Institute: Bodø, Norway, 2020.
5. Kirst, E.; Lang, D.J. Perspectives on Comprehensive Sustainability-Orientation in Municipalities: Structuring Existing Approaches. *Sustainability* **2019**, *11*, 1040. [CrossRef]
6. Valencia, S.C.; Simon, D.; Croese, S.; Nordqvist, J.; Oloko, M.; Sharma, T.; Buck, N.T.; Versace, I. Adapting the Sustainable Development Goals and the New Urban Agenda to the city level: Initial reflections from a comparative research project. *Int. J. Urban Sustain. Dev.* **2019**, *11*, 4–23. [CrossRef]
7. Krellenberg, K.; Bergsträßer, H.; Bykova, D.; Kress, N.; Tyndall, K. Urban Sustainability Strategies Guided by the SDGs—A Tale of Four Cities. *Sustainability* **2019**, *11*, 1116. [CrossRef]

8. Geertman, S.; Stillwell, J. Planning support systems: An inventory of current practice. *Comput. Environ. Urban Syst.* **2004**, *28*, 291–310. [CrossRef]
9. Geertman, S.; Stillwell, J. *Handbook of Planning Support Science*; Edward Elgar Publishing: Northampton, UK, 2020.
10. Flacke, J.; Shrestha, R.; Aguilar, R. Strengthening Participation Using Interactive Planning Support Systems: A Systematic Review. *ISPRS Int. J. Geo Inf.* **2020**, *9*, 49. [CrossRef]
11. Pettit, C.; Shi, Y.; Han, H.; Rittenbruch, M.; Foth, M.; Lieske, S.; Nouwelant, R.V.D.; Mitchell, P.; Leao, S.; Christensen, B.; et al. A new toolkit for land value analysis and scenario planning. *Environ. Plan. B Urban Anal. City Sci.* **2020**, *47*, 1490–1507. [CrossRef]
12. Klosterman, R.E. Planning support systems: A new perspective on computer-aided planning. *J. Plan. Educ. Res.* **1997**, *17*, 45–54. [CrossRef]
13. Geertman, S.; Stillwell, J. *Planning Support Systems in Practice*; Springer: Berlin, Germany, 2003.
14. Couclelis, H. "Where has the future gone?" Rethinking the role of integrated land-use models in spatial planning. *Environ. Plan. A* **2005**, *37*, 1353–1371. [CrossRef]
15. Te Brömmelstroet, M. Towards a pragmatic research agenda for the PSS domain. *Transp. Res. Part A Policy Pract.* **2017**, *104*, 77–83. [CrossRef]
16. Geertman, S.; Stillwell, J. Planning support science: Developments and challenges. *Environ. Plan. B Urban Anal. City Sci.* **2020**, *47*, 1326–1342. [CrossRef]
17. Te Brömmelstroet, M. Equip the warrior instead of manning the equipment: Land use and transport planning support in the Netherlands. *J. Transp. Land Use* **2010**, *3*, 25–41. [CrossRef]
18. Vonk, G.; Geertman, S. Improving the Adoption and Use of Planning Support Systems in Practice. *Appl. Spat. Anal. Policy* **2008**, *1*, 153–173. [CrossRef]
19. Pidd, M. *Tools for Thinking: Modelling in Management Science*, 2nd ed.; Wiley: Chichester, UK, 2003.
20. Rittel, H.W.J.; Webber, M.M. Dilemmas in a general theory of planning. *Policy Sci.* **1973**, *4*, 155–169. [CrossRef]
21. Region Ålesund. *Planprogram. Plan for Areal, Klima og Transport i Ålesundsregionen (Planning Program. Plan for Area, Climate and Ransportation in the Ålesund Region)*; Region Ålesund: Ålesund, Norway, 2020.
22. Ministry of Local Government and Modernisation. Municipal Planning. 2014. Available online: https://www.regjeringen.no/en/topics/plan-bygg-og-eiendom/plan--og-bygningsloven/planning/engelsk-test---planning-in-norway/engelsk-test----2/id710310/ (accessed on 20 January 2021).
23. INCOSE. *Systems Engineering Handbook: A Guide For System Life Cycle Processes and Activities*; Ringgold Inc.: Beaverton, OR, USA, 2006.
24. Haskins, C. *Systems Engineering Analyzed, Synthesized, and Applied to Sustainable Industrial Park Development*; Norges Teknisk-Naturvitenskapelige Universitet: Trondheim, Norway, 2008.
25. Blanchard, B.S.; Fabrycky, W.J. *Systems Engineering and Analysis*; Pearson: Boston, MA, USA, 2011.
26. Palmer, E.; Burton, R.; Haskins, C. A Systems Engineering Framework for Bioeconomic Transitions in a Sustainable Development Goal Context. *Sustainability* **2020**, *12*, 6650. [CrossRef]
27. Fotland, G.; Haskins, C.; Rølvåg, T. Trade study to select best alternative for cable and pulley simulation for cranes on offshore vessels. *Syst. Eng.* **2020**, *23*, 177–188. [CrossRef]
28. Aspen, D.M.; Haskins, C.; Fet, A.M. Application of systems engineering to structuring acquisition decisions for marine emission reduction technologies. *Syst. Eng.* **2018**, *21*, 388–397. [CrossRef]
29. Mitchell, R.K.; Agle, B.R.; Wood, D.J. Toward a Theory of Stakeholder Identification and Salience: Defining the Principle of Who and What Really Counts. *Acad. Manag. Rev.* **1997**, *22*, 853–886. [CrossRef]
30. Clarkson, M.B.E. A Stakeholder Framework for Analyzing and Evaluating Corporate Social Performance. *Acad. Manag. Rev.* **1995**, *20*, 92–117. [CrossRef]
31. Keeney, R.L. *Value-Focused Thinking: A Path to Creative Decisionmaking*; Harvard University Press: Cambridge, MA, USA, 1992.
32. Belton, V.; Stewart, T.J. *Multiple Criteria Decision Analysis: An Integrated Approach*; Kluwer Academic: Boston, MA, USA, 2002.
33. von Winterfeldt, D.; Fasolo, B. Structuring decision problems: A case study and reflections for practitioners. *Eur. J. Oper. Res.* **2009**, *199*, 857–866. [CrossRef]
34. Ministry of the Environment. *Planning and Building Act (2008). Act of 27 June 2008 No. 71 relating to Planning and the Processing of Building Applications (the Planning and Building Act) (the Planning Part)*. 2008. Available online: https://www.regjeringen.no/en/dokumenter/planning-building-act/id570450/ (accessed on 20 January 2021).
35. Ministry of Local Government and Modernisation. *National Expectations Regarding Regional and Municipal Planning 2019–2023.35*; Ministry of Local Government and Modernisation: Oslo, Norway, 2019.
36. Region Ålesund. *PAKT. Plan for Areal, Klima og Transport i Ålesundsregionen*; Region Ålesund: Ålesund, Norway, 2021.
37. Smiciklas, J.; Prokop, G.; Stano, P.; Sag, Z. *Collection Methodology for Key Performance Indicators for Smart Sustainable Cities*; U4SSC: Geneva, Switzerland, 2017.
38. International Telecommunication Union (ITU). *Verification Report-Ålesund, Norway*; International Telecommunication Union: Geneva, Switzerland, 2020.
39. Keeney, R.L.; Raiffa, H. *Decisions with Multiple Objectives: Preferences and Value Tradeoffs*; Cambridge University Press: Cambridge, UK, 1993.

sustainability

MDPI

Article

Applying System-Oriented Sustainability Scoring for Cruise Traffic Port Operators: A Case Study of Geiranger, Norway

Børge Heggen Johansen [1,*], Dina Margrethe Aspen [2], Magnus Sparrevik [3] and Vilmar Æsøy [1]

1 Department of Ocean Operations and Civil Engineering, Norwegian University of Science and Technology, 6009 Ålesund, Norway; vilmar.aesoy@ntnu.no
2 Department of International Business, Norwegian University of Science and Technology, 6009 Ålesund, Norway; dina.aspen@ntnu.no
3 Department of Industrial Economics and Technology Management, Norwegian University of Science and Technology, 7491 Trondheim, Norway; magnus.sparrevik@ntnu.no
* Correspondence: borge.a.h.johansen@ntnu.no; Tel.: +47-91564208

Abstract: Balancing the positive and negative impacts of cruise tourism is a challenging task for port operators. Necessary information for cruise port planning and decision making may be laborious to acquire and further combine for holistic decision support. The current study applies a system-oriented sustainability scoring model to the port of Geiranger, Norway. The aim is to provide a practical and low-threshold approach for appraising sustainability aspects in cruise port planning and decision making. The scoring model provides an estimate of performance on sustainability indicators based on cruise call itinerary information and readily available ship data. Results demonstrate how using the scoring model can prove useful for both port management, planning, stakeholder communication and scenario evaluation.

Keywords: cruise ship; sustainability; indicators; tourism; systems thinking; port management

Citation: Johansen, B.H.; Aspen, D.M.; Sparrevik, M.; Æsøy, V. Applying System-Oriented Sustainability Scoring for Cruise Traffic Port Operators: A Case Study of Geiranger, Norway. *Sustainability* **2021**, *13*, 6046. https://doi.org/10.3390/su13116046

Academic Editor: Manuel Rodríguez-Díaz

Received: 31 March 2021
Accepted: 4 May 2021
Published: 27 May 2021

1. Introduction

Global cruise tourism passenger numbers increased by nearly 70% from 2009 to 2019 [1]. Although the coronavirus disease 2019 (COVID-19) pandemic had a significant impact on the industry in 2020, growth is expected to pick up again in the next few years [2]. Port cities and vicinities near popular cruise destinations benefit from jobs and revenue enabled by the industry. Conversely, there are strains that need to be dealt with. The environmental and social impacts of cruise ships are often concentrated in areas of high natural and cultural value [3]. Klein [4] points out "people pollution" (crowding), homogenization of the port experience, and the reduced authenticity of cultural experience as potential negative social impacts from cruise activities. The economic gains are also challenged as the expansive offering of onboard cruise ship activities may negatively influence the cruise passengers' local spending at cruise destinations. This may lead to an uneven distribution of costs and benefits between the cruise lines and the community ashore at destinations [5–8]. Balancing these positive and negative impacts of cruise tourism is a challenging task for local decision makers and the information available to support decisions for managing visits from cruise ships is often limited.

Through system development, the cruise ship-at-port system is explored according to the feasibility to support decisions on which cruise ships to admit as well as invoking measures countering strains to the port and its vicinity. Building upon the prior works of utilizing systems to understand tourism systems, the current study aims to develop and apply a sustainability scoring model for cruise ships. The purpose of the proposed model aims to provide port operators with a practical approach for evaluating how cruise calls perform on indicators representing environmental, social and economic dimensions of sustainability analogue to the triple bottom line framework [9]. The proposed model will further be applied to a case study on the port of Geiranger, Norway.

Section 2 of this paper describes the theoretical framework for the study drawing experience on previous studies. Section 3 describes the methodology in the proposed systems-based scoring model approach and Section 4 demonstrates how the scoring model could be applied in planning and management, stakeholder communication and scenario evaluation. Section 5 concludes the work and provides ideas on further development and application of the scoring model.

2. Background

Cruise tourism in the context of sustainability in research is a relatively recent and small field of science, with few publications before Ritter and Schafer [10] called for more academic scrutiny of the industry. Since then, several scholars have focused on the environmental impacts of cruise tourism [3,11]. This includes in particular emissions to air in port cities [7,12–19] and addressing them in terms of economic externalities [20–22]. Johnson [23] points not only to the cruise operators and destinations but also calls for the tourist to use their market influence to invoke changes towards more sustainable cruise tourism. Brida and Zapata-Aguirre [24], look at the economic, environmental and socio-cultural impacts of cruise tourism asking if there is certainty that the benefits of cruise outweigh the costs. Although they do not fully answer the question, they invite decision makers in cruise tourism destinations to not blindly implementing efforts to increase cruise tourism without thorough consideration of potential gains and challenges. Klein [4] looks at cruise tourism through the "responsible tourism lens", focusing on "(a) tourism's impact on the environment, (b) the equitable distribution of economic benefits to all segments of a tourist destination, and (c) minimizing negative sociocultural impacts" and the purpose of this framework is to consider the sustainability of cruise tourism in the perspective of the port community. The studies mentioned above are important conceptual papers, but they lack practical means of providing and structuring sustainability information to port operators.

Hritz and Cecil [25] looked at a case study on Key West, Florida, to construct sustainability indicators for cruise tourism focusing on current and potential resources, demographics of local residents and cruise passengers, economic structure and tourism demand–supply relationships. The study uncovered opinions on the indicators from different stakeholder groups and made up a structure to monitor future policy changes to cope with rising tourism. A similar approach, focusing on local stakeholders, was undertaken by Pomeranz, Needham and Kruger [26] regarding a coastal protected area in Alaska, US. Other studies on stakeholder perspectives of cruise tourism have been conducted since focusing on the perspectives of local residents [27,28]. Stefanidaki and Lekakou [29] introduce their conceptual approach "cruise-carrying capacity" based on the drivers, pressures, social processes/state, impacts and responses (DPSIR) relevant for cruise tourism. Based on an extensive review Stefanidaki and Lekakou [29] also provide a comprehensive set of sustainability indicators for cruise destinations in their framework based on a literature review and dialogue with experts. The indicators were derived from literature review and discussion with peers within cruise tourism research. The set of indicators are comprehensive and broad in scope and would provide to a high degree an exhausting set of parameters for societal planning and policy making on a national or regional governmental level. Wu, Chen and Min [30] built upon the work of previous studies building a set of indicators for cruise tourism sustainability for assessing regional sustainability conditions and cruise-industry development. Albeit being comprehensive and focusing on broad aspects of sustainability related to cruise tourism from the perspective of ports, the latter two indicator schemes would be too comprehensive for most port authorities to implement. In order to acquire useful information for sustainability decision support it is useful to comprehend the tourism industry as a complex and dynamic system of close and interdependent ties between components [31] p. 34. A port city—with its inhabitants, businesses, infrastructure and nature—being visited by a cruise ship and its passengers could be considered as such a system. Arnold and Wade [32] define systems thinking as " ... a set of synergistic analytic skills used to improve the capability of identifying and understanding systems, predicting their behaviors, and devising modifications to them in order to produce desired effects ... ". This includes understanding the system structure and its dynamic behavior. Systems thinking approaches have been applied in tourism focusing on e.g., organizational learning [33], future simulation [34] and mapping the effects on climate change from marine tourism [35]. Other recent examples of a systems thinking approach for establishing sustainability indicators are e.g., Stanitsas et al. [36]—in the construction industry and Steiniger et al. [37]—for urban sustainability indicators. More specific to the topic of the current study are Coccossis and Koutsopolou [38] for monitoring sustainability of coastal tourism; Schianetz and Kavanagh [39] for making sustainability indicators for tourism destinations; and a more broad-reaching scheme by Roxas et al. [40] using causal loop diagrams.

3. Materials and Methods

The current study evaluates a model for establishing sustainability indicators for decision support in cruise ports based on a systems thinking approach—seen from the perspective of the port. In this context, multiple systems are interacting. Firstly, there is the cruise ship with the purpose of transport, accommodation and entertainment of tourists which in itself embodies the characteristics of a system. In the context of sustainability, cruise ships interact with natural surroundings and socio-technical structures of the port cities, which also make up a new system that results in desired effects and the opposite. Examples of these sub-systems are the cruise ship–emission/discharge to nature system; cruise ship–road traffic system; and cruise ship–tourist distribution system; opposed to the cruise ship—local business system. These systems encompass both spatial and temporal boundaries together with interactions on other systems and stakeholders that also should be addressed in a cruise call sustainability scoring approach.

The literature assessed in Section 2 does not provide a practical and low threshold approach for appraising sustainability considerations in cruise port planning and decision making. The authors of the current study therefore propose a model for establishing sustainability indicators for decision support in cruise ports as shown in Figure 1. The model builds on a five-step procedure to analyze and evaluate sustainability impacts of cruise calls to port based on cruise ship characteristics and to use the output for decision support. The main focus of this paper is, hence, to demonstrate and evaluate the suitability of the model by using it on a test case of the port of Geiranger, Norway.

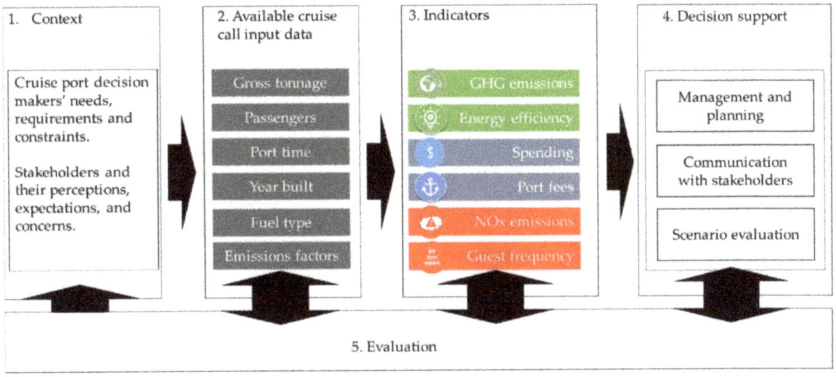

Figure 1. Proposed cruise call sustainability scoring and decision support model.

Firstly, the context of the cruise destination system is defined, including cruise port decision makers' needs, requirements and constraints together with stakeholders and their perceptions, expectations and concerns. The concerns and wants of the stakeholders should be addressed in the indicators used to explain phenomena in the system. Cruise tourism destinations have great variation based on both demography and society. This provides different profiles of stakeholders. Popular cruise tourism destinations span from large cities like Barcelona and Miami to rural societies like the Norwegian fjords or small Caribbean island societies. The relative impact and focus of cruise tourism to stakeholders vary according to this as it in big cities is one of many systems affecting sustainability while it at smaller societies may be a dominant activity regarding economic development and hence the accompanying externalities. The distribution between gains and externalities is also an important aspect. Although stakeholders want to influence, the reality is often that economic and societal structures make up boundaries for involvement and many are aware and accept this but expect at least to be informed and to be heard [41]. Choosing the correct indicators for communication of impacts is an important part of this as they should represent the information requirement.

In the second step, the port operator selects data to compute indicators. The minimum required data are the number of passengers (PAX), gross tonnage (GT), year built and duration of the stay. This is information that significantly correlates to environmental impacts [16]. The second step also consists of informing the model with other available ship data and the planned itineraries or port calls. The approach proposed in this paper is also open for further information to be added, as it will inform the model better and provide more precise impact estimates.

In step three, the indicators for the cruise call sustainability scoring system should reflect the needs uncovered in the first step, while still being quantifiable through the available data mapped in step 2. According to Waas et al. [42] an indicator represents an attribute of a system that is given by a variable and its value. The attribute may be a quality, characteristic or property while the variable may be expressed using a quantitative or qualitative value. System stakeholders should be involved in the development of indicators. Additional literature and resources could also be utilized in the indicator synthesis process. The output of the exercise is a description of the indicator system properties and ultimately its synthesis. The indicators in the sustainability scoring model presented in this paper were established together with professionals at the Port Authority through a series of workshops. The literature discussed in Section 2 of this paper was used to establish a predefined set of indicators, which were presented and discussed among workshop participants. The indicators were evaluated with regards to how they provide valuable information for sustainability decision support, if the required data for calculating values are at hand and whether they are possible to implement. The approach is iterative, meaning that the practitioner should do adjustments on the indicator based on feedback from the Port Authority between workshops.

The fourth step includes taking the information provided by the indicators at the given context into account when making decisions. It is important to firstly understand the dynamics behind the elements of the systems being managed or considered for some act of change-making. How will management or change influence the different stakeholders of the systems? Will the decision have the desired effects on the impacts of sustainability? Will the decision provide a balanced solution within all three dimensions of sustainability? The scope of decision making could vary between day-to-day operations like management and planning to more long-term uses like scenario evaluation, e.g., Pesce et al. [43] utilized indices of different sustainability aspects in the decision making about different routing options for cruise ships at the port of Venice, Italy. The indicators could also be used for communicating to stakeholders "the state of sustainability" at the cruise tourism destination.

Lastly, each step includes evaluating the indicators with the requirements and desired attributes stated by the decision makers. In this step one also reviews what has been done in the other steps and check if improvements could be achieved before implementation of the indicator set. This step is inspired by other circular Systems Engineering approaches like the ones by Fet [44] and Haskins [45]. Analysts using this approach need to evaluate whether the data, models and indicators being used to estimate impacts to sustainability are in line with the desires and concerns of stakeholders. It must also be considered whether the output is understood by stakeholders and gives them valuable input. Either it is decision makers who need the background to make informed choices or stakeholders who require and/or deserve to be kept in the loop.

4. Results and Discussion

4.1. Step 1—Context

The sustainability scoring model is applied within the context of the port of Geiranger, Norway (Figure 2). The first step was taken through document analysis, participating in relevant meetings with stakeholders for the port and working meetings with Stranda Port Authority. The primary systems identified were the cruise ships calling to port and the port destinations with which they interact. A key functional requirement of these indicators is to determine how cruise calls positively and negatively influence the port destination. Critical constraints encompass the time and other resources needed for accessing and modeling data to compute sustainability indicators.

In order to provide meaningful information for decision support for the three purposes given in Figure 1, step four—an overview of the given ports present and future challenges is needed. The quay is situated in a small village with 240 permanent residents, surrounded by steep mountains ranging over 1400 m above fjord level. Geiranger admitted 402,335 cruise passengers in 2019, ranking 3rd in the country [46]. The rising influx of tourists to Geiranger up towards 2019 has led to a debate on the negative effects of mass-tourism and particularly cruise tourism and its accompanying emissions is under much public scrutiny [17,47,48] and perceptions about tourism from local stakeholders are mixed [49], while tourists are generally quite satisfied [50]. The scrutiny has led to actions taken by both local and national government from capping the total daily number of cruise passengers to Geiranger at 6000 by the local port authority to the National Parliament passing a bill requesting that only zero-emission ships shall operate in the World Heritage Fjords by 2026 [51]. Before this, amendments on the shipping emissions for the same area were laid down by the Norwegian Maritime Authority (NMA) to only admit ships with IMO tier III emission standards for NOx by 2025 [52].

Figure 2. Port of Geiranger, Norway. Photo: First author.

4.2. Step 2—Available Cruise Call Input Data

For the second step, data from the port of Geiranger, Norway, from the years 2015 to 2019 were made available from Stranda Municipality Port Authority. This includes the time of arrival and departure of the cruise ships. Based on the cruise ship name as an identifier—data on key ship characteristics were extracted from the IHS-seaweb [53]. Two indicators from each of the three dimensions of sustainability are proposed for the sustainability scoring model (Table 1). The indicators follow the taxonomy given by the triple bottom line framework [9] and Kleins [4] responsible cruise tourism lens, being respectively well known and frequently cited works in the literature on general sustainability and cruise tourism sustainability. Establishing indicators within these dimensions gives a clear structure both on how to utilize indicators for decision support and communication. The indicators are calculated from input variables normally found in cruise call itinerary information kept by port authorities. The input variables were also confirmed by Norwegian cruise port authority representatives. Gross tonnage is the weight of the ship. Port time (t_p) is how long the cruise ship is docked at port either on a quay or at anchor as passengers are landed. PAX is the number of passengers onboard the ship. Year built (YB) is the year the cruise ship was built. The idea is to keep the indicators in the proposed scoring model easy to implement and use for local port authorities as they seldom have the resources to undertake sustainability assessments of the cruise calls they receive. With the scoring model, they can do rough sustainability assessments of each cruise call and be able to use the information when planning itineraries for future cruising seasons as port callings often are booked a couple of years in advance.

Table 1. Input variables needed to determine the scoring model indicators.

Input Variable	Unit	Indicator	Reference Aspect	Area of Interest
GT, t_p	kg	Green house gas emissions	Environmental impact and resource use	Environmental
GT, t_p, PAX	kWh/PAX/h	Energy efficiency		
PAX	NOK	Spending potential	Equitable distribution of economic benefits	Economic
PAX, GT	NOK	Port fee		
GT, t_p, YB	kg	NOx emissions	Minimizing negative social/health impacts	Social
PAX, t_p	PAX/h	Guest frequency		

4.3. Step 3—Testing the Indicators

4.3.1. Environmental Impact

Under the critical constraints encompassing the time and other resources needed for accessing and modeling data to compute sustainability indicators, the scoring model proposed in the current study comes with limitations in this regard as only the hotel phase of cruise ship operations in port are covered by the sustainability scoring system. The combined variable port time and GT is found through regression modelling to be the best independent variables to predict emissions from cruise ship hoteling at port [16]. The proposed indicator for greenhouse gas (GHG) emissions per cruise call presented in the scoring model is hence based on a regression model for hotel power demand (E) for cruise ship hotel power provided by the authors [54]. E is further multiplied by t_p and the emissions factor for greenhouse gas emissions given as kg CO_2 per kWh from the third IMO greenhouse gas study [55]. Greenhouse gas emissions were estimated for all cruise calls at Geiranger for the years 2015 to 2019. The interquartile range has remained steady in the period and the median has increased (Figure 3). E is also suitable for the energy efficiency indicator proposed in the scoring model, but rather than multiplying with an emissions factor it is divided by PAX and t_p. This gives an indicator presenting how much energy is allocated to each of the passengers in the cruise ship. The energy efficiency indicator (Figure 4) gives many more outliers than any of the other indicators, yet the median is on the lower band of the interquartile range. This is due to ships with a high passenger-to-space ratio, meaning that the ships are relatively large compared to the number of passengers. Looking at the details behind the statistical summaries, it is apparent that there is a trend of ships with the highest passenger count are the most energy efficient.

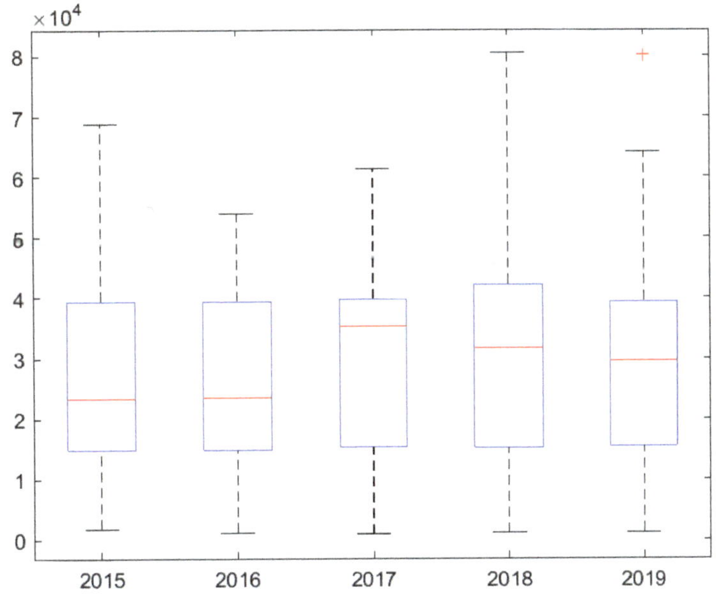

Figure 3. Estimated greenhouse gas emissions [kg] per cruise ship call in Geiranger from 2015–2019, $n = 937$.

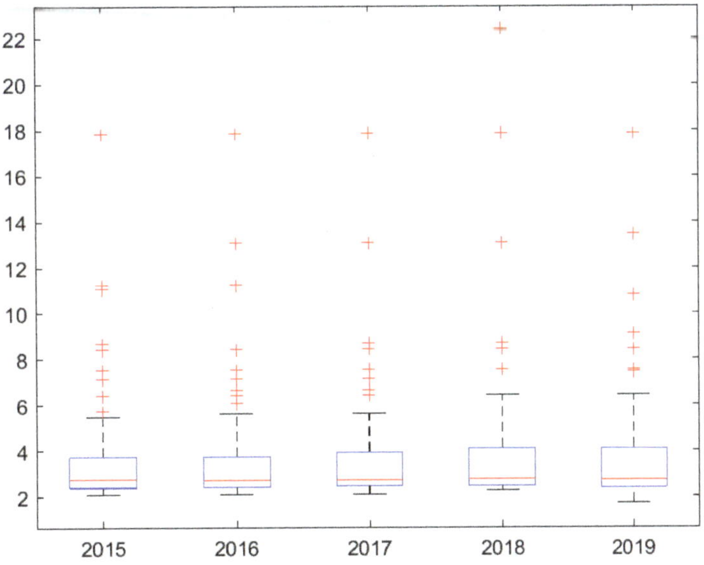

Figure 4. Estimated energy efficiency [kWh/PAX/h] per cruise ship call in Geiranger from 2015–2019, $n = 937$.

4.3.2. Equitable Distribution of Economic Benefits

When cruise calls are ordered by operators the port authority at least knows the passenger capacity of the given ship. Spending potential (SP) is given as the number of passengers multiplied by the average cruise passenger spending factor related to a cruise call for the given port, region or country. The spending factor (SF) gives how much money each cruise passenger on average spends when they go ashore at cruise destinations. The spending factor is normally gathered through surveying cruise passengers to a specific geographical area. The PAX data given for a cruise ship itinerary is commonly in the planning phase of a cruise call based on the passenger capacity of the vessel. It is not common that all passengers disembark the cruise ship at each port call. When calculating spending potential, PAX is multiplied by the landing factor (LF), giving how large percentage of total PAX is going ashore on average. The specific landing factor used when calculating the case data for Geiranger is provided by Stranda Port Authority as between 0.5–0.8 for this specific port. The value for SF for the port of Geiranger is for the data presented in this article LF = 0.7 and SF = 460 NOK [56]. The study also tried to see if the longer the guests stay, the higher is the possibility to spend money on local shops, cafes, etc. [56]. This effect, however, is not quantified in the scoring model in other ways than that cruise calls which have $p_t < 2$ are assumed not landing passengers due to time restrictions, hence giving a spending potential of zero. The port fee (PF) indicator gives how much the cruise ships must pay the local port authorities to use the quay. How the port fee is calculated could differ among ports. For the case of Geiranger, the rules for equating the port fee are extracted from the port authority web pages. Both spending potential and port fee are considered benefits in the scoring model and are hence considered variables that should be maximized.

4.3.3. Minimizing Negative Social/Health Impacts

Social indicators are considered as indicators impacting the well-being of both visitors and local inhabitants. Indicators dealing with emissions to air are considered core indicators, suggesting indicators stating the amount of emissions (SO_x, NO_x, PM) per cruise passenger and the number of days where air quality standards are exceeded for the given pollutants [29]. At the same time, one also could consider the air pollutants with local effects as environmental indicators. Exposure to NO_x may cause impaired lung function, increased susceptibility to respiratory infections and development or worsening of asthma and bronchitis [57]. Although air quality measurements for Geiranger show few occurrences of emissions exceeding the threshold levels for air pollution in national and international regulations, the studies show that the emissions are frequently above the Norwegian Institute of Public Health's air quality criteria during the tourist season [58]. Although the tourist season does not span over the whole year it is still a significant period of the year with significant pollution levels. The purpose of air quality criteria is to prevent negative health effects due to air pollution. The emissions concentrations in the criteria are set so low that most people can be exposed to these levels without suffering harmful health effects. Based upon these arguments the authors of this paper propose that NOx emissions are considered as a social indicator as they affect the health and well-being of local inhabitants and visitors. The guest frequency indicator alludes to the notion of crowding or "people pollution" [4]. The guest frequency (GF) factor is calculated by PAX onshore/port hours. It gives an expression of how many people to be landed/berthed per time unit. Less time also means less time to disperse, meaning more traffic and more crowding.

4.4. Step 4—Decision Support

The fourth step in the methodology proposed in Figure 1 requires that using the indicators is dependent on the management and decision-making context. This includes performing planning and management operations such as assessing port call requests on both a long-term and day-to-day basis, as well as invoking abatement measures to reduce strains or bolster opportunities. Other identified needs were information to support long-term strategic planning as well as the means to communicate towards stakeholders such as actors in the local community as well as cruise corporations and government bodies.

4.4.1. Port Management and Planning

Spending potential and port fee was estimated for all cruise calls coming to Geiranger from 2015 to 2019. These indicators are directly linked to the benefits of having cruise ships visiting the port for local business and the port owner, respectively. The trade-off between benefits and externalities is a central challenge for the port authority. From Table 2 one could observe that both the estimated cumulative spending per cruise call and port fee has increased over the five years, but so have the greenhouse gas emissions as well. This information could be useful for port operators in order to see both the benefit and cost of having more or less cruise tourists going ashore at their destination. Also, it could be useful to see whether, if ships were only to dock and do not let tourists ashore, the port fee might still keep up at high levels, but that would weaken the potential for businesses to generate income. The guest frequency indicator provides a means for port operators to say how many tourists the area needs to absorb. The average number of cruise passengers ashore has remained stable over the five years. A high value on this indicator means that many tourists disembark, are distributed, and board the ship within a specific time window. Decision makers could use this information to observe over time if there are correlations between the indicator and what is observed outside considering the destination's ability to absorb the number of tourists ashore. In this way, anecdotal reports of "crowding" could be related to the guest frequency number and, with many observations, the port operators could potentially gain an idea of whether there are levels of guest frequency when there is an increased risk of experiencing oversaturation of people in the area. This information could further be used to forecast such events and plan measures to abate the issue in advance.

Table 2. Cumulative scoring system indicators for cruise calls to Geiranger 2015–2019.

Year	Environmental		Economic		Social	
	GHG Emissions [ktonnes]	Hotel Energy Use [GWh]	Spending Potential [MNOK]	Port Fee [MNOK]	NOx Emissions [Tonnes]	Number of Tourists '000 [PAX]
2019	5.47	7.74	132	20.9	71.3	288
2018	5.06	7.16	114	18.1	64.3	250
2017	4.97	7.02	117	18.3	71.4	255
2016	4.52	6.39	106	16.7	70.1	230
2015	4.34	6.14	106	16.4	69.7	229

4.4.2. Scenario Evaluation

Considering strategic planning, quantifying greenhouse gas emissions could be used for setting targets and goals and to evaluate the impacts of future policies. Setting a target for cumulative and median greenhouse gas emissions over a season are examples of goals for which this indicator could be used (Tables 2 and 3). Furthermore, policies could be implemented, such as incentivizing cold-ironing or LNG ships and the indicator could be used to see whether these policies are having an effect. Utilizing this information could be valuable in a decision support sense as it provides port authorities information about which ships bring the most spending and port income per energy usage, which is directly proportional to emissions of both global and local pollutants.

The NMA is uncertain whether the cruise industry will be ready for zero emissions in 2026 and, therefore, suggest postponing to 2030, while still keeping Tier III for 2025 [59]. The year 2025 was selected for the scenario, using the Tier III compliant cruise ships scheduled to come to Geiranger in 2020 (before the COVID-19 pandemic brought cruise tourism to a halt) as a proxy. After discussing with local port authorities, it was concluded that making any good prognosis for future cruise tourist visits would be very uncertain due to the unknown short to medium-term effects of the pandemic. The case of emulating 2025 through 2020 was therefore accepted as sufficient for the sake of illustrating the scoring system for scenario evaluation.

Table 3. Average scoring system indicators per cruise calls in Geiranger 2015–2019.

	Environmental		Economic		Social	
	GHG Emissions [tonnes]	**Energy Efficiency [kWh/PAX/h]**	**Spending Potential [MNOK]**	**Port Fee [MNOK]**	**NO_x Emissions [kg]**	**GF [PAX Onshore/Port Hours]**
2019	25.9	3.06	0.63	0.10	338	183
2018	27.5	3.25	0.62	0.10	350	178
2017	27.7	3.07	0.65	0.10	399	188
2016	24.3	3.00	0.57	0.09	377	163
2015	24.5	2.96	0.60	0.09	394	177

Both NO_x emissions and guest frequency are estimated on parameters explaining the size of the ship. Gross tonnage and number of passengers respectively. Due to this one should expect that since greenhouse gas emissions have increased together with the number of passengers the same should apply for NO_x emissions as they are both estimated based on the energy demand. However, this is not the case for the historical data from 2015 to 2019 (Table 2). A possible explanation for this are the policies related to NO_x Tier-standards, requiring ships with lower NO_x emissions. One could based on this observation speculate that the NO_x Tier standard policies have been successful over the five-year period as average spending has increased while median NO_x pollution has stayed still. Regarding the future scenario, results from Table 4 show that there will be a significant reduction in the total number of tourists and spending potential, both with a decrease of 55% from 2019. Cumulative greenhouse gas emissions will decrease even more by 63% and NOx by 92%. These estimates would suggest that the reduction in environmental impact would be greater than the loss in income. When looking at the results from Table 5 one could see that although the cumulative greenhouse gas emissions are decreasing, the emissions per cruise call have increased. The average energy efficiency has also worsened. The explanation of why this might be the case could be due to the more uneven configuration coming to Geiranger under the assumptions made for the case study, as it is either very large or relatively small ships, with the ships in the mid-size class not being as represented as they were historically. One could also see this in the average guest frequency, meaning that although the cumulative number of tourists has drastically declined, the flux of tourists when ships call to port will increase. This could be valuable information when planning how to organize for receiving tourists in the future or considering new regulations. In this specific case the average capacity to absorb tourists through landing boats and distribution with buses etc. needs to remain the same even though the total number and hence possibility for income decreases. Effects like this will be very interesting to study in forthcoming work related to the 2026 zero-emissions regulations being addressed by the authors of the current study.

Table 4. Cumulative scoring system indicators for cruise calls to Geiranger 2019 and 2025.

	Environmental		Economic		Social	
Year	GHG Emissions [ktonnes]	Hotel Energy Use [GWh]	Spending Potential [MNOK]	Port Fee [MNOK]	NO_x Emissions [Tonnes]	Number of Tourists '000 [PAX]
2025	2.01	3.31	59.4	9.18	5.70	129
2019	5.47	7.74	132	20.9	71.3	288

Table 5. Average scoring system indicators per cruise call to Geiranger 2015 and 2025.

	Environmental		Economic		Social	
	GHG Emissions [Tonnes]	Energy Efficiency [kWh/PAX/h]	Spending Potential [NOK]	Port Fee [NOK]	NO_x Emissions [kg]	GF [PAX Onshore/Port Hours]
2025	30.5	3.52	0.90	0.14	86.4	230
2019	25.9	3.06	0.63	0.10	338	183

4.4.3. Communication with Stakeholders

In terms of stakeholder communication, the output from indicators could be valuable to use for informing with quantifiable information on sustainability [29]. Daily or per-cruise call values of the indicators could prove useful information to actors dependent or affected by cruise tourism. The spending potential indicator could prove useful for local business to forecast their revenue and enable better staffing for the given cruise call days. Together with information on GF for a given day, personnel and service level could be tailored based upon this information. In Geiranger, cruise calls are often booked a year or two in advance and therefore the indicators from the proposed scoring system could be communicated to local businesses in advance providing forecast information on the market and a better foundation to choose strategies and investments. Communicating openly on these benefits could also provide more approval of cruise tourism from stakeholder groups with a negative view of the industry. Conversely, the GF indicator could be valuable for people wanting to visit Geiranger on more calm days, avoiding the busiest days and allowing for a more even distribution of visitors. Communicating indicators of local air pollution, like the NO_x indicator, could provide useful information for people sensitive to aerosols. Knowing that there will be potentially large emissions, one could choose to stay out of Geiranger if there are e.g., are weather conditions that tend to worsen air pollution. Using historical information on ship activity and pollutant levels can provide a way of forecasting air pollution levels [60]. The results from the indicator could also be communicated to governing bodies and other stakeholders like greenhouse gas accounting initiatives.

4.5. Step 5—Evaluation

The resulting scoring model provides indicators that may evaluate various sustainability aspects related to cruise calls in both short and long-term perspectives using openly available data. The proposed sustainability scoring model for cruise ships at port has several technical and practical properties that are valuable for cruise port operators and owners. Firstly, the approach is practical, with a low barrier for implementation. This assertion is based upon the fact that the basic input variables to the scoring model are basic cruise call itinerary data and supplementary data that are openly available. Secondly, the scoring model provides estimates on six indicators representing the social, environmental and economic aspects of sustainability and also could be mapped in accordance with the relevant UN Sustainable Development Goals (SDGs) [61]. The indicators are useful when port operators and/or owners seek decision support with aim of enabling sustainability as a control parameter as well as utilizing them for communication with stakeholders. Thirdly, the scoring model is open to assigning weights among the criteria expressed through the indicators. Regarding matters which could be improved, the model could benefit from also including indicators addressing the effect of cruise tourism on onshore traffic strains as tourists are further distributed in the area by bus [62]. The relative importance and/or severity among the criteria expressed in the scoring model could be perceived differently between ports depending on their unique characteristics. Optimizing only one of the criteria does not guarantee the most advantageous solution.

5. Conclusions and Further Work

Balancing the impact of cruise traffic for port operators is challenging and often requires access to extensive tools and methods, while a practical/screening approach is missing. The information gained through the cruise call sustainability scoring model as demonstrated in this paper will, therefore, be a valuable contribution to support decisions in further management. The model provides data on six sustainability indicators based on calculations from readily available cruise call data. This information is then used to determine which measures to invoke for countering environmental and social strains to the port and its vicinity, at the same time balancing it with the economic gains from the activity. The case as demonstrated for the port of Geiranger, Norway, indicates that the cruise call sustainability scoring and decision support model is valuable in practice and beneficial for the cruise traffic port operators in the implementation of the SDGs in operational management procedures.

Scholars investigating sustainability in management and decision support may find the current study relevant as it includes a systems approach to establish and implement the SDGs for practical use. However, we acknowledge there are other models which are more comprehensive like the sustainability indicator sets by e.g., Stefanidaki and Lekakou [29] or Wu, Chen and Min [30], while the model proposed in this study uses an alternative approach, thus being more focused on practicality and low threshold for implementation than the currently available models. To increase the comprehensiveness of the model, more indicators can be included such as other emission factors, shore power availability, discharge to water, visual pollution and other means of local inhabitants and visitors' perceptions of a cruise call. The cumulative strain from multiple and simultaneous cruise calls could also be considered an improvement to the model.

The model could further be extended in its application by adding a multi-criteria decision making (MCDM) component. Instead of being static in the decision support stage, exploring how different stakeholder group perceptions give preference to different ship types through weighting schemes may further increase the possibility of making a balanced decision with credibility among a broad range of stakeholders [63]. Knowledge of how difference in opinions may lead to possible trade-offs may also aid the future strategic decision on the development of tourism destinations and cruise port operations. This will be the topic in a forthcoming study where the sustainability scoring model will be implemented with MCDM methodology for the appraisal of future scenarios of port infrastructure development in the Geirangerfjord area.

Author Contributions: Conceptualization, B.H.J., V.Æ. and D.M.A.; methodology, B.H.J., D.M.A. and M.S.; validation, B.H.J. and D.M.A.; formal analysis, B.H.J.; data curation, B.H.J.; writing—original draft preparation, B.H.J. and D.M.A.; writing—review and editing, B.H.J., D.M.A., M.S. and V.Æ.; supervision, V.Æ.; project administration, D.M.A.; funding acquisition, D.M.A. All authors have read and agreed to the published version of the manuscript.

Funding: This research was funded by the Norwegian Research Council, grant number 267887.

Acknowledgments: The authors are grateful for the participation and knowledge sharing from local actors in the Geiranger area, including Stranda Port Authority in particular.

Conflicts of Interest: The authors declare no conflict of interest. The funders had no role in the design of the study; in the collection, analyses, or interpretation of data; in the writing of the manuscript; or in the decision to publish the results.

References

1. Cruise Lines International Association. *State of the Cruise Industry Outlook 2020*; CLIA: London, UK, 2020.
2. Cruise Lines International Association. *State of the Cruise Industry Outlook 2021*; CLIA: London, UK, 2021.
3. Copeland, C. Cruise ship pollution: Background, laws and regulations, and key issues. In *Maritime Law: Issues, Challenges and Implications*; Harris, J.W., Ed.; Nova Science Publishers, Inc.: New York, NY, USA, 2011.
4. Klein, R.A. Responsible Cruise Tourism: Issues of Cruise Tourism and Sustainability. *J. Hosp. Tour. Manag.* **2011**, *18*, 107–116. [CrossRef]
5. Carić, H. Cruising Tourism Environmental Impacts: Case Study of Dubrovnik, Croatia. *J. Coast. Res.* **2011**, *61*, 104–113. [CrossRef]
6. Maragkogianni, A.; Papaefthimiou, S. Evaluating the social cost of cruise ships air emissions in major ports of Greece. *Transp. Res. Part. D Transp. Environ.* **2015**, *36*, 10–17. [CrossRef]
7. Papaefthimiou, S.; Maragkogianni, A.; Andriosopoulos, K. Evaluation of cruise ships emissions in the Mediterranean basin: The case of Greek ports. *Int. J. Sustain. Transp.* **2016**, *10*, 985–994. [CrossRef]
8. Scarfe, B.L. *Victoria as a Port-of-Call: The Costs and Benefits of Cruise Ship Visits*; James Bay Neighbourhood Association: Victoria, BC, Canada, 2011.
9. Elkington, J. *Cannibals with Forks: The Triple Bottom Line of 21st Century Business*; Capstone Publishing Ltd.: Oxford, UK, 1997.
10. Ritter, W.; Schafer, C. Cruise-Tourism: A Chance of Sustainability. *Tour. Recreat. Res.* **1998**, *23*, 65–71. [CrossRef]
11. Carić, H.; Mackelworth, P. Cruise tourism environmental impacts—The perspective from the Adriatic Sea. *Ocean. Coast. Manag.* **2014**, *102*, 350–363. [CrossRef]
12. Miola, A.; Ciuffo, B. Estimating air emissions from ships: Meta-analysis of modelling approaches and available data sources. *Atmos. Environ.* **2011**, *45*, 2242–2251. [CrossRef]
13. Poplawski, K.; Setton, E.; McEwen, B.; Hrebenyk, D.; Graham, M.; Keller, P. Impact of cruise ship emissions in Victoria, BC, Canada. *Atmos. Environ.* **2011**, *45*, 824–833. [CrossRef]
14. Mölders, N.; Gende, S.; Pirhalla, M. Assessment of cruise–ship activity influences on emissions, air quality, and visibility in Glacier Bay National Park. *Atmos. Pollut. Res.* **2013**, *4*, 435–445. [CrossRef]
15. Carić, H. Challenges and prospects of valuation—cruise ship pollution case. *J. Clean. Prod.* **2016**, *111*, 487–498. [CrossRef]
16. Rodríguez, G.D.M.; Martin-Alcalde, E.; Murcia-González, J.; Saurí, S. Evaluating air emission inventories and indicators from cruise vessels at ports. *WMU J. Marit. Aff.* **2017**, *16*, 405–420. [CrossRef]
17. Simonsen, M.; Walnum, H.J.; Gössling, S. Model for Estimation of Fuel Consumption of Cruise Ships. *Energies* **2018**, *11*, 1059. [CrossRef]
18. Murena, F.; Mocerino, L.; Quaranta, F.; Toscano, D. Impact on air quality of cruise ship emissions in Naples, Italy. *Atmos. Environ.* **2018**, *187*, 70–83. [CrossRef]
19. Ruiz-Guerra, I.; Molina-Moreno, V.; Cortés-García, F.J.; Núñez-Cacho, P. Prediction of the impact on air quality of the cities receiving cruise tourism: The case of the Port of Barcelona. *Heliyon* **2019**, *5*, e01280. [CrossRef] [PubMed]
20. Dragović, B.; Tzannatos, E.; Tselentis, V.; Meštrović, R.; Škurić, M. Ship emissions and their externalities in cruise ports. *Transp. Res. Part. D Transp. Environ.* **2018**, *61*, 289–300. [CrossRef]
21. Paoli, C.; Vassallo, P.; Dapueto, G.; Fanciulli, G.; Massa, F.; Venturini, S.; Povero, P. The economic revenues and the emergy costs of cruise tourism. *J. Clean. Prod.* **2017**, *166*, 1462–1478. [CrossRef]
22. McArthur, D.P.; Osland, L. Ships in a city harbour: An economic valuation of atmospheric emissions. *Transp. Res. Part. D Transp. Environ.* **2013**, *21*, 47–52. [CrossRef]
23. Johnson, D. Environmentally sustainable cruise tourism: A reality check. *Mar. Policy* **2002**, *26*, 261–270. [CrossRef]
24. Brida, J.G.; Zapata-Aguirre, S. Cruise Tourism: Economic, Socio-Cultural and Environmental Impacts. *Int. J. Leis. Tour. Mark.* **2009**, *1*, 22. [CrossRef]
25. Hritz, N.; Cecil, A.K. Investigating the Sustainability of Cruise Tourism: A Case Study of Key West. *J. Sustain. Tour.* **2008**, *16*, 168–181. [CrossRef]
26. Pomeranz, E.F.; Needham, M.D.; Kruger, L.E. Stakeholder Perceptions of Indicators of Tourism Use and Codes of Conduct in a Coastal Protected Area in Alaska. *Tour. Mar. Environ.* **2013**, *9*, 95–115. [CrossRef]
27. James, L.; Olsen, L.S.; Karlsdóttir, A. Sustainability and cruise tourism in the arctic: Stakeholder perspectives from Ísafjörður, Iceland and Qaqortoq, Greenland. *J. Sustain. Tour.* **2020**, *28*, 1425–1441. [CrossRef]
28. McCaughey, R.; Mao, I.; Dowling, R. Residents' perceptions towards cruise tourism development: The case of Esperance, Western Australia. *Tour. Recreat. Res.* **2017**, *43*, 403–408. [CrossRef]
29. Stefanidaki, E.; Lekakou, M. Cruise carrying capacity: A conceptual approach. *Res. Transp. Bus. Manag.* **2014**, *13*, 43–52. [CrossRef]
30. Wu, X.; Chen, H.; Min, J. Sustainability assessment of cruise-industry development: A case study of Xiamen, China. *Marit. Policy Manag.* **2020**, 1–12. [CrossRef]

31. Gunn, C.A. *Tourism Planning*, 4th ed.; Routlege: New York, NY, USA, 2002.
32. Arnold, R.D.; Wade, J.P. A Definition of Systems Thinking: A Systems Approach. *Procedia Comput. Sci.* **2015**, *44*, 669–678. [CrossRef]
33. Schianetz, K.; Kavanagh, L.; Lockington, D. The Learning Tourism Destination: The potential of a learning organisation approach for improving the sustainability of tourism destinations. *Tour. Manag.* **2007**, *28*, 1485–1496. [CrossRef]
34. Walker, P.A.; Greiner, R.; McDonald, D.; Lyne, V. The Tourism Futures Simulator: A systems thinking approach. *Environ. Model. Softw.* **1998**, *14*, 59–67. [CrossRef]
35. Dawson, J.; Maher, P.T.; Slocombe, S.D. Climate Change, Marine Tourism, and Sustainability in the Canadian Arctic: Contributions from Systems and Complexity Approaches. *Tour. Mar. Environ.* **2007**, *4*, 69–83. [CrossRef]
36. Stanitsas, M.; Kirytopoulos, K.; Leopoulos, V. Integrating sustainability indicators into project management: The case of construction industry. *J. Clean. Prod.* **2021**, *279*, 123774. [CrossRef]
37. Steiniger, S.; Wagemann, E.; De La Barrera, F.; Molinos-Senante, M.; Villegas, R.; De La Fuente, H.; Vives, A.; Arce, G.; Herrera, J.-C.; Carrasco, J.-A.; et al. Localising urban sustainability indicators: The CEDEUS indicator set, and lessons from an expert-driven process. *Cities* **2020**, *101*, 102683. [CrossRef]
38. Coccossis, H.; Koutsopoulou, A. Measuring and monitoring sustainability of coastal tourism destinations in the Mediterranean. *Tourism* **2020**, *68*, 482–498. [CrossRef]
39. Schianetz, K.; Kavanagh, L. Sustainability Indicators for Tourism Destinations: A Complex Adaptive Systems Approach Using Systemic Indicator Systems. *J. Sustain. Tour.* **2008**, *16*, 601–628. [CrossRef]
40. Roxas, F.M.Y.; Rivera, J.P.R.; Gutierrez, E.L.M. Framework for creating sustainable tourism using systems thinking. *Curr. Issues Tour.* **2020**, *23*, 280–296. [CrossRef]
41. Kerswill, M.; Mair, H. Big Ships, Small Towns: Understanding Cruise Port Development in Falmouth, Jamaica. *Tour. Mar. Environ.* **2015**, *10*, 189–199. [CrossRef]
42. Waas, T.; Hugé, J.; Block, T.; Wright, T.; Benitez-Capistros, F.; Verbruggen, A. Sustainability Assessment and Indicators: Tools in a Decision-Making Strategy for Sustainable Development. *Sustainability* **2014**, *6*, 5512–5534. [CrossRef]
43. Pesce, M.; Terzi, S.; Al-Jawasreh, R.I.M.; Bommarito, C.; Calgaro, L.; Fogarin, S.; Russo, E.; Marcomini, A.; Linkov, I. Selecting sustainable alternatives for cruise ships in Venice using multi-criteria decision analysis. *Sci. Total Environ.* **2018**, *642*, 668–678. [CrossRef]
44. Fet, A.M. Systems Engineering Methods and Environmental Lifecycle Performance within Ship Industry. Ph.D. Thesis, Norwegian University of Science and Technology, Trondheim, Norway, 1997.
45. Haskins, C. Systems Engineering Analyzed, Synthesized, and Applied to Sustainable Industrial Park Development. Ph.D. Thesis, Norwegian University of Science and Technology, Trondheim, Norway, 2008.
46. Cruise Norway, 2019 Cruise Calls and Pax to Norway. 2020. Available online: https://www.cruise-norway.no/Public-info/Statistics/ (accessed on 13 October 2020).
47. Shlopak, M.; Bråthen, S.; Svendsen, H.J.; Oterhals, O. *Grønn Fjord. Bind II. Beregning av Klimagassutslipp i Geiranger*; Møreforsking Molde AS: Molde, Norway, 2014.
48. Norwegian Maritime Authority. *Utslipp til Luft og sjø fra Skipsfart i fjordområder med stor cruisetrafikk, Report ver*; Norwegian Maritime Authority: Haugesund, Norway, 2017.
49. Yttredal, E.R.; Homlong, N. Perception of Sustainable Development in a Local World Heritage Perspective. *Sustainability* **2020**, *12*, 8825. [CrossRef]
50. Yttredal, E.R.; Homlong, N. *Besøkendes Opplevelse av Geirangerområdet*; Volda University College: Volda, Norway, 2019.
51. Norwegian Parliamient. *Klimastrategi for 2030—Norsk Omstilling i Europeisk Samarbeid*; E.-A.E. Committee: Oslo, Norway, 2018.
52. Norwegtian Minsitry of Climate and Environment. *Forskrift om Miljømessig Sikkerhet for Skip og Flyttbare Innretninger*; Norwegtian Minsitry of Climate and Environment: Oslo, Norway, 2020.
53. IHS-Markit. Sea-Web Ships. 2018. Available online: https://maritime.ihs.com/ (accessed on 20 November 2020).
54. Johansen, B.A.H.; Aspen, D.M.; Sparrevik, M.; Æsøy, V. Air Emission Screening Approach for Sustainable Cruise Port Management. Manuscript submitted for publication. 2020.
55. Smith, T.; Smith Jalkanen, J.; Anderson, B.; Corbett, J.; Faber, J.; Hanayama, S.; O'Keeffe, E.; Parker, S.; Johansson, L.; Aldous LRaucci, C.; et al. *Third IMO GHG Study 2014: Executive Summary and Final Report*; IMO: London, UK, 2014.
56. Yttredal, E.R.; Homlong, N. *Forbruk Blant Besøkende til Geirangerområdet*; Volda University College: Volda, Norway, 2019.
57. World Health Organization. *Ambient (Outdoor) Air Quality and Health Fact. Sheet, Updated September 2016*; World Health Organization: Geneva, Switzerland, 2016.
58. Löffler, J. *Long-Term Air Quality Monitoring Program. UNESCO World Natural Heritage "Geiranger Fjord", Norway Annual Scientific Report 2020*; Universität Bonn: Bonn, Germany, 2020.
59. Cruise Industry News. Norway Extends Zero-Discharge for Heritage Fjords to 2030. Available online: https://www.cruiseindustrynews.com/cruise-news/22871-norway-extends-zero-discharge-for-heritage-fjords-to-2030.html (accessed on 4 May 2020).
60. Aspen, D.M.; Löffler, J.; Hasle, E.; Ellefsen, A.L.; Kader, A.M.H.A.; Cheng, X.; Johansen, B.A.H. *Prediction Modeling of Air Quality in Geiranger*; Norwegian University of Science and Technology: Ålesund, Norway, 2019.

61. United Nations. *Transforming our World: The 2030 Agenda for Sustainable Development*; United Nations: New York, NY, USA, 2015.
62. Díez-Gutiérrez, M.; Babri, S. Explanatory variables underlying the route choice decisions of tourists: The case of Geiranger Fjord in Norway. *Transp. Res. Part. A Policy Pract.* **2020**, *141*, 398–409. [CrossRef]
63. Belton, V.; Stewart, T. *Multiple Criteria Decision Analysis: An. Integrated Approach*; Springer Science Business Media: Berlin, Germany, 2002.

sustainability

MDPI

Article

Patterns for Resilient Value Creation: Perspective of the German Electrical Industry during the COVID-19 Pandemic

Konstantin Neumann [1,*], Tim van Erp [2,3], Erik Steinhöfel [1], Felix Sieckmann [1] and Holger Kohl [1,4]

1 Fraunhofer Institute for Production Systems and Design Technology, 10587 Berlin, Germany; mail@eriksteinhoefel.de (E.S.); felix.sieckmann@ipk.fraunhofer.de (F.S.); holger.kohl@ipk.fraunhofer.de (H.K.)
2 German Electrical and Electronic Manufacturers' Association (ZVEI), 60528 Frankfurt am Main, Germany; Tim.vanErp@zvei.org
3 Department of Technology and Innovation, University of Southern Denmark, 5230 Odense, Denmark; tve@iti.sdu.dk
4 Chair for Sustainable Corporate Development, Technische Universität Berlin, 10587 Berlin, Germany
* Correspondence: konstantin.neumann@ipk.fraunhofer.de; Tel.: +49-30-39006-139

Abstract: The COVID-19 pandemic represents a massive, often unanticipated, external disruption for many companies. As a concept for responding to such disruption, organizational resilience has recently received great attention. In the organizational context, the overriding question is how companies can become more resilient. This study aims to contribute to answering this question by identifying, categorizing, and providing specific business model patterns for achieving resilience on the corporate level. For this purpose, a review of publications by major consulting firms was conducted. Patterns were extracted from publications until a convergence criterion indicated that no new pattern could be identified considering further publications. The 110 extracted unique patterns were clustered into 13 objectives, and additionally categorized according to resilience phases, as well as business model elements, to support the application in practice. The final catalog of patterns was validated through expert interviews and thus provides organizations, such as those in the electrical industry, with an overview and specific approaches on how to tackle industrial resilience through the adaption of their business model.

Keywords: organizational resilience; business model; resilience measure; business model pattern; COVID-19; pandemic; crisis; risk management; business continuity; electrical industry

Citation: Neumann, K.; van Erp, T.; Steinhöfel, E.; Sieckmann, F.; Kohl, H. Patterns for Resilient Value Creation: Perspective of the German Electrical Industry during the COVID-19 Pandemic. *Sustainability* **2021**, *13*, 6090. https://doi.org/10.3390/su13116090

Academic Editor: Ali Bahadori-Jahromi

Received: 31 March 2021
Accepted: 24 May 2021
Published: 28 May 2021

1. Introduction

1.1. Motivation

Against the backdrop of increasing volatility in business environments, organizational resilience is receiving significant attention from research and practice. The current pandemic is accompanied by changing consumer patterns and behavior, disrupted supply chains and the need for effective health measures, among other factors. This underscores the importance of being able to anticipate potential threats, cope effectively with adverse events and adapt to changing conditions.

The concept addressing this ability is resilience. Existing literature covers the concept [1–3] of resilience and, based on literature reviews [1,3–7] or case studies [2,8,9], general principles and attributes of resilient organizations, or parts of organizations like supply chains [10], are discussed. To help companies cope with volatility in the short term and achieve resilience in the long run, it is essential to provide them with practical and effective tools [9]. For example, Mallak [11] provides principles and suggests how to apply them. Fiksel et al. [12] also provide a practitioner-focused description; however, their analysis is on supply chains. Although a variety of general and sustainability-oriented business model pattern collections exist (e.g., [13]), as well as approaches for developing resilient business models (e.g., [14]), no pattern collection focusing on resilience is yet available. Therefore, this study aims at closing this research gap by developing such a pattern collection against the background of the COVID-19 pandemic. The German electrical industry was chosen for validation as they faced several impacts of the pandemic situation.

For this purpose, it is first pointed out how resilience fits into the concept of systems engineering and sustainability. Afterward, using the example of the German electrical industry, the results of a current survey illustrate the impact of the pandemic at various levels. Subsequently, the concept of resilience and the concept of the business model are outlined and brought together to guide the following developments. Based on a review of contributions on resilience by the main players in the consulting industry, measures or so-called business model patterns are identified, classified, and validated through expert interviews. The insights gained are used to provide recommendations for practice and research. In conclusion, the results of the paper are summarized and an outlook for further research and development is given.

1.2. Academic Relevance: Contributions to Systems Engineering and Sustainable Development

The logic of how companies create value is a key element of value creation systems. Understanding value and considering customers, users, other beneficiaries, and suppliers is an essential activity of model-based systems engineering [15]. Business models provide a common, conceptual modeling framework for describing this logic, by defining key elements and their relationships. Developing or innovating business models is thus an important aspect supporting the engineering of complex industrial systems. For example, it supplements the hardware and software of technical innovations [16]. Resilient business model patterns can serve as principles and recommendations as well as best-practice examples for designing complex value creation systems, and thus contribute to research in systems engineering, specifically with relevance to business engineering and to designing innovation systems.

The Sustainable Development Goals (SDGs) of the United Nation aim to enable transformation to move toward a global state of sustainability. This state can change dynamically over time, and is described by a current global consensus, i.e., the SDGs [17]. Sustainable transformation can be divided into two main components: mitigation and adaptation [18]. Mitigation aims at eliminating the underlying negative causes for global trends and their connected risks, whereas adapting aims at building up resilience against global risks by increasing the capability to appropriately react to shocks. Thus, resilience is a possible answer for organizations to adjust to climate change, for instance, with its uncertain consequences or other unknown disturbances. This directly contributes to SDG targets 13.1 and 13.3 [19]. In the literature, it is argued that there is a relationship between the different areas of resilience, e.g., that resilient organizations can lead to resilient communities [9]. Therefore, resilience also facilitates SDG targets 9.1 and 11.b [20,21]. Furthermore, it is shown that social and environmental practices have a positive effect on long-term metrics, and researchers have argued that they therefore support resilience [22]. Hence, this paper intends to contribute directly to research approaches supporting sustainability and the SDGs. It provides a holistic view on how to achieve a higher state of resilience in value creation systems, helping industrial stakeholders to adapt to global sustainability risks.

1.3. Industrial Relevance: Impacts of the COVID-19 Crisis on the Electrical Industry

The electrical industry in Germany is represented by ZVEI, the German Electrical and Electronic Manufacturers' Association. ZVEI has more than 1600 member companies, which account for approximately 90% of the employees of the electrical industry in Germany. Its member structure comprises global players as well as medium-sized and family-owned companies. ZVEI's members employ 867,500 people in Germany, and an additional 790,000 people globally. The sector generated a turnover of approximately 181.9 billion Euros in 2020. The electrical industry accounts for 10% of industrial turnover, 14% of total employees, as well as 3% of the gross domestic product of Germany in 2020 [23]. Its companies employ 26% of all employees in the area of research and development, are responsible for 23% of innovation expenditures of the manufacturing industry, and realize one-third of their turnover with product innovations.

The electrical industry is an essential supplier for the health, building and mobility industry as well as for the energy sector. Typical products are automation components and systems, batteries, consumer electronics, home appliances, microelectronic components and systems, electric power tools, medical systems, and electrical installation systems, as well as electric traction systems and vehicles, among others [24]. Consequently, most of the business models of the electrical industry are B2B-oriented, with comparatively little B2C orientation addressing electrical consumer products such as home appliances. Value creation of the electrical industry is characterized by highly interconnected global networks [25].

In April, during the first wave of the COVID-19 pandemic, ZVEI conducted two surveys on the impact of the crisis on the electrical industry in Germany. 128 member companies participated in the survey at the beginning of April (4–7 April 2020). The participants comprised 24% (45 billion Euros) of the sector's turnover. 114 member companies participated in the second survey at the end of April (23–27 April 2020). The participants of this second survey comprised 41% (78 billion Euros) of the sector's turnover. Comparisons of selected results from both surveys are presented in Figures 1 and 2.

Figure 1. Impact of the COVID-19 crisis on orders in April. (Multiple answers possible, answers weighted by turnover, own illustration data from [26,27]).

Figure 2. Impact of the COVID-19 crisis on the supply chain at the end of April 2020. (Multiple answers possible, answers weighted by turnover, own illustration data from [27].)

In the first ZVEI survey, more than half of the participating companies (55%) had already received fewer orders than before the crisis. 26% reported a slump in orders (see Figure 1). In the same survey, the companies expected sales to fall by an average of 14%. Only 3% considered it realistic to completely make up for these losses within the near future. The rest expected to recover only half of them or even less in the foreseeable future [26]. In the second survey, 12% expected to make up for the loss [27].

In the second survey, 86% of the companies continued to report slight problems with their suppliers. For 7%, the problems had taken on a serious level (see Figure 2). As far as the supply chain was concerned, 68% of companies were experiencing slight difficulties, and 26%, serious difficulties. For around 90% of companies, sales of their own products and services had been impaired. In half of the cases, orders had collapsed. In just under a third of the companies, sales had also been disrupted due to their own production stoppages (see Figure 1) [27].

At the beginning of April 2020, the COVID-19 crisis had already hit the companies of Germany's electrical industry quite hard. In particular, a decrease in orders, as well as production downtime and limited supply chain functionality, hindered the production systems of the electrical industry. The impact of the crisis further increased over April, and the production conditions for the companies continued to deteriorate.

This snapshot of the situation in April is also evident when looking into the economic development of the electrical industry from January to October 2020. According to the ZVEI economic barometer in January 2021, orders were down 6.9% compared to the same period in the previous year. In the first ten months of 2020, production was 8.1% lower than in the first ten months of 2019. In those ten months, the aggregate sector turnover amounted to 146.8 billion Euros, 7.2% short of the previous year's figure. In November 2020, 26% of the sector's companies rated their current economic situation as good, 49% as stable, and 25% as poor. As far as expectations for the coming six months were concerned, 24% of the companies expected business to pick up, 61% expected it to remain stable, and 15% to slow down [28].

The COVID-19 crisis showed the vulnerability of global value creation networks, and led to negative impacts for the electrical industry in Germany with its complex, globally connected manufacturing systems. Resilient value creation approaches can enable industrial stakeholders to cope with future crises in a more efficient and effective manner. Resilient value creation is expected to reduce the negative impacts of internal and external shocks and contribute to the competitiveness of companies in a future that is increasingly shaped by the direct effect of climate change. In conclusion, the identification of specific patterns for resilient value creation, as intended by this paper, can be of substantial industrial relevance.

2. Literature Review

The aforementioned results exemplify the effect of an unexpected crisis on companies. This is where resilient value creation comes into play as an answer to the posed challenges. The concept of value creation, laid out in the following text, is based on a manufacturing-oriented perspective as described by Stock et al. [29], and Stock and Seliger [30]. Companies create value by linking value creation modules of different aggregation levels, e.g., factories, production lines or assembly workspaces, in horizontally and vertically integrated global networks. Individual business models define and frame the activities of stakeholders involved throughout the value creation network, and lead to cooperation and competition among them.

2.1. *Resilient Value Creation*

Organizations must continuously prove themselves against adverse events that can have a negative impact on them. Enterprise risk management and business continuity management are established approaches that support businesses in acting proactively and reactively. These approaches are based on risk identification, which is difficult because they depend on knowledge about the nature of the adverse events, as well as their occurrence probability and impact [12]. Additionally, these methods usually assume the independence of risks, which in today's complex systems and networks is questionable [31]. Furthermore, Fiksel et al. [12] argue that the process of risk management is too simplified, and learning opportunities are not considered, as risk management always assumes a return to its initial state. They conclude that, while building resilience might not be a substitute for other risk management approaches, it is an approach that continuously adapts to changing circumstances and can give a competitive advantage by building capabilities [12].

The term "technological sovereignty" is strongly related to the topic of resilience. It is defined as "the ability of governing bodies to choose and take political, economic, and scientific measures of their own within the operational framework of their international commitments. The aim is to protect well-defined system-critical infrastructures, including the competence to control risks associated with the use of certain technologies" [32] (p. 3). The COVID-19 crisis, especially at the beginning of the pandemic when global supply chains were massively disrupted, demonstrated the dependency of European society on products imported from non-European countries, leading to increasing global competition and tensions in international trade [32]. The capability to provide key technologies without relying on third countries is now an ongoing political discussion. Technological sovereignty thus describes the actions of a government on a macroeconomic level to ensure independence in clearly defined core areas of great social-economic relevance, whereas resilience covers the actions of individual companies on a microeconomic level for increasing the ability to cope with shocks of different sorts.

The overview of the development of the term, "resilience", by Linnenluecke [33] identifies the origins of the concept in publications as early as 1981. The term is used in several contexts such as psychology, ecosystems, infrastructure, etc. Bhamra et al. [1] see the common definition across all fields of resilience as the "capability and ability of an element to return to a stable state after disruption" [1] (p. 5376), although the context might change from an ecological, individual or organizational perspective. McManus et al. [9] point out that there is a connection between the different fields and perspectives of resilience, as resilient organizations can lead to resilient communities.

Previous research on the topic of resilience has shown that there are different definitions of the term [3], and the research itself is divided into several research streams, one of them being the "adaptability of business models" [33]. Duchek [3] identifies three scholarly groups. The first one views resilience as the ability to resist and recover, and the second one focuses on the development "of organizational processes and capabilities", while the last group also brings anticipation into its scope.

Collections of definitions of resilience in the organizational context (and others) can be found, for instance, in Bhamra et al. [1], Kamalahmadi and Parast [5], Linnenluecke [33] and Duchek [3]. So far, there is a lack of consistent conceptualization [3,33].

The measurement of resilience is an ongoing research topic; however most studies measure resilience after specific events, predictive factors are missing [33]. Ruiz-Martin et al. [6] suggest the four levels of fragile, robust, resilient, and antifragile to classify the resilience level, and note that these can change over time. The measurement itself is not the topic of this article. However, it is assumed that achieving higher states of resilience is possible.

To conceptualize resilience, and for ease of understanding the concept, Thoma [34] suggests a resilience cycle consisting of five phases—prepare, prevent, protect, respond and recover—based on approaches from classical disaster management. Duchek [3] uses a three-phase approach that includes anticipation, coping and adaptation. During the first phase, which Duchek [3] defines as an offensive one, the organization should observe its environment and prepare for unexpected events. This is followed by the coping phase, where actions are conducted as a response to the event and adaptation after the event, where the organization should transform and adapt based on learning from the crisis it has undergone [3]. Following this approach, in this study, three phases are applied for ease of usage and increased practical applicability. However, within the first phase, we also include resistance to an unexpected event as a possible reaction and not only adaptation.

Based on the existing definitions from Ruiz-Martin et al. [6], and Vogus and Sutcliffe [7], for the purpose of the paper we define resilient value creation as follows: the ability, capacity or capability of value creation systems, including its stakeholders, to return to a competitive state (which can be based on a different business model than before) by preparing for, responding to, and recovering from internal or external and known or unknown disturbances.

Thus, measures of resilient value creation can be characterized by combining different business model activities throughout the phases of anticipation, coping and adaptation of the resilience cycle.

2.2. Business Models and Business Model Patterns

As mentioned before, "working constructs that are practical and effective in the short and long term" [9] are essential for achieving corporate resilience. Business models can be understood as such constructs as they allow, depicting and thereby capturing organizations with all their elements and respective characteristics in their entirety [35–37]. Furthermore, they allow the adequate analysis and design of organizations [38,39]. In this context, generating value for the organization and its stakeholders represents the overall goal [35–37,40], through which competitive advantage can be realized and organizations' survival can be secured [35,37,41].

In the context of this paper, resilience can thus also be understood as an organization's capability to always have an adequate business model at hand for both short-term and long-term competitiveness, and thereby, survive. For the generation of business models, a variety of approaches exist [36,42].

A particularly promising, efficient and effective approach is the application of so-called business model patterns [13,43,44]. In the relevant literature, business model patterns are generally understood as design options for the configurations of a business model which has proven successful in practice, and represents solutions for recurring problems [44–46]. In addition to the term "business model pattern", different terminologies are used: business model analogies [47]; atomic business models [48]; operating business models [49]; profit models [50]; business models [51]; business model configurations [52]; and business model archetypes [53].

Business model patterns can refer to both a business model as a whole and to individual elements of a business model [47,54]. A business model can thus represent a combination of different patterns. According to Gassmann et al. [46], business model patterns are of particular importance, because their analysis of disruptive business models over the last 50 years shows that more than 90% of all business model innovations are merely recombinations of known ideas, concepts, and elements of business models from other industries. Against this backdrop, leveraging business model patterns that have proven successful in other industries and companies offer an efficient approach for generating business models and for decision support [13,44,55]. To date, a variety of business model pattern collections with different focuses have been created [13,42,43,56,57]. In this context, the focus has been put on different industries and businesses such as banking [58], car-sharing [56], destination management [59], e-business [38,60–62], e-mobility [63], open data [64], the internet of things [65], insurance [66], social business [67,68] and utility computing [69], as well as industry-independent generic patterns [43,52,57,70]. In addition, a variety of pattern collections have been dedicated to the field of sustainability [13,71–75], focusing on the areas of economic, environmental and social dimensions in the frame of business models and their development. Regardless of the respective focus, the development of pattern collections is carried out both by means of literature analyses, taking into account scientific and non-scientific articles (e.g., [43]), and based on empirical research (e.g., [46]), as well as by means of conceptual work (e.g., [62]), and combinations of the different approaches (e.g., [58]).

To the best of the authors' knowledge, so far, no specific collection that focuses on business model patterns with regard to resilience is yet available. Complementary to existing general and sustainability-focused pattern collections, as well as further approaches for designing resilient and sustainable business models (e.g., [14,76]), a generic pattern collection dedicated to resilience seems promising, and thus illustrates a promising approach for supporting organizations in becoming and remaining resilient.

3. Methodology

To investigate which business model patterns are relevant for resilient value creation during COVID-19, and what can be learned for future crises, this research focused on the recommendations and findings of globally operating business consultancies. At the beginning of the pandemic, only a few academic publications on resilience during the COVID-19 crisis were available. In contrast, consultancies were able to publish their experiences ad hoc without the need of going through academic review procedures. Therefore, this study's methodology focuses on the consultancies' insights in order to gain practical-oriented knowledge with high relevance for the industry. Thus, the research results provide an overview of the resilience experiences from global consultancies during the beginning of the COVID-19 crisis.

The research methodology followed the algorithm shown in Figure 3. To support the procedure, Microsoft Excel was used. First, relevant sources were identified by analyzing resilience-related open-access publications of established business consultancies, sourced from their websites. For starting the algorithm, a consultancy was randomly selected from a list of the top 10 consultants by revenue [77], and searched for resilience-related content on their websites. For the respective consultancy under consideration, the relevant content, i.e., individual sources, was then extracted. Subsequently, in each of these publications, relevant statements were identified. A statement represents a self-contained approach for resilient value creation. From each single statement, specific objectives and patterns were condensed and transferred into one list. In the next step, the novelty of the identified patterns was checked by comparing them all before capturing anonymized patterns. After processing each publication, it was evaluated regarding a pre-defined convergence criterion. The convergence criterion was defined as the number of new patterns of a publication divided by all patterns of that publication. For the threshold, 10% was determined. This allowed monitoring of the knowledge gain during the research process. In the case that newly sourced statements repeatedly did not contribute new knowledge in terms of unique patterns, and convergence was reached, the review was stopped. Eventually, four different consultancies (BCG, McKinsey, Deloitte, EY; k = 4) and 10 related publications (p = 10) were analyzed, leading to 171 statements (m = 171). From these statements, 226 objectives (n = 226) and 245 patterns (z = 245) were derived. The convergence criterion for the final source equaled 0.077, meaning that only 8% (or 1 out of 13) of the patterns derived from this source could be classified as novel. The two previous iterations reached 15% (8 out of 54) and 25% (4 out of 16) regarding the convergence criterion. By reaching a value below 0.1 in the final iteration, the algorithm was exited.

Subsequently, the list of objectives and patterns, i.e., the "Bucket of Patterns", was processed. This step included a further clustering of the objectives and patterns in terms of similarity. This resulted in 13 objectives. Under each objective, similar patterns were then grouped together and reformulated as an actionable measure. At the same time, patterns that were too broad or general, or articulate an objective itself, were omitted. In total, this resulted in 110 unique patterns. In the next step, each of the remaining patterns was assigned to one or more resilience phases according to Duchek [3], based on its content. Furthermore, each pattern was assigned to one or more elements of the business model according to Osterwalder and Pigneur [36]. In the case where patterns affected the business model as a whole, they were assigned to an overarching business model level.

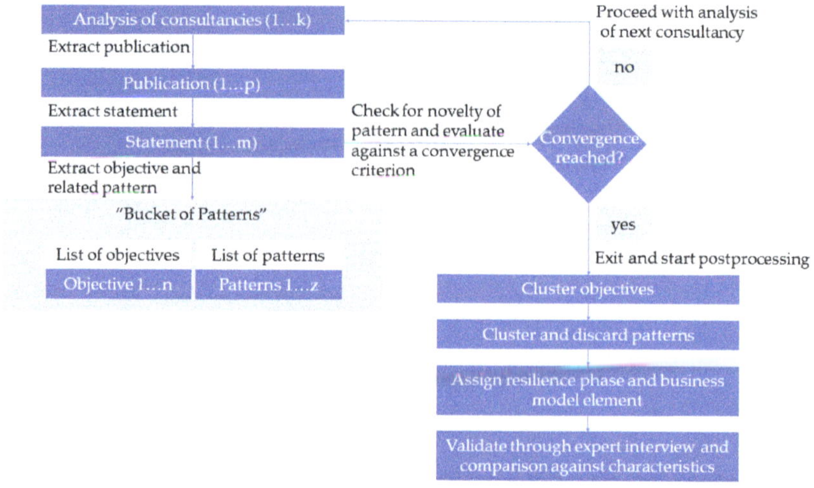

Figure 3. Research methodology.

As a result of this procedure, the pattern catalog comprised 13 objectives and 110 patterns, and was subsequently validated through semi-structured interviews with 3 experts from the electrical industry. The identified patterns were also compared with the aforementioned characteristics of the German electrical industry. In conclusion, the main result of this research is a qualitative meta-study of validated patterns for industrial resilience, based on the practical experiences of well-established business consultancies. The results of this approach are presented in the following section.

4. Results and Validation

This section takes a closer look at the objectives, the patterns and their assignment to resilience phases and business model elements, before describing how the patterns were validated.

4.1. Results

The provided structure aimed at enabling practitioners to choose suitable patterns from the catalog according to their individual needs. All patterns are listed in Supplementary Table S1, which shows the 13 objectives, a short explanation and the number of patterns within each objective. As shown in Table 1, most patterns were assigned to the objective "ensure liquidity" while a few are assigned to "empower people and create culture" and "increase transparency of the value creation system". However, the number of patterns per objective does not allow any conclusions about importance. For a discussion of the objectives against the literature and their relevance, see Section 5.

Table 1. Identified objectives of resilient value creation and number of patterns per objective.

Objective	Brief Explanation	Number of Patterns
Adapt and align leadership	All changes which are necessary in the leadership style of the organization and the strategic orientation.	10
Anticipate change in customer demand	Identify at an early stage how customer needs and distribution channels are changing, e.g., towards digital sales.	10
Anticipate moment of rebound	Identify early on when the disturbance is over, and then prepare accordingly to ensure a smooth and fast ramping-up of operations.	6
Create awareness on different scenarios and risks	Facilitate better understanding about the occurrence and effects of possible adverse events, as well as feasible countermeasures.	12

Table 1. *Cont.*

Objective	Brief Explanation	Number of Patterns
Design a flexible system	Create a system and its components, be it individual organizations, supply chain or technical systems, to be easily modified.	10
Empower people and create culture	Build competencies and motivation to create a higher individual resilience, as well as facilitate a cultural framework in which people can act resiliently.	5
Enable agile actions	Facilitate quick and easy response to any changes when necessary.	8
Ensure employees' safety (in a health crisis)	Guarantee the physical and emotional wellbeing of the employees, especially in a health crisis.	8
Ensure liquidity	Stay on top of finances to have sufficient cash available or be able to convert assets quickly into cash.	15
Identify revenue opportunities	Question product and service listing and reorganize if necessary.	6
Increase robustness	Create stability and reduce vulnerability to severe impacts by increasing redundancy and diversity.	6
Increase transparency of the value creation system	Share information with suppliers and collect information, e.g., by "control towers", and create visibility of material stocks.	5
Use digital processes and work tools	Digitalize the complete system from E2E, and use digital tools to support employees. The basis is IT security, which is a top priority.	9
Total		110

With regard to the distribution of patterns in relation to phases and objectives, different insights can be gained. As shown in Table 2, most patterns are assigned to the coping phase (65). This is followed by 48 patterns assigned to the adaptation phase, and 47 to the anticipation phase, as shown in Table 2. A possible explanation for this distribution among the phases might be the time frame of the study. It was conducted when most companies had to deal with the consequences of COVID-19, and consultancies tried to offer related solutions. Regarding the relationship between objectives and resilience phases, the data show that nearly all objectives are represented in more than one phase, except "anticipate moment of rebound". This objective is only present during adaptation. Contrary to the general focus on coping and adaptation, there is a clear relationship between the "creation of awareness" and the anticipation phase, which seems logical as it is most important before the disturbance affects organizations. In this way, organizations can react early and quickly. The objective "ensure liquidity" has a high relevance during the coping phase, as it is most crucial for organizations to stay on top of their finances during this phase. The objective "ensure employees' safety" was not represented during the anticipation phase, but it shows high representation during the coping phase. Building a robust operation needs a clear focus on anticipation, with no patterns assigned to adaptation. The advancement in digital processes shows little relevance during the adaptation phase. When discussing this topic, it is important to consider that the process of becoming resilient is regarded as a cycle. Therefore, each adaptation phase is followed by an anticipation phase. However, if a company is well prepared, it might require fewer or no activities during the crisis itself.

The distribution of patterns according to the business model elements is illustrated in Table 3. By far the most frequent patterns are those assigned to key resources (52), followed by key activities (28). The explanation of this concentration relates to the high prevalence of patterns in the coping phase. When companies were forced to cope with the initial effects of the pandemic, existing resources and processes were the obvious starting point. A significant, yet only approximately half as large, number was assigned to key partners (23), which might be explained through supply chain difficulties arising. Sixteen patterns showed a relationship to the business model as a whole (row: "meta") signaling the need for broader measures to deal with the pandemic. In general, it shows that, across all patterns, the internal view of the business model (cost structure, key activities, key resource) is more in the foreground than is the external view (revenue stream, customer segments, channels, customer relationships, key partners).

Table 2. Objectives in the respective phases of resilience. Dark shades highlight the highest number of patterns in the respective objective cluster.

Objective	Resilience Phases		
	Number of Patterns in Anticipation Phase	Number of Patterns in Coping Phase	Number of Patterns in Adaption Phase
Increase transparency of the value creation system	3	3	2
Increase robustness	6	2	-
Create awareness on different scenarios and risks	10	4	4
Design a flexible system	6	4	5
Empower people and create culture	2	4	4
Enable agile actions	4	6	1
Ensure employees' safety (in a health crisis)	-	7	1
Ensure liquidity	2	12	6
Adapt and align leadership	4	7	3
Anticipate change in customer demand	2	9	3
Use digital processes and work tools	7	4	9
Identify revenue opportunities	-	3	4
Anticipate moment of rebound	-	-	6
Total	47	65	48

Table 3. Distribution of identified patterns across business model elements.

Business Model Element	Number of Patterns
Meta	16
Key Partners	23
Key Activities	28
Key Resources	52
Value Proposition	9
Customer Relationships	5
Channels	9
Customer Segments	8
Cost Structure	17
Revenue Streams	12

4.2. Validation

The validation of the identified business model patterns, and the related classification according to resilience phases, objectives and business model elements, was based on a joint workshop with three company representatives from the electrical industry, as well as two representatives of ZVEI.

In the beginning, a short questionnaire for assessing the current and future relevance of industrial resilience for Germany's electrical industry was used. All participants agreed that issues related to industrial resilience gained importance during the COVID-19 pandemic and will become an even more important topic in the future.

Afterward, a qualitative discussion of the overall research approach, as well as of individual objectives and patterns related to the electrical industry, was conducted. Most of the patterns were assessed as very important or important by the participants. Some patterns assigned to the objectives "ensure revenue opportunities" and "ensure liquidity" were neither assessed as important nor as unimportant, as the representatives felt more confident about discussing patterns related to manufacturing and supply chain management due to their specific backgrounds. During the discussion, it was stated that the catalog of patterns is an important means for improving industrial resilience in the value creation of companies. However, it was pointed out that many patterns are rather generic and are applicable to any industry. Thus, it was noted that specific challenges of the electrical industry were not being sufficiently addressed. A core statement of the participants was that dual/second sourcing strategies often cannot be effectively realized, since only a few suppliers are globally capable of manufacturing complex electrical or electronic parts and products. This fact is currently also highlighted in the media, as a shortage of microelectronics and chips is hindering the manufacturing process of many companies. The shortage shows that manufacturing capacities of sufficient quality cannot be scaled up quickly, due to the high complexity of the required processes and produced products. Thus, the implementation of redundancies by emphasizing dual-sourcing strategies in value creation networks seems not to be a favorable approach for the electrical industry.

Since the identified objectives and patterns are non-industry-specific for the time being, it is necessary to interpret them against the background of the challenges faced by the electrical industry as described in Section 1. The electrical industry as a whole shows a very high degree of connectivity with other sectors, and therefore covers a very diverse spectrum of products and services. This ranges from end-customer business for building automation and digital services in the healthcare sector, to highly specialized plant technology in industrial production. With this in mind, a collection of generic patterns (see Supplementary Table S1) is generally necessary to cover the wide range of potential challenges faced by individual companies. Nevertheless, several characteristics of the electrical industry must be considered to align the pattern catalog accordingly (see Table 4).

Table 4. Characteristics of the electrical industry in Germany and the resulting focus for the resilience patterns.

Characteristic of Electrical Industry and Relevance for Resilience	Related Resilience Objectives	Related Business Model Elements	Resulting Focus for Resilience Patterns (Reference for Pattern ID in Supplementary Table S1)
Mainly B2B-oriented: long-term and more rational business relationships but limited overview of quickly changing end-user markets	• Anticipate change in customer demand • Anticipate moment of rebound	• Customer relationship • Customer segment • Channels	• Consider long-term effects of prioritization and pricing on relationship with key customers (2.5; 2.10) • Monitor behavior of end-customers of customers to anticipate demand changes early and dampen bullwhip effect (2.4; 3.3; 3.5)
High share of turnover with capital goods: higher susceptibility to delayed investments by customers	• Anticipate change in customer demand • Design flexible systems	• Value proposition • Revenue streams	• Expand products with supporting services to generate continuous revenue (2.1; 2.9; 5.1) • Utilize product-service systems to provide functionality instead of ownership, to accommodate flexible demand and reduce dependency regarding financial investment barriers (2.2; 2.3; 2.4; 5.1)

Table 4. *Cont.*

Characteristic of Electrical Industry and Relevance for Resilience	Related Resilience Objectives	Related Business Model Elements	Resulting Focus for Resilience Patterns (Reference for Pattern ID in Supplementary Table S1)
High portion of digital products and services: ability to support customers' increased transparency and flexibility needs to address crises	• Identify revenue opportunities • Use digital processes and work tools • Increase transparency of value creation system	• Key activities • Value proposition • Channels	• Leverage and advance the technology portfolio to serve upcoming digitization demands created by the crisis (10.2; 10.3; 10.4) • Accelerate internal digitization projects to ensure effective working capabilities (13.1, 13.7; 13.8; 13.9) • In own products, enhance data generation and usage to as basis for transparency needs (12.2; 12.4)
High degree of innovation: ability to realize new technologies and applications for changing needs	• Identify revenue opportunities	• Value proposition • Key activities	• Create new features and services quickly, concentrating on essential features (10.2; 10.3) • Transfer changes into new long-term business models (10.1; 10.2; 10.4)
High number of SME: faster reaction capability but limited financial and personnel redundancies	• Adapt and align leadership • Empower people and create culture • Ensure liquidity	• Key resources	• Utilize closer management proximity from owner/CEO to employees to make and implement decisions, considering the strategic perspective as well as current operative problems (1.3; 1.5; 6.4) • Secure availability and skills of key personnel to reduce dependency on individuals (1.2; 6.2) • Utilize better process overview and autonomy of employees (1.4; 6.1) • Focus on ensuring liquidity with internal measures, as bank loans are less accessible (9.1; 9.3; 9.10)
Highly globally connected supply chains with limited supply alternatives: higher susceptibility to impairments, even locally.	• Create awareness on different scenarios and risks • Design flexible systems • Increase robustness	• Key partners • Key resources	• Monitor events and risks on global (political) landscape (4.6; 4.7) • When redundancies and alternatives cannot be achieved by dual-sourcing, adapt inventory allocation and prioritize customers (5.1; 5.8; 5.10; 11.3)

Most business models in the electrical industry are B2B-oriented, which simplifies communication, as customer contacts are mostly well defined and requirements are more explicit compared to the B2C market. However, when prioritizing individual customers in the event of shortages, greater attention must be paid to the long-term effects on the supplier–customer partnership. As the electrical industry is a driver for digitization and innovations, with digital products and services being elements of many business models, existing capabilities and activities can be leveraged to serve external demands quicker and translate new features into long-term post-crisis business models. At the same time, this provides opportunities to advance internal digitization roadmaps. Small and medium-sized companies (SMEs) make up the majority of the electrical industry. As SMEs have a limited access to external loans from the financial market for investments, in any case, patterns to ensure liquidity without relying on external sources need to be focused. On the other hand, the closer relationship between management and employees can be utilized to establish clear communication and decision making in a quicker and more credible manner. The high level of connectivity in global supply chains represents a major risk for electronics companies, as multiple regions have to be continuously monitored both from a demand and a supply perspective. At the same time, the global networks often cannot be used to create supply alternatives via dual-sourcing, as the necessary resources and components are often only available through very few suppliers. To compensate for this, companies need to resort more often to adapted inventory strategies and the prioritization of customers.

5. Discussion and Future Recommendations

The catalog is based on found statements in consulting publications, and therefore provides a picture of how to achieve resilience from a practical point of view, giving managerial recommendations. At the same time, it provides the basis for comparison with scientific literature. Practitioners can use the catalog in different ways. First, it gives a holistic overview of objectives and assigned patterns for reaching a state of higher resilience. Second, by combining resilience phases and patterns, it helps to shift the focus to what time each objective should be pursued. Third, by combining business model elements and patterns, the catalog on the one side helps to identify patterns that are suited to making single business model elements more resilient, and on the other side, the categorization simplifies the allocation of patterns within the organization. At the same time, this still is a general catalog; the suitability of each pattern for the specific question still needs to be checked every time. A company-specific solution also needs to be worked out by the practitioners. The interrelationship of the patterns was not investigated. It was assumed that each pattern in itself increases resilience.

There are, however, limitations to this study. First, only a few well-known consultancies were considered. Second, the approach concluded the search when a convergence criterion was reached. Continuing might deliver more insights; however, the number is expected to be very small. Third, in the course of this study, validation took place in a qualitative way in a single workshop, with experts from different companies from the German electrical industry. While it is unlikely, as the experts came from multinational organizations, it is possible that in other countries there will be a different result. Thus, applying the catalog to companies to gain insights on its applicability and usability constitutes a logical next step. Lastly, some patterns are certainly a reaction to the type of crisis. As COVID-19 was a health crisis, some patterns and objectives might be less relevant during other types of crisis, especially the cluster of ensuring employees' health and safety, which might be less relevant during non-health-related crises. The cluster regarding a change in customer demand might also be less oriented toward a digital channel during other crises.

Comparing the objectives against existing literature can give more insight into whether the proposed patterns extracted from consultancies improve resilience. As the objectives fit with resilience in the literature, and the patterns fit with the objectives, this can be seen as confirmation of the patterns.

The ability to come to a good decision quickly, among other factors, is influenced by leadership, decision-making structures and knowledge management, and the flexibility of an organization [9]. This confirms the general relevance of the objectives "adapt and align leadership" and "empower people and create culture". Defining a long-term vision (1.9) is seen as a positive factor for resilience [4]. According to the literature, including other opinions into the decision-making process, as well as decentralizing control [4], supports resilience. This confirms patterns 1.4 and 1.7. However, it contradicts pattern 1.3. A possible explanation is that, according to the analyzed literature, at different times, different leadership styles should be utilized. The focus on finding solutions quickly, rather than finding the perfect solution (1.6), is found in the literature as the term "bricolage" [78], which is seen as an important aspect of resilience [11]. Creating a culture where disruptions are seen as an opportunity, and which is supporting creativity [4,11], is a driver of resilience, yet this is not explicitly mentioned in the patterns. Having enough available personnel (6.2) and ensuring that they are motivated and aligned (6.4), e.g., by ensuring their wellbeing (6.5), can also be found in the literature [4].

The clusters "anticipate change in customer demand" and "anticipate moment of rebound" are not found in the literature with this particular focus. However, the implications of this focus on manufacturing and on fulfilling customer demands (2.8, 2.9, 3.2, 3.4, 3.5) are important topics when addressing supply chain resilience [12].

The objective "create awareness on different scenarios and risks" can be found in many sources. Existing literature addresses the awareness of the environment in which the organization is operating, e.g., by building scenarios [9], and at the same time addresses managing potential negative impacts on the organization [9]. Organizations need to be able to observe the environment, identify possible threats, and prepare accordingly [3]. In this context, it is also important to recognize the problem [3]. Most of the patterns in this cluster are placed into the anticipation phase (see Table 2). This is in line with Duchek [3], where "observation and identification" [3] is placed in the anticipation phase. This capability in turn is closely related to the objective "increase transparency of the value creation system", as it is also related to the information within the system. Patterns can therefore be seen as relevant. The patterns related to the supply chain (12.2, 12.4, 12.5) can also be found as visibility [12] in previous literature.

The objective "ensure liquidity" is focused on financial resource availability, as that was a particular focus within the analyzed publications. While that factor can also be found in the scientific literature [3,4,6], there it is not only understood in terms of financial resources but also as the availability of other resource types.

The cluster "design a flexible system" can be found in the literature regarding supply chains [12]. Patterns within the objective "enable agile actions" are focused more on the ability of individuals to perform agile actions [4,9,11], which has been covered above.

Redundancies to build robustness are mentioned especially in the literature [3,4,6,12], and are therefore seen as relevant. Furthermore, the literature mentions the importance of networks and collaborations [4,6], which is also represented by the pattern "preserve supplier networks on which you depend" in the objective of robustness.

The cluster "ensure health" is not mentioned explicitly in the literature. Nevertheless, the well-being of staff is mentioned in the literature [4,12]. The frequent mention in the analyzed sources of this objective is certainly a consequence of the COVID-19 pandemic being a health crisis. Therefore, it might be less relevant in other types of crisis; however, as it is also generalized in the literature it seems to be relevant to a certain point.

The patterns of "identify revenue opportunities" are not prevalent in the scientific literature. In contrast, liquidity and revenue generation are frequently mentioned in the grey literature of consultancies. This can be seen as too narrow a focus on short-term and reactive measures in the field of resilience management. On the other hand, it can also indicate a research gap regarding the criticality of available cash as a basis to survive a crisis.

The objective of "digital processes and work tools" is not directly represented in the scientific literature. The aspect of cybersecurity (13.4, 13.5) is, however, mentioned [12]. The analyzed literature stems mainly from the context of the COVID-19 pandemic, during which capabilities for remote working by utilizing digital tools were highly relevant. It can be assumed that digitization is not an unlimited measure for improving resilience and may also have negative effects in specific crisis situations.

An aspect that is highlighted by the literature is learning from the previous crisis and changing accordingly afterward [3]. This aspect was not present in the analyzed consulting literature; pattern 4.4 only partly includes it.

Further insights can be gained by comparing the results of this study with existing pattern collections from the business model domain. When matching the patterns with collections focusing on sustainability, the main differences persist with regard to the content of patterns and the structure or classification of patterns. In collections such as those by Lüdeke-Freund et al. [13] and by Lüdeke-Freund et al. [72], and Bocken et al. [73], content wise focus is predominantly put on the environmental as well as the social dimension of business models. This aims to improve regular business performance with regard to these dimensions, while economic performance is also considered. When being applied to specific industries, as for example in the case of the banking industry, the patterns are adapted to that industry's characteristics [58]. Regarding the results of this study, this would entail adapting the pattern catalog to the electronic industry, which has not been realized at present. Crises in terms of single events that are needed to be dealt with are not explicitly focused on or taken into consideration in the mentioned collections.

The predominant focus on sustainability dimensions is also reflected in the structuring or classification of patterns in these collections. Patterns are clustered in topics related to the environmental and social dimension, e.g., creating value from waste, substituting with renewables and natural processes, and encouraging sufficiency [58], or are assigned to the three dimensions of sustainability or subdimensions of the sustainability triangle [13]. On the other hand, a similarity with regard to the current study persists in the fact that business model elements and underlying elements are partly applied to structuring or classifying patterns [72].

The comparison of the pattern catalog of this study to generic pattern collections draws a different picture. In contrast to the aforementioned pattern collections with a focus on sustainability, the economic performance seems to be in the foreground. Patterns that are related to social or environmental aspects are partly considered, but seemingly play a subordinate role here. Crises, on the other hand, are also not explicitly considered [56,57]. In relation to this study, it shows that although patterns of similar content are included in the generic pattern collections, especially those related to increasing revenue and decreasing costs, they are also related to sourcing and other aspects [43]. A further similarity with this study is the structuring or classification of patterns, which also takes into account business model elements and an overarching business model perspective [43]. Regarding the structuring of patterns, the contribution by Remane et al. [56] provides some insights on how structuring can be conducted in the case of specific industries. Here, the patterns are assigned to specific characteristics of the car-sharing business, which is in general a promising option for structuring existing collections according to specific industries.

Against the backdrop of the preceding explanations, it can be concluded that resilience has until now not been taken thoroughly into consideration in the context of business model patterns, although overlaps between the content of this study's catalog and existing collections exist. Thus, bringing the different collections together and structuring them according to the respective focus of business model development, e.g., improving resilience, seems promising. In this context, the structuring of patterns plays a very important role, as it determines the access to and the usability of collections.

Besides the differences and commonalities between the different collections, their application requires a structured approach for target-oriented business model development. Therefore, the focus should be put on synergies between existing collections and further pattern-independent approaches. In the context of resilience, bringing this study's collection and, for example, the approaches by Carayannis et al. [14] and Biloslavo et al. [76] together, is a promising next step towards a systematic business model development for resilience.

6. Conclusions and Outlook

This study made a first attempt to fill the research gap related to the lack of pattern collections, with a focus on resilience. Based on publications by large consultancies, 110 unique patterns were identified and compiled in a catalog. For structuring the catalog and thereby enabling target-oriented access to the patterns, these were assigned to 13 objectives related to resilience, developed through clustering, and assigned to the three phases of resilience as well as to business model elements. The structured, yet generic, catalog was validated based on interviews with representatives of the German electrical industry. Additionally, the pattern catalog was analyzed and interpreted with regard to the characteristics of the German electrical industry. This resulted in the selection of patterns that are of high relevance for the industry, with industry-specific interpretations. These steps showed the catalog's general relevance, as well as its usefulness in the context of the electrical industry.

With regard to practice, the catalog thus provides companies with an overview and specific starting points for becoming more resilient, which can be selected according to company-specific needs. The assignment to the resilience phases and business model elements supports this selection. By improving the resilience capabilities of the value-creation system through the identified patterns, companies can continue to contribute social and economic value, especially to local communities, while maintaining economic, ecological, and social resources. In this context, practitioners could be further assisted by catalogs that combine different targets like sustainability and resilience.

Furthermore, the structured catalog was analyzed with regard to relevant scientific literature, and differences as well as commonalities were identified, providing a variety of starting points for further research and development. In this regard, it showed that not all of the 13 objectives are represented in scientific literature, but that a general consistency exists. The analyzed publications showed a stronger focus on health concerns, revenue opportunities and digital processes, while the aspect of learning was less prevalent. The timing of the analysis, as well as the nature of the current crisis, might be causal for these circumstances, and future research should analyze the underlying reasons in more detail. In addition, subsequent research should repeat the approach chosen in this study after the pandemic, and see whether the identified patterns change.

Another starting point for further research persists in the analysis of existing pattern collections from the literature with a focus on patterns that are potentially relevant for resilience, to extend the structured catalog of this study. In addition, applying the catalog to specific industries and extending it by industry-specific patterns through case studies and literature analysis represents a promising opportunity to increase its relevance and usefulness for specific organizations. Further research could also consist in reviewing scientific and further literature independently of the current pandemic. In doing so, the catalog could be supplemented by scientific findings, as well as by experiences from other crises, such as the financial crisis and the ongoing disruptions between China and the USA. In addition to the pattern-based approach outlined in this article, further pattern-independent approaches exist that also support companies in becoming more resilient, and that can be combined with this study's results to unlock synergy potential. Apart from the focus on resilience, merging existing pattern collections independent of their focus, e.g., sustainability, creating industry-specific or general collections, and developing classifications or taxonomies that enable demand-oriented or occasion-related access on this basis, represent great research and development potential.

Following up on the approach of the literature review as chosen in this contribution, the next logical step consists in the practical application of the catalog, e.g., in a case study approach to validate its effectiveness either within the electrical industry or outside. In the long run, this could result in the development and deployment of a maturity model that helps companies to follow a predefined path to resilience, based on their current status quo.

Supplementary Materials: The following are available online at https://www.mdpi.com/article/10.3390/su13116090/s1, Table S1: Catalog of patterns.

Author Contributions: Conceptualization, K.N., T.v.E. and F.S.; methodology, K.N., T.v.E., E.S. and F.S.; validation, K.N., T.v.E., E.S. and F.S.; formal analysis, K.N., T.v.E., E.S. and F.S.; investigation, K.N., T.v.E., E.S. and F.S.; data curation, K.N., T.v.E., E.S. and F.S.; writing—original draft preparation, K.N., T.v.E., E.S. and F.S.; writing—review and editing, K.N., T.v.E., E.S., F.S. and H.K.; funding acquisition, K.N.; visualization, T.v.E.; project administration, F.S. All authors have read and agreed to the published version of the manuscript.

Funding: This research received no external funding.

Institutional Review Board Statement: Not applicable.

Informed Consent Statement: Not applicable.

Data Availability Statement: The data presented in this study are available in Supplementary Material Table S1.

Acknowledgments: The authors thank the German Electrical and Electronic Manufacturers' Association (ZVEI) and the participants of the joint workshop as well as the reviewers for their reviews and valuable comments.

Conflicts of Interest: The authors declare no conflict of interest.

References

1. Bhamra, R.; Dani, S.; Burnard, K. Resilience: The concept, a literature review and future directions. *Int. J. Prod. Res.* **2011**, *49*, 5375–5393. [CrossRef]
2. Burnard, K.; Bhamra, R. Organisational resilience: Development of a conceptual framework for organisational responses. *Int. J. Prod. Res.* **2011**, *49*, 5581–5599. [CrossRef]
3. Duchek, S. Organizational resilience: A capability-based conceptualization. *Bus. Res.* **2020**, *13*, 215–246. [CrossRef]
4. Barasa, E.; Mbau, R.; Gilson, L. What Is Resilience and How Can It Be Nurtured? A Systematic Review of Empirical Literature on Organizational Resilience. *Int. J. Health Policy Manag.* **2018**, *7*, 491–503. [CrossRef]
5. Kamalahmadi, M.; Parast, M.M. A review of the literature on the principles of enterprise and supply chain resilience: Major findings and directions for future research. *Int. J. Prod. Econ.* **2016**, *171*, 116–133. [CrossRef]
6. Ruiz-Martin, C.; Lopez-Paredes, A.; Wainer, G. What we know and do not know about organizational resilience. *Int. J. Prod. Manag. Eng.* **2018**, *6*, 11–28. [CrossRef]
7. Vogus, T.J.; Sutcliffe, K.M. Organizational resilience: Towards a theory and research agenda. In Proceedings of the 2007 IEEE International Conference on Systems, Man and Cybernetics, Montreal, QC, Canada, 7–10 October 2007; IEEE Service Center: Piscataway, NJ, USA, 2007; pp. 3418–3422, ISBN 978-1-4244-0990-7.
8. Gittell, J.H.; Cameron, K.; Lim, S.; Rivas, V. Relationships, Layoffs, and Organizational Resilience. *J. Appl. Behav. Sci.* **2006**, *42*, 300–329. [CrossRef]
9. McManus, S.; Seville, E.; Vargo, J.; Brunsdon, D. Facilitated Process for Improving Organizational Resilience. *Nat. Hazards Rev.* **2008**, *9*, 81–90. [CrossRef]
10. Christopher, M.; Peck, H. Building the Resilient Supply Chain. *Int. J. Logist. Manag.* **2004**, *15*, 1–14. [CrossRef]
11. Mallak, L. Putting Organizational Resilience to Work. *Ind. Manag. Chic. Atlanta* **1998**, *40*, 8–13.
12. Fiksel, J.; Polyviou, M.; Croxton, K.L.; Pettit, T.J. From Risk to Resilience Learning to Deal with Disruption. *MIT Sloan Manag. Rev.* **2015**, *56*, 79–86.
13. Lüdeke-Freund, F.; Carroux, S.; Joyce, A.; Massa, L.; Breuer, H. The sustainable business model pattern taxonomy—45 patterns to support sustainability-oriented business model innovation. *Sustain. Prod. Consum.* **2018**, *15*, 145–162. [CrossRef]
14. Carayannis, E.G.; Grigoroudis, E.; Sindakis, S.; Walter, C. Business Model Innovation as Antecedent of Sustainable Enterprise Excellence and Resilience. *J. Knowl. Econ.* **2014**, *5*, 440–463. [CrossRef]
15. Dori, D. Conceptual Modeling: Purpose and Context. In *Model-Based Systems Engineering with OPM and SysML*, 1st ed.; Dori, D., Ed.; Springer: New York, NY, USA, 2016; pp. 75–96, ISBN 978-1-4939-3295-5.

16. Stock, T.; Seliger, G. Methodology for the Development of Hardware Startups. *Adv. Mater. Res.* **2016**, *1140*, 505–512. [CrossRef]
17. United Nations. United Nations Sustainable Development: 17 Goals to Transform Our World. Available online: https://www.un.org/sustainabledevelopment/ (accessed on 21 April 2021).
18. Denton, F.; Wilbanks, T.; Burton, I.; Chandani, A.; Gao, Q.; Lemos, M.C.; Masui, T.; O'Brien, K.; Warner, K.; Dickinson, T.; et al. Climate-resilient pathways: Adaptation, Mitigation, and Sustainable Development. In *Climate Change 2014: Impacts, Adaptation, and Vulnerability. Part A: Global and Sectoral Aspects. Contribution of Working Group II to the Fifth Assessment Report of the Intergovernmental Panel on Climate Change*; Field, C.B.V.R., Barros, D.J., Dokken, K.J., Mach, M.D., Mastrandrea, T.E., Bilir, M., Chatterjee, K.L., Ebi, Y.O., Estrada, R.C., Genova, B., et al., Eds.; Cambridge University Press: Cambridge, UK; New York, NY, USA, 2014; pp. 1101–1131. Available online: https://www.ipcc.ch/site/assets/uploads/2018/02/WGIIAR5-Chap20_FINAL.pdf (accessed on 21 April 2021).
19. United Nations. Goal 13 | Department of Economic and Social Affairs. Available online: https://sdgs.un.org/goals/goal13 (accessed on 21 April 2021).
20. United Nations. Goal 11 | Department of Economic and Social Affairs. Available online: https://sdgs.un.org/goals/goal11 (accessed on 21 April 2021).
21. United Nations. Goal 9 | Department of Economic and Social Affairs. Available online: https://sdgs.un.org/goals/goal9 (accessed on 21 April 2021).
22. Ortiz-de-Mandojana, N.; Bansal, P. The long-term benefits of organizational resilience through sustainable business practices. *Strat. Mgmt. J.* **2016**, *37*, 1615–1631. [CrossRef]
23. ZVEI—German Electrical and Electronic Manufacturers' Association. German Electric Industry—Facts & Figures. Available online: https://www.zvei.org/fileadmin/user_upload/Presse_und_Medien/Publikationen/Regelmaessige_Publikationen/Daten_Zahlen_und_Fakten/Die_deutsche_Elektroindustrie_Daten_Zahlen_Fakten/Fact-Sheet-March-2021_V2.pdf (accessed on 19 April 2021).
24. ZVEI. Divisions. Available online: https://www.zvei.org/en/association/divisions (accessed on 19 April 2021).
25. ZVEI—German Electrical and Electronic Manufacturers' Association. ZVEI-Foreign-Trade-Report-March-2021. Available online: https://www.zvei.org/fileadmin/user_upload/Presse_und_Medien/Publikationen/2021/Maerz/ZVEI-Aussenhandelsreport_Maerz_2021/ZVEI-Foreign-Trade-Report-March-2021.pdf (accessed on 19 April 2021).
26. ZVEI. Ergebnisse der ZVEI-Umfrage. Available online: https://www.zvei.org/themen/maerkte-recht/konjunktur-analysen/ergebnisse-der-zvei-umfrage (accessed on 16 January 2021).
27. Jürgen Polzin. Ergebnisse der Dritten Ad-hoc-Umfrage zur Corona-Krise. Available online: https://www.zvei.org/presse-medien/newsletter/auf-den-punkt-3/2020-zvei-newsletter/ergebnisse-der-dritten-ad-hoc-umfrage-zur-corona-krise (accessed on 16 January 2021).
28. ZVEI—Zentralverband Elektrotechnik-und Elektronikindustrie e. V. ZVEI-Konjunkturbarometer-Dezember-2020. Available online: https://www.zvei.org/fileadmin/user_upload/Presse_und_Medien/Publikationen/2020/Dezember/ZVEI-Konjunkturbarometer_Dezember_2020/ZVEI-Konjunkturbarometer-Dezember-2020.pdf (accessed on 16 January 2021).
29. Stock, T.; Obenaus, M.; Kunz, S.; Kohl, H. Industry 4.0 as enabler for a sustainable development: A qualitative assessment of its ecological and social potential. *Process. Saf. Environ. Prot.* **2018**, *118*, 254–267. [CrossRef]
30. Stock, T.; Seliger, G. Opportunities of Sustainable Manufacturing in Industry 4.0. *Procedia CIRP* **2016**, *40*, 536–541. [CrossRef]
31. Dalziell, E.P.; McManus, S.T. Resilience, Vulnerability, and Adaptive Capacity: Implications for System Performance. In Proceedings of the 1st International Forum for Engineering Decision Making (IFED), Stoos, Switzerland, 5–8 December 2004.
32. ZVEI—German Electrical and Electronic Manufacturers' Association. Technological-Sovereignty-Industrial-Resilience-and-European-Competences-Discussion-Paper 2020. Available online: https://www.zvei.org/fileadmin/user_upload/Presse_und_Medien/Publikationen/2020/Oktober/Technologische_Souveraenitaet__Resilienz_der_Industrie_und_europaeische_Kompetenzen/Technological-Sovereignty-Industrial-Resilience-and-European-Competences-Discussion-Paper.pdf (accessed on 14 March 2021).
33. Linnenluecke, M.K. Resilience in Business and Management Research: A Review of Influential Publications and a Research Agenda. International Journal of Management Reviews. *Int. J. Manag. Rev.* **2017**, *19*, 4–30. [CrossRef]
34. Resilien-Tech. *Resilience-by-Design Strategie für die Technologischen Zukunftsthemen*; Thoma, K., Ed.; Utz: Munchen, Germany, 2014; ISBN 978-3-8316-4375-2.
35. Morris, M.; Schindehutte, M.; Allen, J. The entrepreneur's business model: Toward a unified perspective. *J. Bus. Res.* **2005**, *58*, 726–735. [CrossRef]
36. Osterwalder, A.; Pigneur, Y. *Business Model Generation: A Handbook for Visionaries, Game Changers, And Challengers*; Wiley: Hoboken, NJ, USA, 2010; ISBN 0470876417.
37. Wirtz, B.W.; Pistoia, A.; Ullrich, S.; Göttel, V. Business Models: Origin, Development and Future Research Perspectives. *Long Range Plan.* **2016**, *49*, 36–54. [CrossRef]
38. Afuah, A.; Tucci, C.L. *Internet Business Models and Strategies: Text and Cases*, 2nd ed.; McGraw Hill: Boston, MA, USA, 2003; ISBN 0-07-115123-0.
39. Pateli, A.G.; Giaglis, G.M. A research framework for analysing eBusiness models. *Eur. J. Inf. Syst.* **2004**, *13*, 302–314. [CrossRef]
40. Teece, D.J. Business Models, Business Strategy and Innovation. *Long Range Plan.* **2010**, *43*, 172–194. [CrossRef]

41. Schallmo, D. *Geschäftsmodell-Innovation: Grundlagen, Bestehende Ansätze, Methodisches Vorgehen und B2B-Geschäftsmodelle*; Springer Gabler: Wiesbaden, Germany, 2013; ISBN 9783658002442.
42. Gassmann, O.; Frankenberger, K.; Csik, M. *The Business Model Navigator: 55 Models that Will Revolutionise Your Business*; Pearson: Harlow, UK, 2014; ISBN 978-1-292-06581-6.
43. Remane, G.; Hanelt, A.; Tesch, J.F.; Kolbe, L.M. The business model pattern database—A tool for systamatic business model innovation. *Int. J. Innov. Mgt.* **2017**, *21*, 1750004. [CrossRef]
44. Abdelkafi, N.; Makhotion, S.; Posselt, T. Business Model Innovations for Electric Mobility—What can be learned from existing business model patterns? *Int. J. Innov. Mgt.* **2013**, *17*, 1–41. [CrossRef]
45. Gausemeier, J.; Amshoff, B. Diskursive Geschäftsmodellentwicklung—Erfolgreiche Positionierung in der Wettbewerbsarena durch integrative Entwicklung von Marktleistung und Geschäftsmodell. *ZwfZ. Für Wirtsch. Fabr.* **2014**, *109*, 428–434.
46. Gassmann, O.; Frankenberger, K.; Choudury, M. *Geschäftsmodelle Entwickeln: 55+ Innovative Konzepte mit dem St. Galler Business Model*; Navigator, 3.; Überarbeitete und Erweiterte Auflage; Hanser: München, Germany, 2021; ISBN 9783446465213.
47. Johnson, M.W. *Seizing the White Space: Business Model Innovation for Growth and Renewal*; Harvard Business Press: Boston, MA, USA, 2010; ISBN 9781422124819.
48. Weill, P.; Vitale, M.R. *Place to Space: Migrating to eBusiness Models*; Harvard Business School Press: Boston, MA, USA, 2001; ISBN 157851245X.
49. Linder, J.; Cantrell, S. Changing Business Models: Surveying the Landscape. Working Paper. Accenture Institute for Strategic Change: Cambridge, MA, USA, 2000.
50. Tuff, G.; Wunker, S. Beacons for Business Model Innovation. Available online: http://cdn2.hubspot.net/hub/88808/file-890913188-pdf/docs/beacons_for_busniess_model_innovation_download.pdf (accessed on 19 October 2018).
51. Rappa, M. Managing the Digital Enterprise: Business Models on the Web. Available online: http://digitalenterprise.org/models/models.pdf (accessed on 26 October 2018).
52. Taran, Y.; Nielsen, C.; Montemari, M.; Thomsen, P.; Paolone, F. Business model configurations: A five-V framework to map out potential innovation routes. *Eur. J. Innov. Manag.* **2016**, *19*, 492–527. [CrossRef]
53. Massa, L.; Tucci, C.L. Business Model Innovation. In *The Oxford Handbook of Innovation Management*; Dodgson, M., Gann, D.M., Phillips, N., Eds.; OUP: Oxford, UK, 2014; pp. 420–441. ISBN 9780199694945.
54. Amshoff, B.; Dülme, C.; Echterfeld, J.; Gausemeier, J. Business model patterns for disruptive technologies. *Int. J. Innov. Manag.* **2015**, *19*, 1540002. [CrossRef]
55. Amshoff, B. *Systematik zur Musterbasierten Entwicklung Technologie-Induzierter Geschäftsmodelle: Approach for a Pattern-Based Design of Technology-Induced Business Models*; Heinz Nixdorf Institut, Universität Paderborn: Paderborn, Germany, 2016; ISBN 3942647761.
56. Remane, G.; Nickerson, R.C.; Hanelt, A.; Tesch, J.F.; Kolbe, L.M. A Taxonomy of Carsharing Business Models. In Proceedings of the 37th International Conference on Information Systems, Dublin, Irland, 11–14 December 2016.
57. Weking, J.; Hein, A.; Böhm, M.; Krcmar, H. A hierarchical taxonomy of business model patterns. *Electr. Mark* **2018**, *17*, 359. [CrossRef]
58. Yip, A.W.; Bocken, N.M. Sustainable business model archetypes for the banking industry. *J. Clean. Prod.* **2018**, *174*, 150–169. [CrossRef]
59. Reinhold, S.; Beritelli, P.; Grünig, R. A business model typology for destination management organizations. *Tour. Rev.* **2019**, *74*, 1135–1152. [CrossRef]
60. Hedman, J.; Kalling, T. The business model concept: Theoretical underpinnings and empirical illustrations. *Eur. J. Inf. Syst.* **2003**, *12*, 49–59. [CrossRef]
61. Zott, C.; Amit, R.; Massa, L. Working Paper. In *The Business Model: Theoretical Roots, Recent Developments, and Future Research*; University of Navarra: Barcelona, Spain, 2010.
62. Wise, R.; Morrison, D. Beyond the exchange—The future of B2B. *Harv. Bus. Rev.* **2000**, *78*, 86–96. [PubMed]
63. Hunke, F.; Schüritz, R.; Kuehl, N. Towards a unified approach to identify business model patterns: A case of e-mobility services. *Lect. Notes Bus. Inf. Process.* **2017**, *279*, 182–196. [CrossRef]
64. Zeleti, F.A.; Ojo, A.; Curry, E. Emerging business models for the open data industry: Characterization and analysis. In Proceedings of the ACM International Conference Proceeding Sereries, Aguascalientes, Mexico, 18–21 June 2014. [CrossRef]
65. Fleisch, E.; Weinberger, M.; Wortmann, F. Business models and the internet of things. In *Interoperability and Open-Source Solutions for the Internet of Things*; Springer: Cham, Switzerland, 2015; pp. 6–10.
66. Kobler, D. *Innovative Geschäftsmodelle: Entwicklung und Gestaltung innovativer Geschäftsmodelle für Schweizer Versicherungsunternehmen im Privatkundensegment Hamburger Schriften zur Marketingforschung*; Rainer Hampp Verlag: München und Mering, Germany, 2005; ISBN 3879889686.
67. Alter, S.K. Social enterprise models and their mission and money relationships. In *Social Entrepreneurship: New Models of Sustainable Social Change*; Oxford University Press: Oxford, UK, 2006; pp. 205–232.
68. Dohrmann, S.; Raith, M.; Siebold, N. Monetizing Social Value Creation—A Business Model Approach. *Entrep. Res. J.* **2015**, *5*, 308. [CrossRef]
69. Rappa, M.A. The utility business model and the future of computing services. *IBM Syst. J.* **2004**, *43*, 32–42. [CrossRef]
70. Schallmo, D.R. *Geschäftsmodelle erfolgreich entwickeln und implementieren: Mit Aufgaben, Kontrollfragen und Templates*, 2nd ed.; Springer Gabler: Berlin, Germany, 2018; ISBN 9783662576045.

71. Albino, V.; Fraccascia, L. The Industrial Symbiosis Approach: A Classification of Business Models. *Procedia Environ. Sci. Eng. Manag.* **2015**, *2*, 217–223.
72. Lüdeke-Freund, F.; Gold, S.; Bocken, N.M. A Review and Typology of Circular Economy Business Model Patterns. *J. Ind. Ecol.* **2019**, *23*, 36–61. [CrossRef]
73. Bocken, N.; Short, S.W.; Rana, P.; Evans, S. A literature and practice review to develop sustainable business model archetypes. *J. Clean. Prod.* **2014**, *65*, 42–56. [CrossRef]
74. Dressler, M.; Paunović, I. Towards a conceptual framework for sustainable business models in the food and beverage industry. *Br. Food J.* **2020**, *122*, 1421–1435. [CrossRef]
75. Reinhardt, R.; Christodoulou, I.; García, B.A.; Gassó-Domingo, S. Sustainable business model archetypes for the electric vehicle battery second use industry: Towards a conceptual framework. *J. Clean. Prod.* **2020**, *254*, 119994. [CrossRef]
76. Biloslavo, R.; Bagnoli, C.; Edgar, D. An eco-critical perspective on business models: The value triangle as an approach to closing the sustainability gap. *J. Clean. Prod.* **2018**, *174*, 746–762. [CrossRef]
77. consultancy.uk. The 10 Largest consulting Firms in the World. Available online: https://www.consultancy.uk/news/14018/the-10-largest-consulting-firms-in-the-world (accessed on 22 March 2021).
78. Weick, K.E. The Collapse of Sensemaking in Organizations: The Mann Gulch Disaster. *Adm. Sci. Q.* **1993**, *38*, 628. [CrossRef]

MDPI
St. Alban-Anlage 66
4052 Basel
Switzerland
Tel. +41 61 683 77 34
Fax +41 61 302 89 18
www.mdpi.com

Sustainability Editorial Office
E-mail: sustainability@mdpi.com
www.mdpi.com/journal/sustainability

www.ingramcontent.com/pod-product-compliance
Lightning Source LLC
Chambersburg PA
CBHW040808250926
51850CB00007BA/1115
* 9 7 8 3 0 3 6 5 2 2 2 1 0 *